ENERGY, ENVIRONMENT, ECOLOGY AND SOCIETY

By

Smriti Srivastava

S.K. KATARIA & SONS®

Publisher of Engineering Books

4885/109, Prakash Mahal, Dr. Subhash Bhargav Lane,
Daryaganj, New Delhi-110002 (India)
Phone: +91-11-43551243; Mobile: +91-9871775858, 7678425209
e-mail: katariabook@gmail.com, katariabooks@yahoo.com
Website : www.skkatariaandsons.com

(e-books, subscriptions and bulk buying also available on our website)

Published By:

S.K. Kataria & Sons®

4885/109, Prakash Mahal, Dr. Subhash Bhargav Lane,
Opposite Delhi Medical Association, Daryaganj, New Delhi-110002
Phone : +91-11-43551243; Mobile: +91-9871775858, 7678425209
e-mail : katariabook@gmail.com, katariabooks@yahoo.com

First Edition: 2010; Second Edition: 2011; Reprint: 2025

Price: ₹ 350/-

Laser Typesetting by :
AISA ENTERPRISES, Delhi

Printed at :
New A.S. Offset Printers, Delhi

PREFACE

A publication like this can only be a beginning. Strategies to conserve natural resources are needed at all national/international levels and should form an integral part of all environment protection plans. These strategies should be turned into action.

This book is intended to meet the academic requirements of the subject Energy, Environment, Ecology and Society for graduate students in Indian universities irrespective of their major subject of study. While meeting the exact syllabus requirements of the university, this book has a holistic approach to the subject for easy and logical comprehension. The contents have been prepared keeping in mind the widest possible variations in the background of the users. A comprehensive and balanced coverage of the entire syllabus is the hallmark of this book.

The entire syllabus is covered in the six chapters. To avoid confusion with the technical jargon associated with Energy, Environment, Ecology and Society the definitions and figures are provided in the concern chapters and this could be of particular use to students with little or no environmental science back-ground.

I express my deep sense of gratitude to my parents for being contant inspiration for me and making me what I am. I owe special thanks to my husband Shubham Kumar Srivastava for his unfailing support and encouragement throughout this book and his keen and inquisitive observation on various aspect of the book. Continuous support of my brother Anand V. Srivastava is greatly acknowledged. Special recognition is due to my son Hardik for his love, patience and understanding during all long hours of work that kept me busy and for making my life more lively and interesting.

I thanks to Sanjeev Kumar Kataria of M/s S. K. Kataria and Sons, New Delhi for their wonderful work in bringing out this book in an attractive manner.

I also thank Mr. Sachin Kumar for his wonderful editing and bringing out this book in an attractive manner.

Smriti Srivastava
Author

PREFACE

Dedicated to

My Motherhood

CONTENTS

CHAPTER

1

Energy

1.1. INTRODUCTION

INSIDE THIS CHAPTER

The word energy derives from Greek word ενεέ ργεια (energeia-"activity, operation"), which appears for the first time in the work Nicomachean Ethics of Aristotle in the 4th century BC. In physics, energy is a scalar physical quantity that describes the amount of work that can be performed by a force, an attribute of objects and systems that is subject to a conservation law. Different forms of energy include kinetic, potential, thermal, gravitational, sound, light, elastic, and electromagnetic energy. The forms of energy are often named after a related force.

Any form of energy can be transformed into another form, but the total energy always remains the same. This principle, the conservation of energy, was first postulated in the early 19th century, and applies to any isolated system. According to Noether's theorem, the conservation of energy is a consequence of the fact that the laws of physics do not change over time.

Although the total energy of a system does not change with time, its value may depend on the frame of reference. For example, a seated passenger in a moving airplane has zero kinetic energy relative to the airplane, but non-zero kinetic energy relative to the Earth.

The concept of energy emerged out of the idea of vis viva, which Leibniz defined as the product of the mass of an object and its velocity squared; he believed that total vis viva was conserved.

1.2. ENERGY IN VARIOUS CONTEXTS SINCE THE BEGINNING OF THE UNIVERSE

The concept of energy and its transformations is useful in explaining and predicting most natural phenomena. The direction of transformations in energy (what kind of energy is transformed to what other kind) is often described by entropy (equal energy spread among all available degrees of freedom) considerations, since in practice all energy transformations are permitted on a small scale, but certain larger transformations are not permitted because it is statistically unlikely that energy or matter will randomly move into more concentrated forms or smaller spaces.

The concept of energy is widespread in all sciences.

- In biology, energy is an attribute of all biological systems from the biosphere to the smallest living organism. Within an organism it is responsible for growth and development of a biological cell or an organelle of a biological organism. Energy is thus often said to be stored by cells in the structures of molecules of substances such as carbohydrates (including sugars) and lipids, which release energy when reacted with oxygen. In human terms, The human equivalent (H-e) (Human energy conversion) energy indicates, for a given amount of energy expenditure, the relative quantity of energy needed for human metabolism, assuming an average human energy expenditure of 12,500kJ per day and a basal metabolic rate of 80 watts. For example, if our bodies run (on average) at 80 watts, then a light bulb running at 100 watts is running at 1.25 human equivalents ($100 \div 80$) *i.e.* 1.25 H-e.

- In geology, continental drift, mountain ranges, volcanoes, and earthquakes are phenomena that can be explained in terms of energy transformations in the Earth's interior. While meteorological phenomena like wind, rain, hail, snow, lightning and cyclones, are all a result of energy transformations brought about by solar energy on the atmosphere of the planet Earth.

- In cosmology and astronomy the phenomena of stars, nova, supernova, quasars and gamma ray bursts are the universe's highest-output energy transformations of matter. All stellar phenomena (including solar activity) are driven by various kinds of energy transformations. Energy in such transformations is either from gravitational collapse of matter (usually molecular hydrogen) into various classes of astronomical objects (stars, black holes, etc.), or from nuclear fusion (of lighter elements, primarily hydrogen).

Energy transformations in the universe over time are characterized by various kinds of potential energy, later being "released" (transformed to more active types of energy such as kinetic or radiant energy), when a triggering mechanism is available.

Familiar examples of such processes include nuclear decay, in which energy is released which was originally "stored" in heavy isotopes (such as uranium and thorium), by nucleosynthesis, a process which ultimately uses the gravitational potential energy released from the gravitational collapse of supernovae, to store energy in the creation of these heavy elements before they were incorporated into the solar system and the Earth. This energy is triggered and released in nuclear fission bombs.

In another similar chain of transformations beginning at the dawn of the universe, nuclear fusion of hydrogen in the Sun releases another store of potential energy.

Fig. 1.1. Lightning

Lightning is the electric breakdown of air by strong electric fields and is a flow of energy. The electric potential energy in the atmosphere changes into heat, light and sound which are other forms of energy.

Fig. 1.2. Burning of Energy in Body

Energy lights our cities, powers our vehicles, and runs machinery in factories. It warms and cools our homes, cooks our food, plays our music, and gives us pictures on television. Energy is defined as the ability or the capacity to do work.

We use energy to do work and make all movements. When we eat, our bodies transform the food into energy to do work. When we run or walk or do some work, we 'burn' energy in our bodies. Cars, planes, trolleys, boats, and machinery also transform energy into work. Work means moving or lifting something, warming or lighting something. There are many sources of energy that help to run the various machines invented by man.

The discovery of fire by man led to the possibility of burning wood for cooking and heating thereby using energy. For several thousand years human energy demands were met only by renewable energy sources—sun, biomass (wood, leaves, twigs), hydel (water) and wind power.

Fig 1.3. Wind Mill

As early as 4000–3500 BC, the first sailing ships and windmills were developed harnessing wind energy. With the use of hydropower through water mills or irrigation systems, things began to move faster. Fuel, wood and dung cakes are even today a major source of energy in rural India. Solar energy is used for drying and heating.

With the advent of the Industrial Revolution, the use of energy in the form of fossil fuels began growing as more and more industries were set up. This occurred in stages, from the exploitation of coal deposits to the exploitation of oil and natural gas fields. It has been only half a century since nuclear power began being used as an energy source. In the past century, it became evident that the consumption of non-renewable sources of energy had caused more environmental damage than any other human activity. Electricity generated from fossil fuels such as coal and crude oil has led to high concentrations of harmful gases in the atmosphere. This has in turn led to problems such as ozone depletion and global warming. Vehicular pollution is also a grave problem.

There has been an enormous increase in the demand for energy since the middle of the last century as a result of industrial development and population growth. World population grew 3.2 times between 1850 and 1970, per capita use of industrial energy increased about twenty-fold, and total world use of industrial and traditional energy forms combined increased more than twelve-fold.

Due to the problems associated with the use of fossil fuels, alternative sources of energy have become important and relevant in today's world. These sources, such as the sun and wind, can never be exhausted and are therefore called renewable. Also known as non-conventional sources of energy, they cause less emission and are available locally. Their use can significantly reduce chemical, radioactive, and thermal pollution. They are viable sources of clean and limitless energy. Most of the renewable sources of energy are fairly non-polluting and considered clean. However, biomass is a major polluter indoors.

Renewable energy sources include the sun, wind, water, agricultural residue, fuel-wood, and animal dung. Fossil fuels are non-renewable sources. Energy generated from the sun is known as solar energy. Hydel is the energy derived from water. Biomass – firewood, animal dung, and biodegradable waste from cities and crop residues – is a source of energy when it is burnt. Geothermal energy is derived from hot dry rocks, magma, hot water springs, natural geysers, etc. Ocean thermal is energy derived from waves and also from tidal waves.

Through the method of co-generation a cleaner and less polluting form of energy is being generated. Fuel cells are also being used as cleaner energy source. In India a number of initiatives have been taken. A good example is the model village of Ralegaon Siddhi.

Total commercial energy consumption has been growing tremendously since the last decade. Per capita commercial energy consumption in low-income countries have more than doubled. About 15 per cent of the world's population living in the wealthy industrialized nations consume over half the energy used in the world. The number of motor vehicles in use worldwide has more than doubled since 1970.

In some respects, the global energy system has evolved in a cleaner direction in the last 25 years. The share of world primary energy derived from natural gas – the cleanest fossil fuel – has increased by more than 25 per cent. So has the use and generation of renewable energy sources.

Still, the overall efficiency of energy production remains extremely low: on average, more than 90 per cent of energy consumed is lost or wasted in the process of conversion from raw materials such as coal to the final energy service such as the light to read a book. The main problem isn't that we use energy, but how we produce and consume energy resources. What we really need are energy sources that will last forever and can be used without polluting the environment. Conserving energy has become the need of the day be it in the transport, household, or industrial sectors.

DO YOU KNOW

How energy is measured : One of the basic measuring blocks for energy is called a Btu or British thermal unit. Btu is defined as the amount of heat energy, it takes to raise the temperature of 1 pound of water by 1 degree Fahrenheit, at sea level. One Btu equals about one black-tip kitchen match. It takes about 2000 Btu to make a pot of coffee.

Energy can also be measured in joules (pronounced the same way as 'jewels'). One joule is the amount of energy needed to lift 1 pound about 9 inches. It takes 1000 joules to equal a Btu. It would take 2 million joules to make a pot of coffee.

Joule is named after an English physicist named James Prescott Joule who lived from 1818 to 1889. He discovered that heat is a type of energy.

1.3. ENERGY TRANSFER

Because energy is strictly conserved and is also locally conserved (wherever it can be defined), it is important to remember that by definition of energy the transfer of energy between the "system" and adjacent regions is work. A familiar example is mechanical work. In simple cases this is written as:

$$\Delta E = W \qquad ...(i)$$

if there are no other energy-transfer processes involved. Here E is the amount of energy transferred, and W represents the work done on the system.

More generally, the energy transfer can be split into two categories :

$$\Delta E = W + Q \qquad ...(ii)$$

where Q represents the heat flow into the system.

There are other ways in which an open system can gain or lose energy. In chemical systems, energy can be added to a system by means of adding substances with different chemical potentials, which potentials are then extracted (both of these process are illustrated by fueling an auto, a system which gains in energy thereby, without addition of either work or heat). Winding a clock would be adding energy to a mechanical system. These terms may be added to the above equation, or they can generally be subsumed into a quantity called "energy addition term E*" which refers to any type of energy carried over the surface of a control volume or system volume. Examples may be seen above, and many others can be imagined (for example, the kinetic energy of a stream of particles entering a system, or energy from a laser beam adds to system energy, without either being either work-done or heat-added, in the classic senses).

$$\Delta E = W + Q + E^* \qquad ...(iii)$$

Where E^* in this general equation represents other additional advected energy terms not covered by work done on a system, or heat added to it.

Energy is also transferred from potential energy (E_p) to kinetic energy (E_k) and then back to potential energy constantly. This is referred to as conservation of energy. In this closed system, energy can not be created or destroyed, so the initial energy and the final energy will be equal to each other. This can be demonstrated by the following :

$$E_{pi} + E_{ki} = E_{pf} + E_{kf}$$

The equation can then be simplified further since $E_p = mgh$ (mass times acceleration due to gravity times the height) and $E_p = mv^2/2$ (half times mass times velocity squared). Then the total amount of energy can be found by adding $E_p + E_k = E_{total}$.

1.4. TYPES OF ENERGY

Energy comes in different forms—heat (thermal), light (radiant), mechanical, electrical, chemical, and nuclear energy. Energy is in everything. We use energy to do everything, from making a jump shot, to baking our favourite cookies, to sending astronauts into space—energy is there, making sure we have the power to do it all. There are two types of energy—stored (potential) energy and working (kinetic) energy. For example, the food we eat contains chemical energy, and our body stores this energy until we release it when we work or play. Energy can be classified into several types based on the following criteria:

- Primary and Secondary energy.
- Commercial and Non commercial energy.
- Renewable and Non-Renewable energy.

1.4.1. Primary and Secondary Energy

Primary energy sources are those that are either found or stored in nature. Primary energy can also be used directly. Some energy sources have non-energy uses, for example coal or natural gas can be used as a feedstock in fertilizer plants. Common primary energy sources are coal, oil, natural gas, and biomass (such as wood). Other primary energy sources available include nuclear energy from radioactive substances, thermal energy stored in earth's interior, and potential energy due to earth's gravity.

Primary energy sources are mostly converted in industrial utilities into *secondary energy* sources; for example coal, oil or gas converted into steam and electricity.

1.4.2. Commercial Energy and Non Commercial Energy

Commercial Energy

The energy sources that are available in the market for a definite price are known as commercial energy. Commercial energy forms the basis of industrial, agricultural, transport and commercial development in the modern world. In the industrialized countries, commercialized fuels are predominant source not only for economic production, but also for many household tasks of general population. Examples: Electricity, lignite, coal, oil, natural gas etc.

Non-Commercial Energy

The energy sources that are not available in the commercial market for a price are classified as non-commercial energy. Non-commercial energy sources include fuels such as firewood, cattle dung and agricultural wastes, which are traditionally gathered, and not bought at a price used especially in rural households. These are also called traditional fuels. Non-commercial energy is often ignored in energy accounting. Example: Firewood, agro waste in rural areas; solar energy for water heating, electricity generation, for drying grain, fish and fruits; animal power for transport, threshing, lifting water for irrigation, crushing sugarcane; wind energy for lifting water and electricity generation.

1.4.3. Renewable and Non-Renewable Energy

Renewable energy is energy obtained from sources that are essentially inexhaustible. The most important feature of renewable energy is that it can be harnessed without the release of harmful pollutants. Renewable energy sources include biomass, geothermal energy, hydropower, solar energy, and wind energy. They are called renewable energy sources because they are replenished in a short time. Day after day, the sun shines, the wind blows, and the rivers flow. We use renewable energy sources mainly to make electricity.

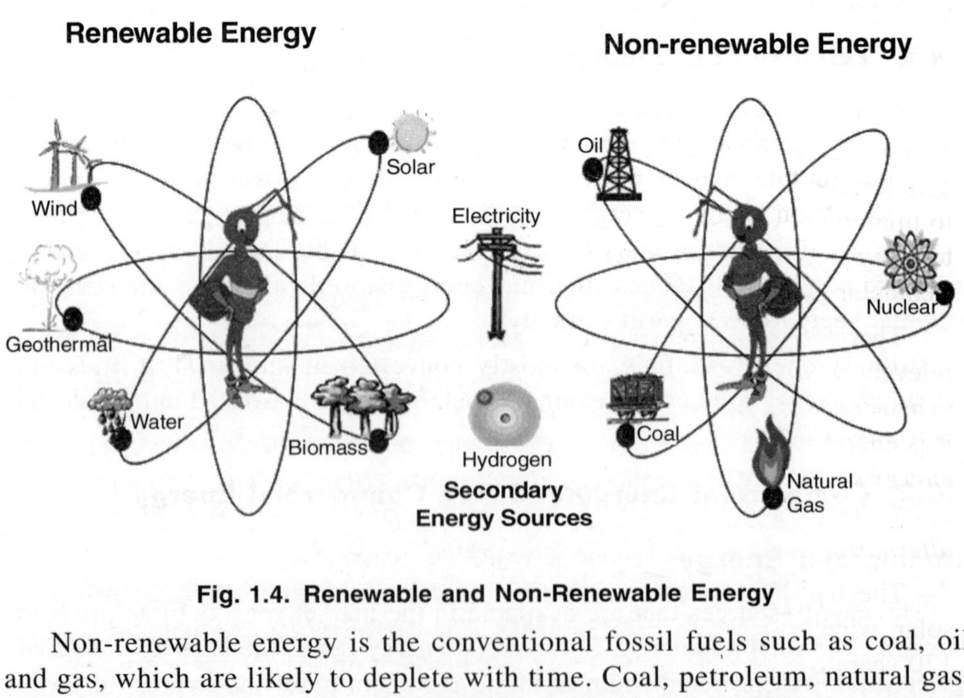

Fig. 1.4. Renewable and Non-Renewable Energy

Non-renewable energy is the conventional fossil fuels such as coal, oil and gas, which are likely to deplete with time. Coal, petroleum, natural gas, propane, and uranium are nonrenewable energy sources. They are used to make electricity, to heat our homes, to move our cars, and to manufacture all kinds of products.

These energy sources are called nonrenewable because their supplies are limited. Petroleum, for example, was formed millions of years ago from the remains of ancient sea plants and animals. We can't make more petroleum in a short time.

It was estimated by the Energy Information Administration that in 2006 primary sources of energy consisted of petroleum 36.8 per cent, coal 26.6 per cent, natural gas 22.9 per cent, amounting to an 86 per cent share for fossil fuels in primary energy production in the world. Non-fossil sources included hydroelectric 6.3 per cent, nuclear 6.0 per cent, and (geothermal, solar, tide, wind, wood, waste) amounting 0.9 per cent. World energy consumption was growing about 2.3 per cent per year.

Electricity and hydrogen are different from the other energy sources because they are **secondary** sources of energy. Secondary sources of energy—energy carriers—are used to store, move, and deliver energy in easily usable form. We have to use another energy source to make electricity or hydrogen.

All forms of energy are stored in different ways, in the energy sources that we use every day. These sources are divided into two groups — renewable (an energy source that can be replenished in a short period of time) and nonrenewable (an energy source that we are using up and cannot recreate in a short period of time). Renewable and nonrenewable energy sources can be used to produce secondary energy sources including electricity and hydrogen.

1.5. RENEWABLE SOURCES

In the past century, it has been seen that the consumption of non-renewable sources of energy has caused more environmental damage than any other human activity. Electricity generated from fossil fuels such as coal and crude oil has led to high concentrations of harmful gases in the atmosphere. This has in turn led to many problems being faced today such as ozone depletion and global warming. Vehicular pollution has also been a major problem.

Therefore, alternative sources of energy have become very important and relevant to today's world. These sources, such as the sun and wind, can never be exhausted and therefore are called renewable. Renewable or regenerative energy it is energy from sources which on human scales are inexhaustible. ***Renewable energy sources*** can be replenished in a short period of time. Renewable energy is an alternative to fossil fuels and nuclear power, and was commonly called *alternative energy* in the 1970s and 1980s.

The five renewable sources used most often include hydropower (water), solar, wind, geothermal, and biomass. The use of renewable energy is not new. 120 years ago, wood supplied up to 90 per cent of our energy needs.

Fig. 1.5. Wind Mill

Fig. 1.6. Solar Energy

They cause less emission and are available locally. Their use can, to a large extent, reduce chemical, radioactive, and thermal pollution. They stand out as a viable source of clean and limitless energy. These are also known as non-conventional sources of energy. Most of the renewable sources of energy are fairly non-polluting and considered clean though biomass, a renewable source, is a major polluter indoors.

Under the category of renewable energy or non-conventional energy are such sources as the sun, wind, water, agricultural residue, firewood, and animal dung. The non-renewable sources are the fossil fuels such as coal, crude oil, and natural gas. Energy generated from the sun is known as solar energy. Hydel is the energy derived from water. Biomass–firewood, animal dung, biodegradable waste from cities and crop residues is a source of energy when it is burnt. Geothermal energy is derived from hot dry rocks, magma, hot water springs, natural geysers, etc. Ocean thermal is energy derived from waves and also from tidal waves.

The main principle of use of renewable energy consists in its extraction from processes constantly occurring in an environment and granting for technical application. Streams of renewed energy which the person can use occur from various primary sources :

- Thermonuclear synthesis of the Sun.

- Radioactive disintegration of terrestrial elements.
- Movement of space bodies, in particular rotation of the Earth about the axis.

On the Earth energy from these primary sources is shown in the form of a wind power, hydro power, a tidal energy, a sunlight, geothermal energy and bio energy.

Historically, low fossil fuel prices, especially for natural gas, have made growth difficult for renewable fuels. In 2006, about 18 per cent of global final energy consumption came from renewable sources, with 13 per cent coming from traditional biomass, like wood-burning. Hydropower was the next largest renewable source, providing 3 per cent, followed by hot water/heating which contributed 1.3 per cent. Modern technologies, such as geothermal, wind, solar, and ocean energy together provided some 0.8 per cent of final energy consumption. The technical potential for their use is very large, exceeding all other readily available sources.

Thermonuclear synthesis of the Sun is a source of the majority of kinds of renewable energy, except for geothermal energy and a tidal energy. By calculations of the astronomers, the remained life expectancy of the Sun makes about five billions years so on human scales of the renewable energy occurring from the Sun, the exhaustion does not threaten.

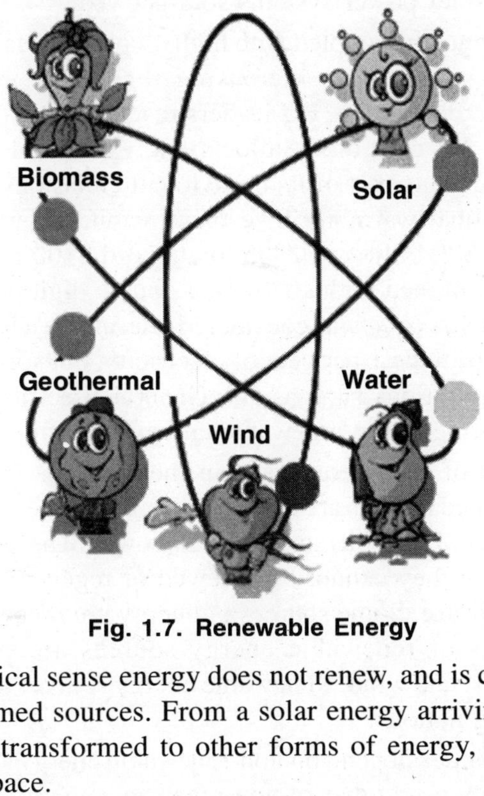

Fig. 1.7. Renewable Energy

In strictly physical sense energy does not renew, and is constantly withdrawn from the above-named sources. From a solar energy arriving to the Earth, only very small part is transformed to other forms of energy, and the biggest part simply leaves in space.

Extraction of mineral energy carriers, such as coal, oil, natural gas or peat is opposed to use of constant processes. In wide understanding they too is renewable, but not to measures of the person as their formation demands hundreds millions years, and their use passes much more quickly.

Renewable energy technologies are sometimes criticized for being unreliable or unsightly, yet the market is growing for many forms of renewable energy. Wind power has a worldwide installed capacity of 74,223 MW and is widely used in several European countries and the USA. The manufacturing output of the photovoltaics industry reached more than 2,000 MW per year in 2006, and PV power plants are particularly popular in Germany. Solar thermal power stations operate in the USA and Spain, and the largest of these is the 354 MW SEGS power plant in the Mojave Desert. The world's largest geothermal power installation is The Geysers in California, with a rated capacity of 750 MW. Brazil has one of the largest renewable energy programs in the world, involving production of ethanol fuel from sugar cane, and ethanol now provides 18 per cent of the country's automotive fuel. Ethanol fuel is also widely available in the USA.

While there are many large-scale renewable energy projects, renewable technologies are also suited to small off-grid applications, sometimes in rural and remote areas, where energy is often crucial in human development. Kenya has the world's highest household solar ownership rate with roughly 30,000 small (20–100 watt) solar power systems sold per year.

Climate change concerns coupled with high oil prices, peak oil and increasing government support are driving increasing renewable energy legislation, incentives and commercialization. EU leaders reached agreement in principle in March 2007 that 20 per cent of the bloc's energy should be produced from renewable fuels by 2020, as part of its drive to cut emissions of carbon dioxide, blamed in part for global warming. Investment capital flowing into renewable energy climbed from $80 billion in 2005 to a record $100 billion in 2006. This level of investment combined with continuing double digit percentage increases each year has moved what once was considered alternative energy to mainstream. Wind was the first to provide 1 per cent of electricity, but solar is not far behind. Some very large corporations such as BP, General Electric, Sharp, and Royal Dutch Shell are investing in the renewable energy sector.

Over 85 per cent of the energy used in the world is from non-renewable supplies. Most developed nations are dependent on non-renewable energy sources such as fossil fuels (coal and oil) and nuclear power. These sources are called non-renewable because they cannot be renewed or regenerated quickly enough to keep pace with their use. Some sources of energy are renewable or potentially renewable. Examples of renewable energy sources are: solar, geothermal, hydroelectric, biomass, and wind. Renewable energy sources are more commonly by used in developing nations.

Industrialized societies depend on non-renewable energy sources. Fossil fuels are the most commonly used types of non-renewable energy. They were formed when incompletely decomposed plant and animal matter was buried in the earth's crust and converted into carbon-rich material that is useable as fuel. This process occurred over millions of years. The three main types of fossil fuels are coal, oil, and natural gas. Two other less-used sources of fossil fuels are oil shales and tar sands.

Renewable energy's impact on the world's energy picture is significant. Many important events have occurred during the history of using renewable sources to generate electricity - but the overall use of these fuels has declined by about 17 per cent from their 1996 peak to about 6 quads in 2003.

The use of renewable energy is not new. Five generations (125 years) ago, wood supplied up to 90 per cent of our energy needs. Due to the convenience and low prices of fossil fuels, wood use has fallen. Now, the biomass which would normally present a disposal problem is converted into electricity (e.g., manufacturing wastes, rice hulls, and black liquor from paper production).

Historically, low fossil fuel prices, especially for natural gas, have made growth difficult for renewable fuels. The deregulation and restructuring of the electric power industry could have a major impact on renewable energy consumption. Demands for cheaper power in the short term would likely decrease demand for renewable energy, while preferences for renewables included in some versions of proposed electricity restructuring legislation would breathe new life into this industry.

Use of renewables in the United States is not currently expected to approach that of the major fuels, and due to their limitations (e.g., their intermittent nature - cloudy days have no solar gain, quiet days mean no wind blows to drive wind turbines, dams are primarily for flood control, so hydroelectricity production varies as dams' water levels change), renewables may never provide "the" answer to all energy problems. Around the world, renewable energy is proving to be of great value.

In 2003, about 6.2 Quadrillion Btu (Quads) of US energy came from renewable fuels. Each of the energy sources we use is measured, purchased, and sold in a different form. Many units of measurement are used to measure the energy we use each day. Learn more about converting energy units in the Units of Measurement section.

1.6. SOLAR ENERGY

Solar energy is the most readily available source of energy. It does not belong to anybody and is, therefore, free. It is also the most important of the non-conventional sources of energy because it is non-polluting and, therefore, helps in lessening the greenhouse effect.

NOTE *India receives solar energy equivalent to over 5000 trillion kWh/year, which is far more than the total energy consumption of the country.*

Solar power involves using solar cells to convert sunlight into electricity, using sunlight hitting solar thermal panels to convert sunlight to heat water or air, using sunlight hitting a parabolic mirror to heat water (producing steam), or using sunlight entering windows for passive solar heating of a building. It would be advantageous to place solar panels in the regions of highest solar radiation.

Solar energy has been used since prehistoric times, but in a most primitive manner. Before 1970, some research and development was carried out in a few countries to exploit solar energy more efficiently, but most of this work remained mainly academic. After the dramatic rise in oil prices in the 1970s, several countries began to formulate extensive research and development programmes to exploit solar energy.

When we hang out our clothes to dry in the sun, we use the energy of the sun. In the same way, solar panels absorb the energy of the sun to provide heat for cooking and for heating water. Such systems are available in the market and are being used in homes and factories.

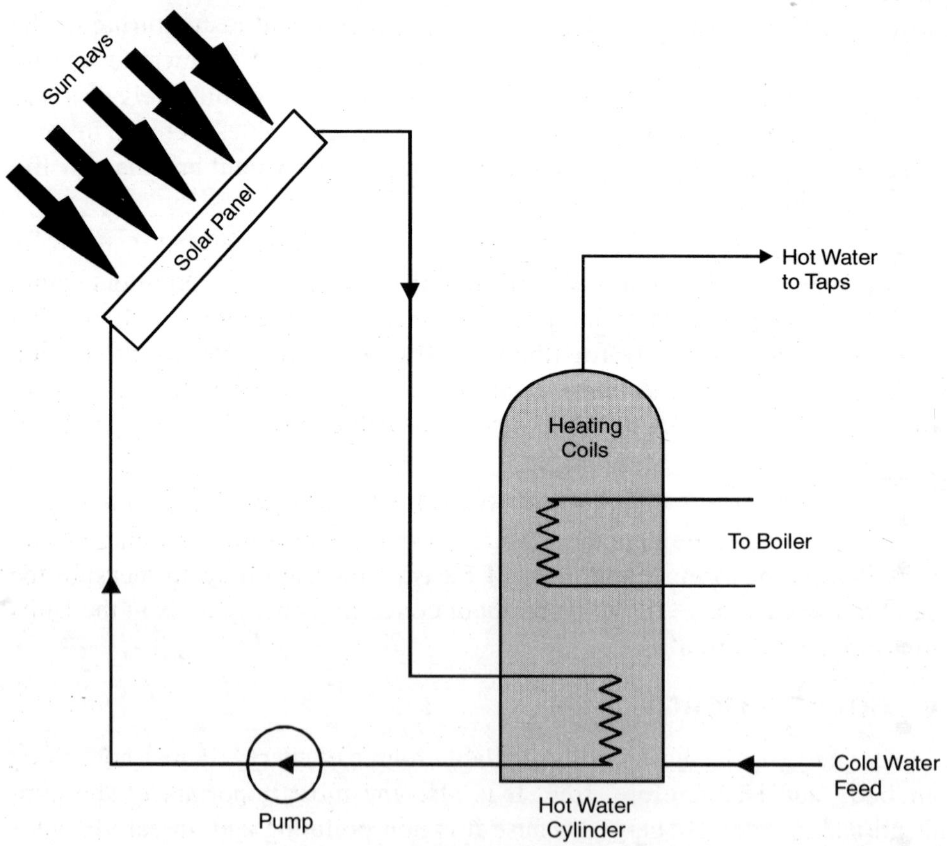

Fig. 1.8. Solar Energy

In the next few years it is expected that millions of households in the world will be using solar energy as the trends in USA and Japan show. In India too, the Indian Renewable Energy Development Agency and the Ministry of Non-Conventional Energy Sources are formulating a programme to have solar energy in more than a million households in the next few years. However, the people's initiative is essential if the programme is to be successful.

India is one of the few countries with long days and plenty of sunshine, especially in the Thar desert region. This zone, having abundant solar energy available, is suitable for harnessing solar energy for a number of applications. In areas with similar intensity of solar radiation, solar energy could be easily harnessed. Solar thermal energy is being used in India for heating water for both industrial and domestic purposes. A 140 MW integrated solar power plant is to be set up in Jodhpur but the initial expense incurred is still very high.

Solar energy can also be used to meet our electricity requirements. Through Solar Photovoltaic (SPV) cells, solar radiation gets converted into DC electricity directly. This electricity can either be used as it is or can be stored in the battery. This stored electrical energy then can be used at night. SPV can be used for a number of applications such as :

(*a*) Domestic lighting

(*b*) Street lighting

(*c*) Village electrification

(*d*) Water pumping

(*e*) Desalination of salty water

(*f*) Powering of remote telecommunication repeater stations and

(*g*) Railway signals.

If the means to make efficient use of solar energy could be found, it would reduce our dependence on non-renewable sources of energy and make our environment cleaner.

Advantages

- Solar power imparts no fuel costs.
- Solar power is a renewable resource. As long as the Sun exists, its energy will reach Earth.
- Solar power generation releases no water or air pollution, because there is no combustion of fuels.
- In sunny countries, solar power can be used in remote locations, like a wind turbine. This way, isolated places can receive electricity, when there is no way to connect to the power lines from a plant.
- Solar energy can be used very efficiently for heating (solar ovens, solar water and home heaters) and daylighting.
- Coincidentally, solar energy is abundant in regions that have the largest number of people living off grid—in developing regions of Africa, Indian subcontinent and Latin America. Hence cheap solar, when available,

opens the opportunity to enhance global electricity access considerably, and possibly in a relatively short time period.

- Photovoltaics are much more efficient in their conversion of solar energy to usable energy than biofuel from plant materials.

Disadvantages

- Solar electricity is currently more expensive than grid electricity.
- Solar heat and electricity are not available at night and may be unavailable due to weather conditions; therefore, a storage or complementary power system is required for off-the-grid applications.
- Solar cells produce DC which must be converted to AC (using a grid tie inverter) when used in currently existing distribution grids. This incurs an energy loss of 4–12 per cent.
- The energy payback time—the time necessary for producing the same amount of energy as needed for building the power device — for photovoltaic cells is about 1–5 years, depending primarily on location.

1.7. WIND ENERGY

Fig. 1.9. Wind Mills

Wind energy is the kinetic energy associated with the movement of atmospheric air. It has been used for hundreds of years for sailing, grinding grain, and for irrigation. Wind energy systems convert this kinetic energy to more useful forms of power. Wind energy systems for irrigation and milling have been in use since ancient times and since the beginning of the 20th century it is being used to generate electric power. Windmills for water pumping have been installed in many countries particularly in the rural areas.

Wind turbines transform the energy in the wind into mechanical power, which can then be used directly for grinding etc. or further converting to electric power to generate electricity. Wind turbines can be used singly or in clusters called 'wind farms'. Small wind turbines called aero-generators can be used to charge large batteries.

Five nations—Germany, USA, Denmark, Spain and India–account for 80 per cent of the world's installed wind energy capacity. Wind energy continues to be the fastest growing renewable energy source with worldwide wind power installed capacity reaching 14,000 MW.

NOTE ***India ranks 5th in the world with a total wind power capacity if 1080 MW out of which 1025 MW have been established in commercial projects.***

Realizing the growing importance of wind energy, manufacturers have steadily been increasing the unit size of the wind electric generators since the late 1980s. Another important development has been the offshore (*i.e.* in the sea) wind farms in some regions of Europe, which have several advantages over the on-shore ones. The third major development has been the use of new techniques to assess the wind resource for techno-commercial viability.

In India the states of Tamilnadu and Gujarat lead in the field of wind energy. At the end of March 2000 India had 1080-MWs capacity wind farms, of which Tamilnadu contributed 770-MW capacity. Gujarat has 167MW followed by Andhra Pradesh, which has 88 MW installed wind farms. There are about a dozen wind pumps of various designs providing water for agriculture, afforestation, and domestic purposes, all scattered over the country.

The design of the Auroville multi-blade windmill has evolved from the practical experience gained in operating these mills over a period of 20 years or so. It has a high tripod tower and its double-action pump increases water output by about 60 per cent compared to the conventional single-action pumps.

1.8. HYDEL ENERGY

1.8.1. Energy from Water Sources

The energy in the flowing water can be used to produce electricity. Waves result from the interaction of the wind with the surface of the sea and represent a transfer of energy from the wind to the sea. Energy can be extracted from tides by creating a reservoir or basin behind a barrage and then passing tidal waters through turbines in the barrage to generate electricity.

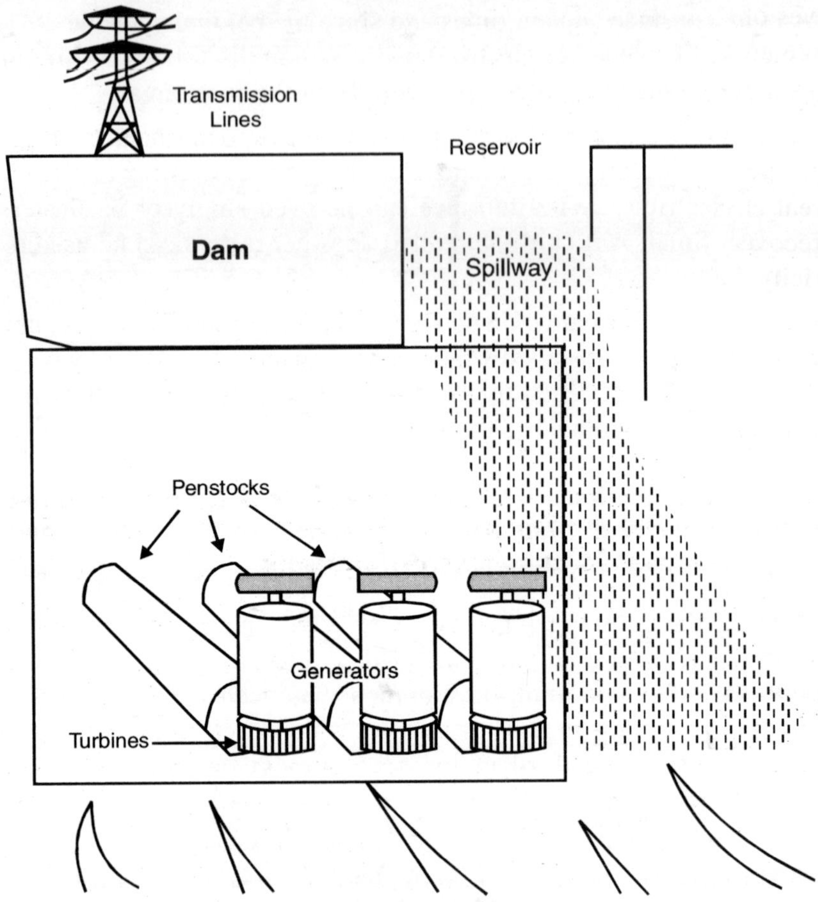

Fig. 1.10. Hydel Energy

> NOTE
> ***On an average, the 60 million sq km of tropical seas absorb solar radiation equal to the heat content of 245 billion barrels of oil.***

1.8.2. Mini or Micro Hydro Power

Hydro power is one of the best, cheapest, and cleanest source of energy, although, with big dams, there are many environmental and social problems as has been seen in the case of the Tehri and the Narmada Projects. Small dams are, however, free from these problems. This is in fact one of the earliest known renewable energy sources, in the country (since the beginning of the 20th century).

In fact, for the last few hundred years, people living in the hills of the

Himalayas have been using water mills, or *chakki*, to grind wheat. The 130 KW small hydropower plant in Darjeeling set up in 1897, was the first in India. Besides being free from the problem of pollution, such plants are also free from issues and controversies that are associated with the bigger projects, namely affecting the lives of thousands of people living along the banks of the rivers, destruction of large areas under forest, and seismological threats.

New environmental laws affected by the danger of global warming have made energy from small hydropower plants more relevant. These small hydropower plants can serve the energy needs of remote rural areas independently. The real challenge in a remote area lies in successful marketing of the energy and recovering the dues. Local industries should be encouraged to use this electricity for sustainable development.

It is a technology with enormous potential, which could exploit the water resources to supply energy to remote rural areas with little access to conventional energy sources. It also eliminates most of the negative environmental effects associated with large hydro projects.

1.9. ENERGY FROM THE SEA—OCEAN THERMAL, TIDAL AND WAVE ENERGY

Tidal power can be extracted from Moon-gravity-powered tides by locating a water turbine in a tidal current, or by building impoundment pond dams that admit-or-release water through a turbine. The turbine can turn an electrical generator, or a gas compressor, that can then store energy until needed. Coastal tides are a source of clean, free, renewable, and sustainable energy.

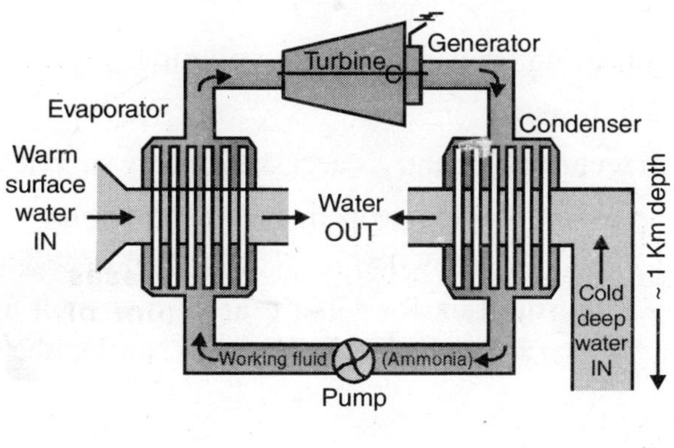

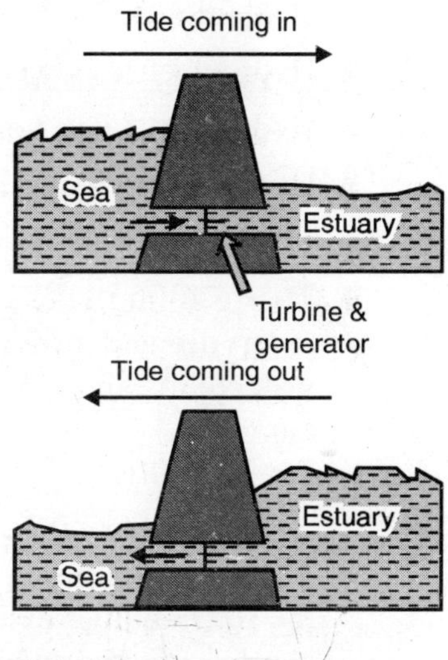

Fig. 1.11. Energy from the Sea

Large amounts of solar energy are stored in the oceans and seas. On an average, the 60 million square kilometre of the tropical seas absorb solar radiation equivalent to the heat content of 245 billion barrels of oil. Scientists feel that if this energy can be tapped a large source of energy will be available to the tropical countries and to other countries as well. The process of harnessing this energy is called OTEC (ocean thermal energy conversion). It uses the temperature differences between the surface of the ocean and the depths of about 1000m to operate a heat engine, which produces electric power.

Energy is also obtained from waves and tides. In India, the first wave energy, project with a capacity of 150MW, has been set up at Vizhinjam near Trivandrum. A major tidal wave power project costing of Rs.5000 crores, is proposed to be set up in the Hanthal Creek in the Gulf of Kutch in Gujarat.

In some countries such as Japan small scale power generators run by energy from waves or the ocean, have been used as power sources for channel marking buoys.

1.10. HYDROELECTRIC ENERGY

In hydro energy, the gravitational descent of a river is compressed from a long run to a single location with a dam or a flume. This creates a location where concentrated pressure and flow can be used to turn turbines or water wheels, which drive a mechanical mill or an electric generator.

Advantages

- Hydroelectric power stations can promptly increase to full capacity, unlike other types of power stations. This is because water can be accumulated above the dam and released to coincide with peak demand.
- Electricity can be generated constantly, so long as sufficient water is available.
- Hydroelectric power produces no primary waste or pollution.
- Hydropower is a renewable resource.
- Hydroelectricity assists in securing a country's access to energy supplies.
- Much hydroelectric capacity is still undeveloped, such as in Africa.
- The resulting lake can have additional benefits such as doubling as a reservoir and providing opportunities for leisure activities such as watersports and fishing as is the case with Kielder Water in Northumberland, UK.

Disadvantages

- The construction of a dam can have a serious environmental impact on the surrounding areas. The amount and the quality of water downstream can be affected, which affects plant life both aquatic, and land-based. Because a river valley is being flooded, the local habitats of many species are destroyed, while people living nearby may have to relocate their homes.

- Hydroelectricity can only be used in areas where there is a sufficient and continuing supply of water.
- Flooding submerges large forests (if they have not been harvested). The resulting anaerobic decomposition of the carboniferous materials releases methane, a greenhouse gas.
- Dams can contain huge amounts of water. As with every energy storage system, failure of containment can lead to catastrophic results, e.g. flooding.
- Dams create large lakes that may have adverse effects on Earth tectonic system causing intense earthquakes.
- Hydroelectric plants rarely can be erected near load centers, requiring long transmission lines.

1.11. GEOTHERMAL ENERGY

We live between two great sources of energy, the hot rocks beneath the surface of the earth and the sun in the sky. Our ancestors knew the value of geothermal energy; they bathed and cooked in hot springs. Today we have recognized that this resource has potential for much broader application.

Geothermal energy harnesses the heat energy present underneath the Earth. Two wells are drilled. One well injects water into the ground to provide water. The hot rocks heat the water to produce steam. The steam that shoots back up the other hole(s) is purified and is used to drive turbines, which power electric generators. When the water temperature is below the boiling point of water a binary system is used. A low boiling point liquid is used to drive a turbine and generator in a closed system similar to a refrigeration unit running in reverse.

The core of the earth is very hot and it is possible to make use of this geothermal (in Greek it means heat from the earth) energy. These are areas where there are volcanoes, hot springs, and geysers, and methane under the water in the oceans and seas. In some countries, such as in the USA, water is pumped from underground hot water deposits and used to heat people's houses.

The utilization of geothermal energy for the production of electricity dates back to the early part of the twentieth century. For 50 years the generation of electricity from geothermal energy was confined to Italy and interest in this technology was slow to spread elsewhere. In 1943, the use of geothermal hot water was pioneered in Iceland.

NOTE ***Geothermal manifestations are wide spread in India in the form of 340 hot spring sites.***

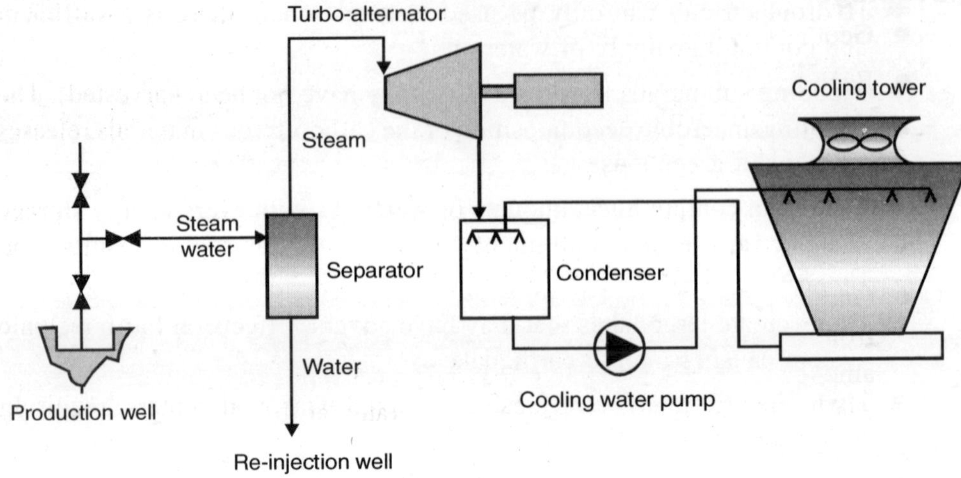

(a)

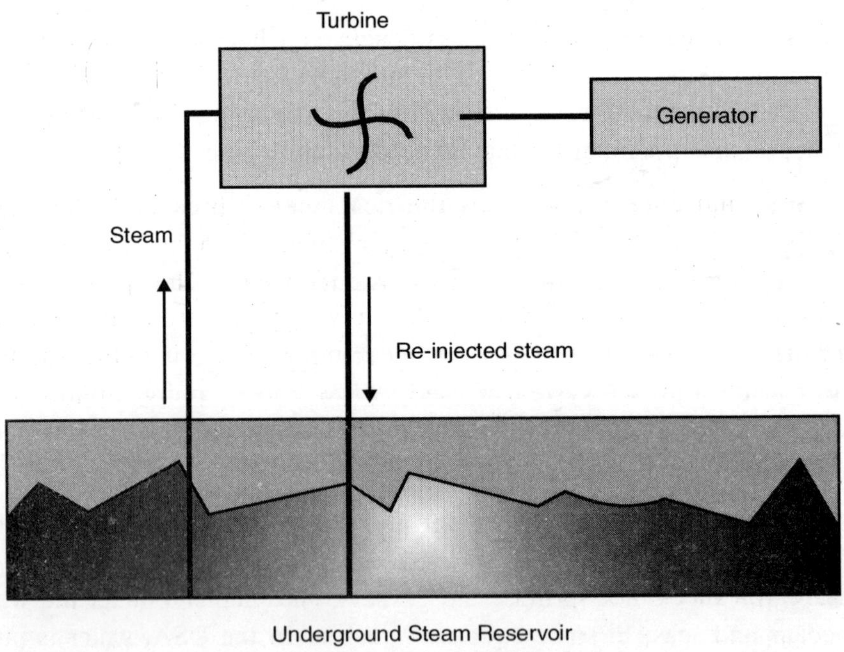

(b)

Fig 1.12. Generation of Geothermal Energy

In India, Northwestern Himalayas and the western coast are considered geothermal areas. The Geological Survey of India has already identified more than 350 hot spring sites, which can be explored as areas to tap geothermal energy. Satellites like the IRS-1 have played an important role, through infrared photographs of the ground, in locating geothermal areas. The Puga valley in the Ladakh region has the most promising geothermal field. An experimental 1-kW generator is already in operation in this area. It is being used mainly for poultry farming, mushroom cultivation, and pashmina-wool processing, all of which need higher temperature.

Advantages

- Geothermal energy is base load power.
- Economically feasible in high grade areas now.
- Low deployment costs.
- Geothermal power plants have a high capacity factor; they run continuously day and night with an uptime typically exceeding 95 per cent.
- Once a geothermal power station is implemented, the energy produced from the station is practically free, minus maintenance costs. A small amount of energy is required in order to run a pump, although this pump can be powered by excess energy generated at the plant.
- Geothermal power stations are relatively small, and have a lesser impact on the environment than tidal or hydroelectric plants. Because geothermal technology does not rely on large bodies of water, but rather, small, but powerful jets of water, like geysers, large generating stations can be avoided without losing functionality.
- Geothermal is now feasible in areas where the earth's crust is thicker. Using enhanced geothermal technology, it's possible to drill deeper and inject water to generate geothermal power.
- Geothermal energy does not produce air or water pollution if performed correctly.

Disadvantages

- Geothermal power extracts small amounts of minerals such as sulfur that are removed prior to feeding the turbine and re-injecting the water back into the injection well.
- Geothermal power requires locations that have suitable subterranean temperatures within 5 km of surface.
- Some geothermal stations have created geological instability, even causing earthquakes strong enough to damage buildings.

1.12. BIOMASS

Biomass is a renewable energy resource derived from the carbonaceous waste of various human and natural activities. It is derived from numerous sources, including the by-products from the timber industry, agricultural crops, raw material from the forest, major parts of household waste and wood.

Biomass is a biological, organic material cultivated from renewable sources such as agricultural waste and trees that can be utilized for fuel. Bio-energy Solutions focuses on wood products and agricultural waste as the most efficient, cost effective sources of biomass for energy production.

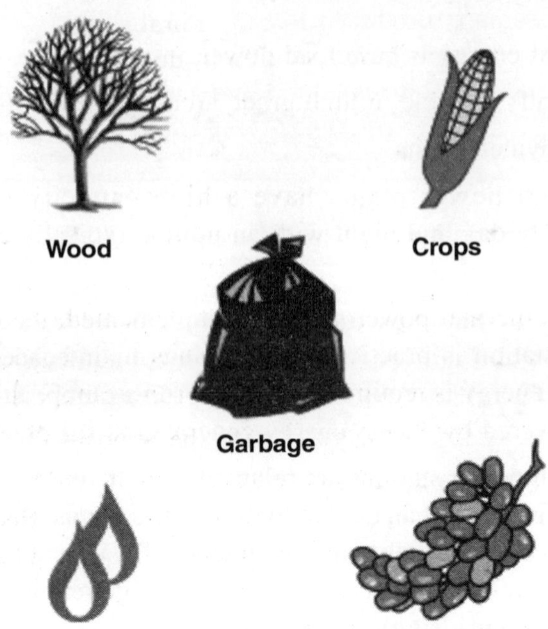

Fig. 1.13. Types of Biomass

Biomass production involves using garbage or other renewable resources such as corn or other vegetation to generate electricity. When garbage

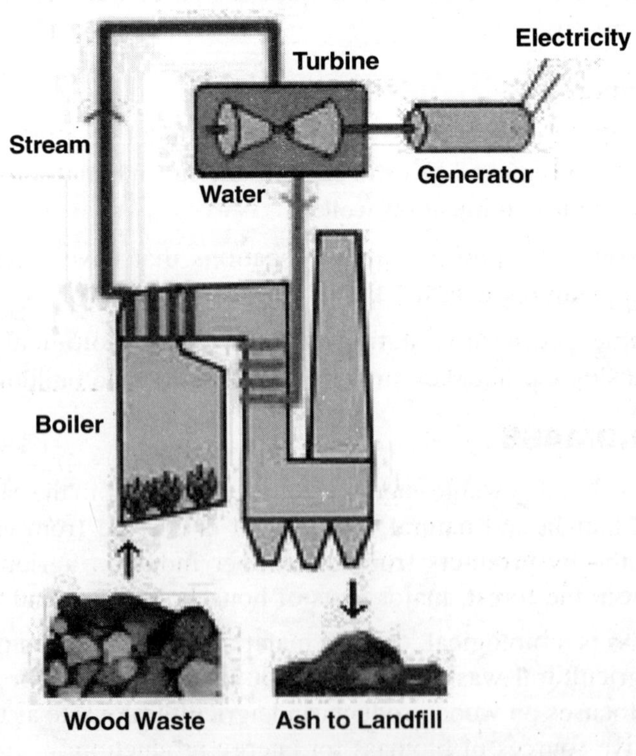

Fig. 1.14. Electricity Generation from Biomass

decomposes, the methane produced is captured in pipes and later burned to produce electricity. Vegetation and wood can be burned directly to generate energy, like fossil fuels, or processed to form alcohols.

When burned, the carbon dioxide that was absorbed while the plant was growing is released back into the atmosphere upon combustion, making it carbon neutral. Additionally, when the biomass is produced and harvested sustainably, there is no net increase in CO_2 as a result of its use.

Vegetable oil is generated from sunlight and CO_2 by plants. It is safer to use and store than gasoline or diesel as it has a higher flash point. Straight vegetable oil works in diesel engines if it is heated first. Vegetable oil can also be transesterified to make biodiesel, which burns like normal diesel.

Biomass does not add carbon dioxide to the atmosphere as it absorbs the same amount of carbon in growing as it releases when consumed as a fuel. Its advantage is that it can be used to generate electricity with the same equipment or power plants that are now burning fossil fuels. Biomass is an important source of energy and the most important fuel worldwide after coal, oil and natural gas.

Traditional use of biomass is more than its use in modern application. In the developed world, biomass is again becoming important for applications such as combined heat and power generation. In addition, biomass energy is gaining significance as a source of clean heat for domestic heating and community heating applications. In fact in countries like Finland, USA and Sweden the per capita biomass energy used is higher than it is in India, China or in Asia.

NOTE ***Half a kilo of dry plant tissue can produce as much as 1890 KCal of heat which is equivalent to the heat available from a quarter of kilogram of coal.***

Biomass fuels used in India account for about one third of the total fuel used in the country, being the most important fuel used in over 90 per cent of the rural households and about 15 per cent of the urban households.

Instead of burning the loose biomass fuel directly, it is more practical to compress it into briquettes (compressing them through a process to form blocks of different shapes) and thereby improve its utility and convenience of use. Such biomass in the dense briquetted form can either be used directly as fuel instead of coal in the traditional chulhas and furnaces or in the gasifier. Gasifier converts solid fuel into a more convenient-to-use gaseous form of fuel called producer gas.

Scientists are trying to explore the advantages of biomass energy as an alternative energy source as it is renewable and free from net CO_2 (carbon dioxide) emissions, and is abundantly available on earth in the form of agricultural residue, city garbage, cattle dung, firewood, etc. Bio-energy, in the form of biogas, which is derived from biomass, is expected to become one of the key energy resources for global sustainable development.

At present, biogas technology provides an alternative source of energy in rural India for cooking. It is particularly useful for village households that have their own cattle. Through a simple process, cattle dung is used to produce a gas, which serves as fuel for cooking. The residual dung is used as manure.

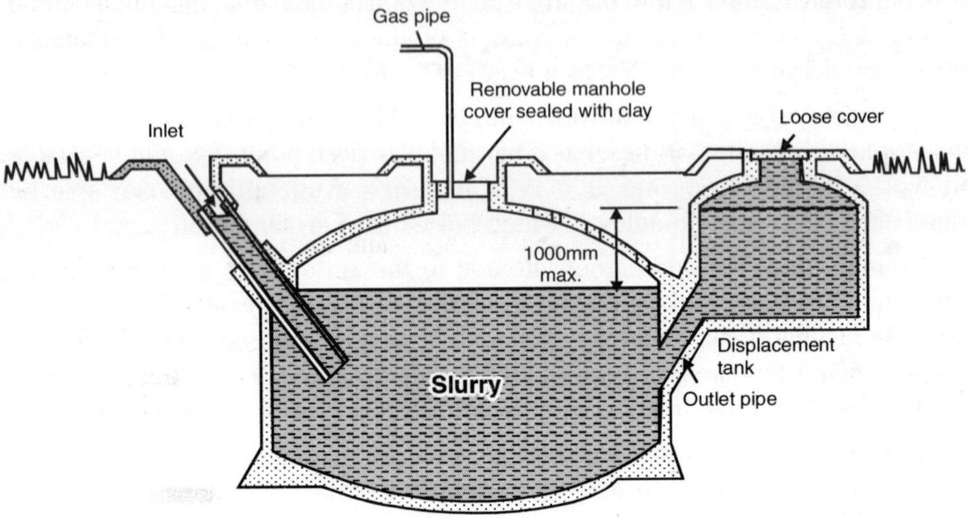

Fig. 1.15. Biogas Digester

Biogas plants have been set up in many areas and are becoming very popular. Using local resources, namely cattle waste and other organic wastes, energy and manure are derived. A mini biogas digester has recently been designed and developed, and is being in-field tested for domestic lighting.

Indian sugar mills are rapidly turning to bagasse, the leftover of cane after it is crushed and its juice extracted, to generate electricity. This is mainly being done to clean up the environment, cut down power costs and earn additional revenue. According to current estimates, about 3500 MW of power can be generated from bagasse in the existing 430 sugar mills in the country. Around 270 MW of power has already been commissioned and more is under construction.

Advantages

- Biomass production can be used to burn organic waste products resulting from agriculture. This type of recycling encourages the philosophy that nothing on this Earth should be wasted. The result is less demand on the Earth's resources, and a higher carrying capacity for Earth because non-renewable fossil fuels are not consumed.
- Biomass is abundant on Earth and is generally renewable. In theory, we will never run out of organic waste products as fuel, because we are continuously producing them. In addition, biomass is found throughout the world, a fact that should alleviate energy pressures in third world nations.

- When methods of biomass production other than direct combustion of plant mass are used, such as fermentation and pyrolysis, there is little effect on the environment. Alcohols and other fuels produced by these alternative methods are clean burning and are feasible replacements to fossil fuels.
- Since CO_2 is first taken out of the atmosphere to make the vegetable oil and then put back after it is burned in the engine, there is no net increase in CO_2.
- By combining the use of biomass with geo-sequestration of CO_2 then this could result in a net decrease of CO_2 in the atmosphere.
- Vegetable oil has a higher flash point and therefore is safer than most fossil fuels.
- Transitioning to vegetable oil could be relatively easy as biodiesel works where diesel works, and straight vegetable oil takes relatively minor modifications.
- The World already produces more than 100 billion gallons a year for food industry, so we have experience making it.
- Algaculture has the potential to produce far more vegetable oil per acre than current plants.
- Infrastructure for biodiesel around the World is significant and growing.

Disadvantages

- Direct combustion of any carbon-based fuel leads to air pollution similar to that from fossil fuels.
- Some researchers claim that when biomass crops are the product of intensive farming, ethanol fuel production results in a net loss of energy after one accounts for the fuel costs of petroleum and natural-gas fertilizer production, farm equipment, and the distillation process.
- There is a long list of reasons why even non-food based cellulosic ethanol cannot solve our energy crisis or global warming problems.
- Direct competition with land use for food production and water use.
- Current production methods would require enormous amounts of land to replace all gasoline and diesel. With current technology, it is not feasible for biofuels to replace the demand for petroleum.
- Even with the most-optimistic current energy return on investment claims, in order to use 100 per cent solar energy to grow corn and produce ethanol (fueling machinery with ethanol, distilling with heat from burning crop residues, using NO fossil fuels at all), the consumption of ethanol to replace only the current U.S. petroleum use would require three quarters of all the cultivated land on the face of the Earth.

1.13. BIODIESEL—A RENEWABLE FUEL

1.13.1. Biodiesel Made From Vegetable Oils and Animal Fats

Biodiesel is a renewable fuel that can be used instead of diesel fuel made from petroleum. Biodiesel can be made from vegetable oils, animal fats, or greases. Most biodiesel today is made from soyabean oil. About half of biodiesel producers are able to make biodiesel from used oils or fats, including recycled restaurant grease.

Biodiesel is most often blended with petroleum diesel in ratios of 2 per cent (B2), 5 per cent (B5), or 20 per cent (B20). It can also be used as pure biodiesel (B100). Biodiesel fuels can be used in regular diesel vehicles without making any changes to the engines. It can also be stored and transported using diesel tanks and equipment.

Fueling engines with biodiesel has just started to catch on, but this isn't a new idea. Before petroleum diesel fuel became popular, Rudolf Diesel, the designer of the diesel engine, experimented with using vegetable oil (biodiesel) as fuel.

1.13.2. Biodiesel as a Transportation Fuel

Most trucks, buses, and tractors in the United States use diesel fuel. Diesel is a nonrenewable fuel made from petroleum. Using biodiesel means that we use a little bit less petroleum. Biodiesel results in less pollution than petroleum diesel. Any vehicle that operates on diesel fuel can switch to biodiesel without changes to its engine.

Because it is so clean burning and easy to use, biodiesel is the fastest growing and most cost efficient fuel for fleet vehicles. Many school districts are switching to biodiesel blends for their school buses. Biodiesel is also being used for fleets of snowplows, garbage trucks, mail trucks, and military vehicles. So far, the use of biodiesel has been limited to fleets of vehicles that have their owner fueling stations. As the number of public fueling stations that offer biodiesel grows, it may become more popular with individual consumers.

B100 and biodiesel blends are sensitive to cold weather and may require special anti-freeze, just like petroleum-based diesel fuel does. Biodiesel acts like a detergent additive, loosening and dissolving sediments in storage tanks. Because biodiesel is a solvent, B100 may cause rubber and other components to fail in older vehicles. This problem does not occur with biodiesel blends.

1.13.3. Biodiesel and The Environment

Biodiesel is renewable, nontoxic, and biodegradable. Compared to diesel, biodiesel, is significantly cleaner burning. It produces fewer air pollutants, like particulates, carbon monoxide, hydrocarbons, and air toxics. It does slightly increase emissions of nitrogen oxides, though. Biodiesel produces less black smoke, and smells better, too. Sometimes biodiesel smells like french fries!

Regular diesel fuel contains sulfur. Sulfur can cause damage to the environment when it is burned in fuels. New environmental laws will require the amount of sulfur in diesel fuel to be dramatically reduced over the next few years. When sulfur is removed from regular diesel fuel, the fuel doesn't work as well. Adding a small amount of biodiesel can fix the problem. Biodiesel has no sulfur, so it can reduce sulfur levels in the nation's diesel fuel supply while making engines run more smoothly.

1.14. FUEL CELLS

Fuel cells are electrochemical devices that convert the chemical energy of a fuel directly and very efficiently into electricity (DC) and heat, thus doing away with combustion. The most suitable fuel for such cells is hydrogen or a mixture of compounds containing hydrogen. A fuel cell consists of an electrolyte sandwiched between two electrodes. Oxygen passes over one electrode and hydrogen over the other, and they react electrochemically to generate electricity, water, and heat.

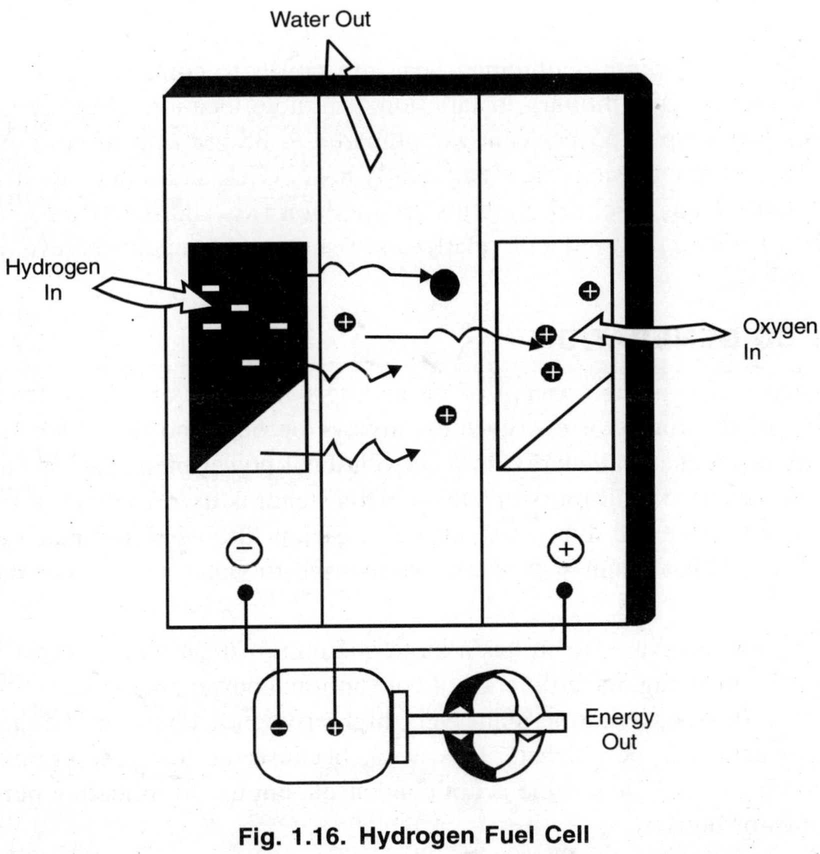

Fig. 1.16. Hydrogen Fuel Cell

Though fuel cells have been used in space flights and combined supplies of heat and power, electric vehicles are the best option available to dramatically reduce urban air pollution. Compared to vehicles powered by the internal combustion engine, fuel-cell powered vehicles have very high energy conversion efficiency, (almost double that of currently used engines) and near-zero pollution, $CO_2$and water vapour being the only emissions. Fuel-cell-powered EV's (electric vehicles) score over battery operated EV's in terms of increased efficiency and easier and faster refuelling.

Canada's Ballad's Power Systems and Germany's Dailmer-Benz are world leaders in the application of fuel cell technology for meeting transportation needs. In India, diesel-run buses are a major means of transport and these emit significant quantities of SPM and SO_2. Thus, fuel-cell powered buses could be introduced with relative ease and yet make a positive impact on urban air quality. Such buses are already in operation in Vancouver in Canada and in Illinois and California in the USA. Though rapid progress has been made, high initial cost is still the biggest hurdle in the widespread commercialization of fuel cells.

Fuel Cells for Power Generation : India has a large gap between the demand for and supply of power. Conventional large-scale power plants use non-renewable fuels with significant adverse ecological and environmental impacts. Fuel cell systems are excellent candidates for small-scale decentralized power generation.

Fuel cells can supply combined heat and power to commercial buildings, hospitals, airports and military installation at remote locations. Fuel cells have efficiency levels up to 55 per cent as compared to 35 per cent of conventional power plants. The emissions are significantly lower (CO_2 and water vapour being the only emissions). Fuel cell systems are modular (i.e. additional capacity can be added whenever required with relative ease) and can be set up wherever power is required.

1.15. CO-GENERATION

Co-generation is the concept of producing two forms of energy from one fuel. One of the forms of energy must always be heat and the other may be electricity or mechanical energy. In a conventional power plant, fuel is burnt in a boiler to generate high-pressure steam. This steam is used to drive a turbine, which in turn drives an alternator through a steam turbine to produce electric power. The exhaust steam is generally condensed to water which goes back to the boiler.

As the low-pressure steam has a large quantum of heat which is lost in the process of condensing, the efficiency of conventional power plants is only around 35 per cent. In a cogeneration plant, very high efficiency levels, in the range of 75–90 per cent, can be reached. This is so, because the low-pressure exhaust steam coming out of the turbine is not condensed, but used for heating purposes in factories or houses.

Since co-generation can meet both power and heat needs, it has other advantages as well in the form of significant cost savings for the plant and reduction in emissions of pollutants due to reduced fuel consumption.

Even at conservative estimates, the potential of power generation from co-generation in India is more than 20,000 MW. Since India is the largest producer of sugar in the world, bagasse-based cogeneration is being promoted. The potential for cogeneration thus lies in facilities with joint requirement of heat and electricity, primarily sugar and rice mills, distilleries, petrochemical sector and industries such as fertilizers, steel, chemical, cement, pulp and paper, and aluminum.

Fig. 1.17. Co-generation

1.16. FOSSIL FUEL

When estimating the world's fossil fuel supply, experts distinguish between *reserves* and *resources*. Reserves are fossil fuel deposits that have already been discovered and are immediately available. Resources are fossil fuel deposits that geologists believe are located in certain sedimentary basins, but have not yet been discovered.

Fossil fuel reserves can be further divided into *proved reserves* and *inferred reserves*. Proved reserves are deposits that have been measured, sampled, and evaluated for production quality. Inferred reserves have been discovered but have not been measured or evaluated.

Fossil fuels are formed from organic material, both plant and animal remains, that have settled to the bottom of oceans. The immense pressure of the water changes this organic matter into liquid and gaseous hydrocarbons. This process is known as catagenesis. Where the bulk of this organic material of plant matter, the process tends to form coal. Many of the coal fields date to one specific in history, which has been named the Carboniferous era.

Fossil fuels can be burned to produce significant amounts of energy. Consequently they have been commercially exploited in the last few hundred years.

Fossil fuels, in terms of energy, involve the burning of coal or hydrocarbon fuels, which are the remains of the decomposition of plants and animals. Fossil fuels or mineral fuels are fuels formed by natural resources such as anaerobic decomposition of buried dead organisms. The age of the organisms and their resulting fossil fuels is typically millions of years, and sometimes exceeds 650 million years. These fuels contain high percentage of carbon and hydrocarbons.

Energy is crucial for sustainable development, poverty eradication and meeting the increasing demand on energy and avoiding gaps between needs and availability; increasing access; enhancing efficiency; strengthening capacity building; the need for adequate predictable new and additional financial resources, technology transfer and political will, as well as commitment to innovative ways of applying energy efficient, environmentally sound, cost effective and socially acceptable technologies and systems; education and awareness-raising, energy regulations and enforcement and changing unsustainable patterns of consumption and production, with developed countries taking the lead.

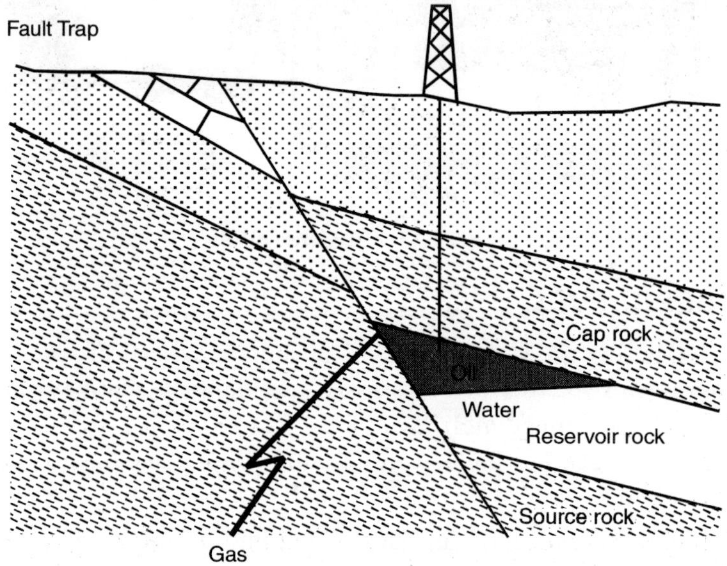

Fig. 1.18. Fossil Fuel

The Industrial Revolution in Europe in the 19th century fired man's search for alternative sources of fuel to meet energy needs of the mushrooming industries. With the realization that fossil fuels could meet this requirement, the energy needs of the world were fulfilled for the time being.

Fossil fuels are called so because they have been derived from fossils, which were formed millions of years ago during the time of the dinosaurs. They are fossilized organic remains that over millions of years have been converted to oil, gas, and coal. Because their formation takes so long, these sources are also called non-renewable.

These fuels are made up of decomposed plant and animal matter. When plants, dinosaurs, and other ancient creatures died, they decomposed and were buried, layer upon layer under the ground. Their decomposed remains gradually changed over the years. It took millions of years to form these layers into a hard, black rock-like substance called coal, a thick liquid called oil or petroleum, and natural gas—the three major forms of fossil fuels.

Fossil fuels are usually found below ground. Coal is either mined or dug out while oil and natural gas are pumped out. Coal is widely distributed and is easier to locate than oil and gas.

The Industrial Revolution was based on the availability of fossil fuels. The invention of the internal combustion engine allowed machines to manufacture consumer goods. Trains, cars and lorries followed with the resulting demand for petrol and diesel. Coal was used for heating and cooking, and then used in power stations to generate electricity.

Fossil fuels are formed by the anaerobic decomposition of remains of organisms including phytoplankton and zooplankton that settled to the sea (or lake) bottom in large quantities under anoxic conditions, millions of years ago. Over geological time, this organic matter, mixed with mud, got buried under heavy layers of sediment. The resulting high levels of heat and pressure caused the organic matter to chemically alter, first into a waxy material known as kerogen which is found in oil shales, and then with more heat into liquid and gaseous hydrocarbons in a process known as catagenesis.

Terrestrial plants, on the other hand, tend to form coal. Many of the coal fields date to the Carboniferous period of Earth's history. Terrestrial plants also form type III kerogen, a source of natural gas.

There are three main types of fossil fuels: coal, petroleum, and natural gas. Another fossil fuel, Liquefied Petroleum Gas (LPG), is principally derived from the production of natural gas. Heat from burning fossil fuel is used either directly for space heating and process heating, or converted to mechanical energy for vehicles, industrial processes, or electrical power generation.

Fossil fuels range from volatile materials with low carbon:hydrogen ratios like methane, to liquid petroleum to nonvolatile materials composed of almost pure carbon, like anthracite coal. Methane can be found in hydrocarbon fields, alone, associated with oil, or in the form of methane clathrates. It is generally accepted that they formed from the fossilized remains of dead plants and animals by exposure to heat and pressure in the Earth's crust over hundreds of millions of years. This biogenic theory was first introduced by Georg Agricola in 1556 and later by Mikhail Lomonosov in the 18th century.

Fossil fuels are non-renewable resources because they take millions of years to form, and reserves are being depleted much faster than new ones are being formed. The production and use of fossil fuels raise environmental concerns. A global movement toward the generation of renewable energy is therefore under way to help meet increased energy needs.

The burning of fossil fuels produces around 21.3 billion tonnes (21.3 gigatonnes) of carbon dioxide per year, but it is estimated that natural processes can only absorb about half of that amount, so there is a net increase of 10.65 billion tonnes of atmospheric carbon dioxide per year (one tonne of atmospheric carbon is equivalent to 44/12 or 3.7 tonnes of carbon). Carbon dioxide is one of the greenhouse gases that enhances radiative forcing and contributes to global warming, causing the average surface temperature of the Earth to rise in response, which climate scientists agree will cause major adverse effects.

Fossil fuels will continue to play a dominant role in the energy mix in the decades to come, therefore, the development and use of advanced and cleaner fossil fuel technologies should be increased. More efforts should go into supporting the further development and dissemination of those technologies.

At the same time, we are cognizant of the need to diversify energy supply and developing advanced, cleaner, more efficient affordable and cost-effective energy technologies, including fossil fuel technologies and renewable energy technologies, hydro included and their transfer to developing countries on concessional terms as mutually agreed. Substantial increase in the global share of renewable energy sources is an urgent imperative; with the objective of increasing it's contribution to total energy supply, recognizing the role of national and regional voluntary targets and initiatives, where they exist.

Fossil fuels take millions of years to make, but burn and disappear in seconds. Once they are used, they cannot be reused. People have irretrievably damaged the planet by extracting and burning these fuels. It is best not to waste fossil fuels as they are not renewable. We have to learn to conserve these sources of energy.

Every year, millions of tonnes of coal is consumed as energy. This has led to global warming (greenhouse effect) and the depletion of resources.

At present, the worldwide burning of coal, oil, and natural gas releases billions of tonnes of carbon dioxide (measured as carbon) into the atmosphere every year. Burning any fossil fuel means pollution of some sort. Even if the fuel is low in sulphur, the atmosphere contains nitrogen, which combines with oxygen at the high burning temperatures found in boilers, jet, or car engines. This yields nitrogen oxides, which like sulphur dioxide, dissolves in rain to form nitric acid. Both gases are poisonous to humans.

Mining and exploration for fossil fuels can cause disturbance to the surrounding ecosystem. The burning of fossil fuels emits oxides of sulphur and nitrogen into the atmosphere.

Importance

Fossil fuels are of great importance because they can be burned (oxidized to carbon dioxide and water), producing significant amounts of energy. The use of coal as a fuel predates recorded history. Coal was used to run furnaces for the melting of metal ore. Semi-solid hydrocarbons from seeps were also burned in ancient times, but these materials were mostly used for waterproofing and embalming.

Commercial exploitation of petroleum, largely as a replacement for oils from animal sources (notably whale oil) for use in oil lamps began in the nineteenth century.

Natural gas, once flared-off as an un-needed byproduct of petroleum production, is now considered a very valuable resource.

Heavy crude oil, which is much more viscous than conventional crude oil, and tar sands, where bitumen is found mixed with sand and clay, are becoming more important as sources of fossil fuel. Oil shale and similar materials are sedimentary rocks containing kerogen, a complex mixture of high-molecular weight organic compounds, which yield synthetic crude oil when heated (pyrolyzed). These materials have yet to be exploited commercially. These fuels are employed in internal combustion engines, fossil fuel power stations and other uses.

Prior to the latter half of the eighteenth century, windmills or watermills provided the energy needed for industry such as milling flour, sawing wood or pumping water, and burning wood or peat provided domestic heat. The wide-scale use of fossil fuels, coal at first and petroleum later, to fire steam engines, enabled the Industrial Revolution. At the same time, gas lights using natural gas or coal gas were coming into wide use. The invention of the internal combustion engine and its use in automobiles and trucks greatly increased the demand for gasoline and diesel oil, both made from fossil fuels. Other forms of transportation, railways and aircraft also required fossil fuels. The other major use for fossil fuels is in generating electricity and the petrochemical industry. Tar, a leftover of petroleum extraction, is used in construction of roads.

Advantages

- The technology and infrastructure already exist for the use of fossil fuels.
- Petroleum energy density in terms of volume (cubic space) and mass (weight) is superior to some alternative energy sources (or energy storage devices, like a battery (electricity)).

Disadvantages

- Petroleum-powered vehicles are very inefficient. Only about 30 per cent of the energy from the fuel they consume is converted into mechanical energy. The rest of the fuel-source energy is inefficiently expended as waste heat. The heat and gaseous pollution emissions harm our environment.
- The inefficient atmospheric combustion (burning) of fossil fuels in vehicles, buildings, and power plants contributes to urban heat islands.
- The combustion of fossil fuels leads to the release of pollution into the atmosphere.

- Dependence on fossil fuels from volatile regions or countries creates energy security risks for dependent countries. Oil dependence in particular has led to war, major funding of radical terrorists, monopolization, and socio-political instability.
- Fossil fuels are non-renewable, un-sustainable resources, which will eventually decline in production and become exhausted, with dire consequences to societies that remain highly dependent on them. (Fossil fuels are actually slowly forming continuously, but we are using them up at a rate approximately 100,000 times faster than they are formed.)
- Extracting fossil fuels is becoming more difficult as we consume the most accessible fuel deposits. Extraction of fossil fuels is becoming more expensive and more dangerous as mines get deeper and oil rigs must drill deeper, and go further out to sea.
- Extraction of fossil fuels results in extensive environmental degradation, such as the strip mining and mountaintop removal of coal.

Since these power plants are thermal engines, and are typically quite large, waste heat disposal becomes an issue at high ambient temperature. Thus, at a time of peak demand, a power plant may need to be shut down or operate at a reduced power level, as sometimes do nuclear power plants, for the same reasons.

1.17. COAL

Coal is by far the most abundant fossil fuel on earth. It is essentially carbon and is mainly used as a combustion fuel. The large-scale use of coal began with the Industrial Revolution in the 19th century. As the number of industries increased, demand for more sources of energy grew.

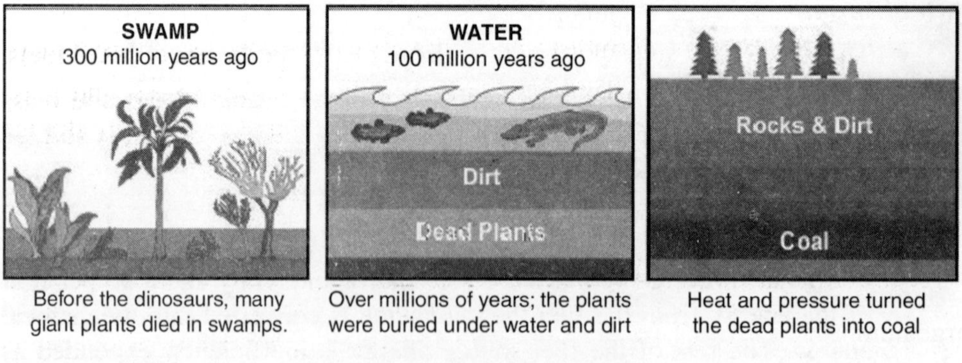

Fig. 1.19. Formation of Coal

Coal is the product of plants, mainly trees, that died tens or hundreds of millions of years ago. Due to water logging in low-lying swampy areas or in slowly sinking lagoons, dead trees and plants did not decompose as they normally would. The dead plant matter was covered with water and protected from the oxidizing effect of air. The action of certain bacteria released the oxygen and hydrogen, making the residue richer and richer in carbon. Thick layers of this carbon-rich substance, called peat, built up over thousands of years. As more material accumulated above the peat, the water was squeezed out leaving just carbon-rich plant remains. Pressure and temperature further compressed the material. This aided the process of producing coal as more gases were forced out and the proportion of carbon continued to increase. The carbon slowly metamorphosed into coal over millions of years.

There are three main types of coal: lignite, bituminous, and anthracite. Lignite and bituminous have a lesser percentage of carbon in them and therefore burn faster. They release a great deal of pollutants into the atmosphere. Anthracite has about 98 per cent carbon and therefore burns slowly and releases much less smoke. Coal of all types contains sulphur to some degree. Sulphur is the worst of the pollutants and causes damage to human health and to vegetation. Though petroleum gained importance over the 20th century and continues to do so, coal remains essential for the industrial sector. It is the principal heat source for electricity generation in most countries and is used directly in such heavy industries as iron and steel making.

Fig. 1.20. Underground Mining

Until recently, most coal came from underground mines. But now there are a large number of open cast mines. Underground coal mines are notorious killers due to roof falls and explosions. Accidents have resulted in the deaths of hundreds of miners. Almost 80 per cent of today's coal comes from surface strip mines (opencast mines), which is much safer. Huge earth-moving equipment strips off the soils and rocks covering the buried coal seams. The land is backfilled and returned to normal after the coal has been removed, thereby repairing the landscape. But most companies do not refill the excavated area and leave it damaged. Most countries have now enforced backfilling by law.

Fig. 1.21. Open Cast Mining

Coal Mining in India

India has about seven per cent of the world's proven coal reserves. Coal supplies more than 50 per cent of the country's total energy requirements. By current estimates, the reserves are enough to meet India's needs for at least another 100 years.

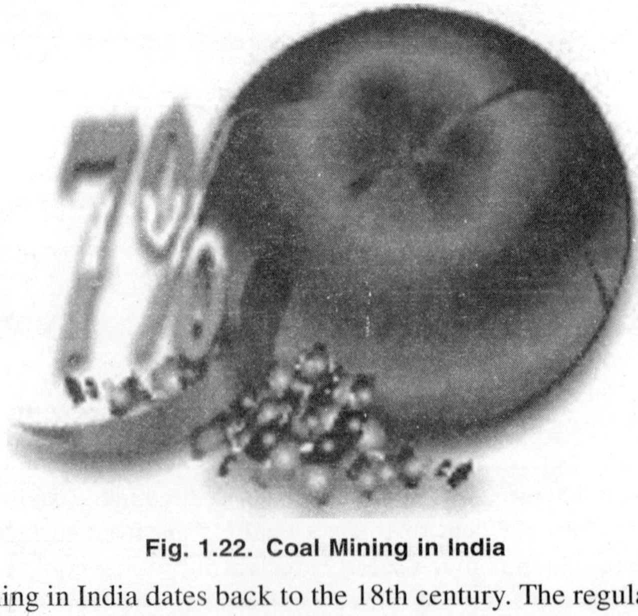

Fig. 1.22. Coal Mining in India

Coal mining in India dates back to the 18th century. The regulation of its use in the industrial sector was conceived in 1923. In 1972-73, the Indian government nationalized the coal industry, primarily to develop the sector, since it was considered to be of strategic importance for rapid industrial development.

India's coal demand is expected to increase several fold within the next 5–10 years due to the completion of ongoing coal-based power projects, and demand from metallurgical and other industries.

Coal is the dominant source of fuel in the industrial sector, with a share of nearly 72.5 per cent in the total energy consumption. The industrial sector is the largest consumer of electricity, with a share of 41 per cent of the total consumption. The transport sector is the largest consumer of petroleum products, and accounts for nearly 50 per cent of the total consumption.

The small amount of coal presently consumed adds atmospheric pollutants, some of which precipitate into the ground and water. This assault on the environment has been the cause of heavy pollution in many areas of the world.

1.18. OIL AND GAS

Almost all oil and natural gas are found deep underground in tiny holes in rocks. Millions of years ago a sea covered much of what is now dry land. In prehistoric times, tiny plants and animals lived in the sea. When these creatures died, they sank to the bottom of the sea, and got buried in layers of mud and sand. As the ages passed, this organic material sank deeper and deeper. The earth's crust changed its shape, and put intense pressure and heat on what was once only plants and tiny animals. Heat from the earth's interior and the weight of the overlying rocks gradually changed the energy-containing substances in the accumulated plants into hydrocarbon liquids and gases. As millions of years passed, these deposits turned into chemicals that are now called 'hydrocarbons'.

Hydrocarbons are simple molecules made up of carbon and hydrogen atoms joined together in chains or in rings. These molecules, being light and mobile, migrated upwards through the rocks but eventually became trapped beneath impermeable rock structures in the earth's crust. That is where oil and natural gas come from. Some were created millions of years ago, some were created thousands of years ago, and some are being created right now!

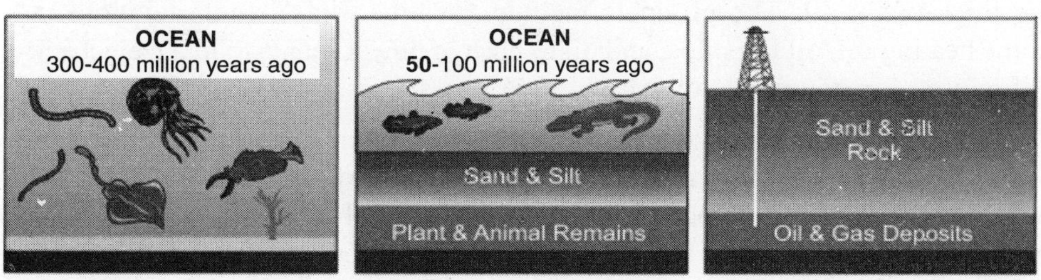

Tiny sea plants and animals died and were buried on the ocean floor. Over time, they were covered by layers of silt and sand.

Over millions of years, the remains were buried deeper and deeper. The enormous heat and pressure turned them into oil and gas.

Today, we drill down through layers of sand, silt, and rock to reach the rock formations that contain oil and gas deposits.

Fig. 1.23. Formation of Oil and Natural Gas

Much of the oil and gas production now comes from underneath the sea-bed. As the technology for extraction continues to advance, production becomes possible from deeper and deeper waters. But the supplies are limited. Every drop of oil burnt adds to the monumental environment problems already created by pumping gases like carbon dioxide into the atmosphere. Many scientists worry that this continual release of carbon dioxide is an important cause of global warming.

Natural gas is usually found underground near an oil source. It is a mixture of light hydrocarbons including methane, ethane, propane, butane, and pentane. Other compounds found in natural gas include carbon dioxide, helium, hydrogen sulphide, and nitrogen. It is found around the world, but the largest reserves are in the former Soviet Union and the Middle East. This gas is lighter than air and is highly flammable, made up mainly of a gas called methane. Methane is a simple chemical compound that is made up of carbon and hydrogen atoms. Natural gas usually has no odour and cannot be seen. Before it is sent to the pipelines and storage tanks, it is mixed with a chemical that gives it a strong odour, almost like rotten eggs. The odour makes it easy to detect a leak.

Natural gas is the cleanest burning fossil fuel. When it is burned, it gives off less carbon dioxide than oil or coal, virtually no sulphur dioxide, and only small amounts of nitrous oxides. Natural gas is mostly composed of methane and other light hydrocarbons. Both the carbon and the hydrogen in methane combine with oxygen when natural gas is burned, giving off heat. Coal and oil contain proportionally more carbon than natural gas, therefore giving off more carbon dioxide per unit of energy produced. Natural gas gives off 50 per cent of the carbon dioxide released by coal and 25 per cent less carbon dioxide than oil, for the same amount of energy produced. Carbon dioxide is the most important greenhouse gas contributing to global warming.

To find oil and natural gas, companies drill through the earth to the deposits deep below the surface. The oil and natural gas are then pumped from below the ground by oil rigs. They then usually travel through pipelines.

At oil refineries, crude oil is split into various types of products by heating the thick black oil. The products include gasoline, diesel fuel, aviation fuel, home heating oil, oil for ships, and oil to burn in power plants to make electricity. Oil is used for transportation—cars, airplanes, trucks, buses, and motorcycles.

Oil is stored in large tanks until it is sent to various places to be used. Oil is also made into many different products—fertilizers for farms, clothes, toothbrush, plastic bottle, and plastic pen. There are thousands of other products that come from oil. Almost all plastic comes originally from oil. Oil is transported in huge pipelines and tanker ships to places where it is made into other products.

The origin of the oil industry in India can be traced back to the last part of the 19th century when petroleum was discovered in Digboi in north-east India. Thereafter large numbers of oil fields have been discovered both inland and off-shore. This has led to the setting up of refineries to process the oil and gas for use in various sectors.

LPG (Liquefied Petroleum Gas)

LPG is the abbreviation or short form for liquefied petroleum gas. Like all fossil fuels, it is a non-renewable source of energy. It is extracted from crude oil and natural gas. The main composition of LPG is hydrocarbons containing three or four carbon atoms. The normal components of LPG thus, are propane (C_3H_8) and butane (C_4H_{10}). Small concentrations of other hydrocarbons may also be present. Depending on the source of the LPG and how it has been produced, components other than hydrocarbons may also be present.

LPG is a gas at atmospheric pressure and normal ambient temperatures, but it can be liquefied when moderate pressure is applied or when the temperature is sufficiently reduced. It can be easily condensed, packaged, stored and utilized, which makes it an ideal energy source for a wide range of applications.

Normally, the gas is stored in liquid form under pressure in a steel container, cylinder or tank. The pressure inside the container will depend on the type of LPG (commercial butane or commercial propane) and the outside temperature.

When we start using LPG, some of the pressure in the container is released. Some of the liquid LPG then boils to produce vapour. Heat is needed to convert the liquid to vapour (known as the latent heat of vaporization). As the liquid boils, it draws the heat energy from its surroundings. This explains why containers feel cold to touch and why, if there is a heavy off-take, water or ice may appear on the container. When we stop using LPG, the pressure will return to the equilibrium value for the surrounding temperature. The pressure of the LPG in the container varies with the surrounding temperature. It is also much higher than is needed by the appliances that use it; it needs to be controlled to ensure a steady supply at constant pressure. This is done by a regulator, which limits the pressure to suit the appliance that is being fuelled. It is a colourless and odourless gas to which foul-smelling mercaptan is added so that leak can be easily detected.

LPG is highly inflammable and must therefore be stored away from sources of ignition and in a well-ventilated area, so that any leak can disperse safely. Another reason why care should be taken during storage is that LPG vapour is heavier than air, so any leakage will sink to the ground and accumulate in low lying areas and may be difficult to disperse. LPG expands rapidly when its temperature rises. So whenever a container is filled, sufficient space is left to allow for such expansion. LPG will cause natural rubber and some plastics to deteriorate. This is why only hoses and other equipment specifically designed for LPG should be used.

Although LPG is non-toxic, its abuse – (like that of solvents) – is highly dangerous. LPG should always be treated with respect and kept away from children whenever possible.

Liquid petroleum gases were discovered in 1912 when Dr. Walter Snelling, an American scientist, realized that these gases could be changed into liquids and stored under moderate pressure. From 1912 and 1920, LPG uses were developed. The first LPG cook stove was made in 1912, and the first LPG - fueled car was developed in 1913. The LPG industry began sometime shortly before World War I. At that time, a problem in the natural gas distribution process popped up. Gradually facilities were built to cool and compress natural gas, and to separate the gases that could be turned into liquids (including propane and butane). LPG was sold commercially by 1920.

Case Study

Digboi: The First Oil Well in India

Digboi in Assam is an oil town that can be traced to the early 18th century, when oil was first discovered here. Digboi can proudly boast of two unique features: a 100-year-old extant oilfield and the world's oldest operating oil refinery.

The earliest recorded reference of oil in Assam can be traced to Lieut. R Wilcox of the 46th Regiment Native Infantry who saw it as '... rising to the surface at Supkhong with great bubbling of gas and green petroleum...' in the *book Memoir of a survey of Assam and the Neighbouring Countries executed in 1825-6-7-8.*

Barely seven years after Edwin L. Drake drilled the world's first oil well in 1859 at Titusville, Pennsylvania, USA, history registered another exploration of the black liquid gold, in the largest continent. In 1867, a large group of men were engaged in laying railway tracks for the Assam Railway and Trading Co. Ltd. in an area very close to what is now the town of Digboi. They had to work in the dense forest where the only visitors were animals, birds and insects. The place is said to have smelt of the rain: soaked forest mingled with a heavy odour of something that smelt like oil.

Legend has it that an elephant hauling logs from the forest returned with distinct traces of oil on its feet and trail. The excited owners of the elephant followed its footprints and found seepage of oil bubbling to the surface. 'Dig, boy, dig!', probably this is what the Englishman cried out to his men, hence the name Digboi. From that day, this tiny habitat, among the rolling hills of Assam found a place in the map of the world's petroleum industry.

The first commercial discovery of crude oil in the country was, however, made in 1889 at Digboi when a group of men erected a 20 meter high thatch covered wooden structure, marking the establishment of the first well at Digboi (Well No. 1 or the Discovery well). Thereafter, systematic drilling began in 1891, and in 1901, Asia's first oil refinery was set. In the meantime the AOC (Assam Oil Company) was formed in 1899 to look after the running of the oil business in this area.

Today, though the production is very low, Digboi still has the distinction of being the world's oldest continuously producing oilfield. Digboi oilfield is not another oilfield, but an oil museum with a history to be proud of.

The Jharia Coal Field Fire

The Jharia coalfield in Bihar is an exclusive storehouse of prime coke coal in the country, consisting of 23 large underground and nine large open cast mines. The mining activities in these coalfields started in 1894 and had really intensified in 1925. The history of coal-mine fire in Jharia coalfield can be traced back to 1916 when the first fire was detected. At present, more than 70 mine fires are reported from this region.

Coal, a non-renewable source of energy, is found in several parts of the world. The coal layers are mined by two methods: open cast mining and underground mining. Coal is formed from organic matter with a high carbon content, which when exposed to certain conditions (temperature, moisture, oxygen etc.) tends to ignite/ burn spontaneously at rather low temperatures. This may occur naturally or the combustion process may be triggered by other causes.

However, once a coal seam catches fire, and efforts to stop it an early stage fail, it may continue to burn for tens to hundreds of years, depending primarily on the availability of coal and oxygen. Coal fires have occurred in nearly all parts of the world like India, the US, Indonesia, South Africa, Australia, China, Germany and many other countries. However, the nature and magnitude of the problem differs from country to country. In India, the fire in the Jharia coalfield has mainly been due to unscientific mining and extraction of coal in the past.

Fires may occur in coal layers that are exposed to the surface of the earth or areas close to it. These are visible to the naked eye. Also, fires erupt in the underground seams, which have large cracks that serve as channels for oxygen to the burning coal. The main cause of natural coal fires are lightening, forest fires, bush fires, etc. Among human causes are accidents, negligent acts, domestic fires, lighting fires in abandoned underground mines for heating or distilling alcohol etc. Besides, burning away of an important energy resource, it creates problems for exploitation of coal, poses danger to humankind, raises the temperature of the area, and when present underground, can cause land to subside.

The pollution caused by these fires affects air, water, and land. Smoke, from these fires contains poisonous gases such as oxides and dioxides of carbon, nitrogen and sulphur, which along with particulate matter are the causes of several lung and skin diseases. High levels of suspended particulate matter increase respiratory diseases such as chronic bronchitis and asthma, while the gases contribute to global warming besides causing health hazards to the exposed population. Methane emission from coal mining depends on the mining methods, depth of coal mining, coal quality and entrapped gas content in the coal seams. These fires also pollute water by contaminating it and increasing its acidity, which is due to a certain percentage of sulphur that is present in coal. These fires lead to degradation of land and does not allow any vegetation to grow in the area.

The measures for controlling coal mine fires, in the case of Jharia coalfields, include bull dozing, leveling and covering with soil to prevent the entry of oxygen and to stabilize the land for vegetation. Fire fighting in this area requires relocation of a large population, which poses to be a bigger problem than the actual fire fighting operations.

1.19. NUCLEAR ENERGY

Nuclear energy is the energy that is trapped inside each atom. The ancient Greeks believed that the smallest part of nature is an atom. But they did not know 2000 years ago that atoms are made up of further smaller particles—a nucleus of protons and neutrons, surrounded by electrons, which swirl around the nucleus much like the earth revolves around the sun.

Fig. 1.24. Nuclear Energy

One of the laws of the universe is that matter and energy can neither be created nor destroyed. But they can be changed in form. Matter can be changed into energy. Albert Einstein's famous mathematical formula $E = mc^2$ explains this. The equation says: E [energy] equals m [mass] times c^2 [c stands for the speed or velocity of light]. This means that it is mass multiplied by the square of the velocity of light. Scientists used Einstein's equation as the key to unlock atomic energy and to create atomic bombs.

An atom's nucleus can be split apart. This is known as fission. When this is done, a tremendous amount of energy in the form of both heat and light is released by the initiation of a chain reaction. This energy, when slowly released, can be harnessed to generate electricity. When it is released all at once, it results in a tremendous explosion as in an atomic bomb.

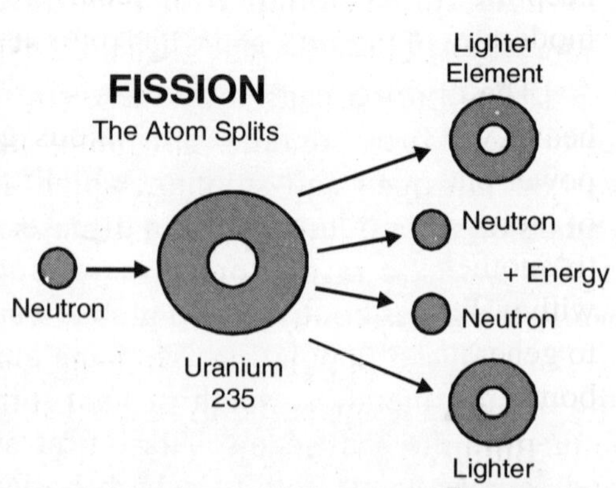

Fig. 1.25. Nuclear Fusion

Nuclear energy can also be harnessed by fusion. A fusion reaction occurs when two hydrogen atoms combine to produce one helium atom. This reaction takes place at all times in the sun, which provides us with the solar energy. This technology is still at the experimental stage and may become viable in future.

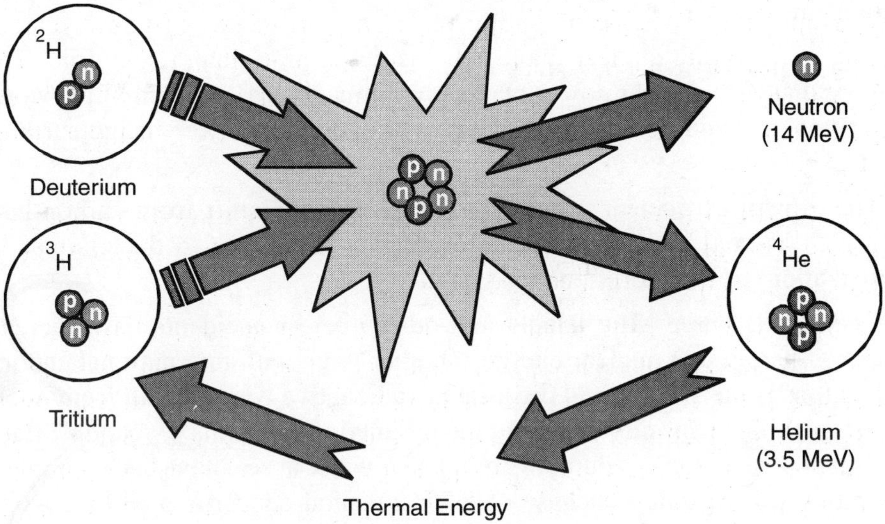

Fig. 1.26. Nuclear Fission

Uranium is the main element required to run a nuclear reactor where energy is extracted. Uranium is mined from many places around the world. It is processed (to get enriched uranium, i.e. the radioactive isotope) into tiny pellets. These pellets are loaded into long rods that are put into the power plant's reactor. Inside the reactor of an atomic power plant, uranium atoms are split apart in controlled chain reaction. Other fissile material includes plutonium and thorium.

In a chain reaction, particles released by the splitting of the atom strike other uranium atoms and split them. The particles released by this further split other atoms in a chain process. In nuclear power plants, control rods are used to keep the splitting regulated, so that it does not occur too fast. These are called moderators.

The chain reaction gives off heat energy. This heat energy is used to boil heavy water in the core of the reactor. So, instead of burning a fuel, nuclear power plants use the energy released by the chain reaction to change the energy of atoms into heat energy. The heavy water from around the nuclear core is sent to another section of the power plant. Here it heats another set of pipes filled with water to make steam. The steam in this second set of pipes rotates a turbine to generate electricity. If the reaction is not controlled, you could have an atomic bomb.

But in atomic bombs, almost pure pieces of uranium-235 or plutonium, of a precise mass and shape, must be brought together and held together with great force. These conditions are not present in a nuclear reactor.

The reaction also creates radioactive material. This material could hurt people if released, so it is kept in a solid form. A strong concrete dome is built around the reactor to prevent this material from escaping in case of an accident.

Experiences with nuclear programmes differ and the future of nuclear power remains uncertain because of public reaction. But in the past few years the capacity of operating nuclear plants has increased more than twentyfold. There are more than 400 nuclear power plants providing about 7 per cent of the world's primary energy and about 25 per cent of the electric power in industrialized nations.

The growth of nuclear power combined with the shift from carbon-heavy fuels such as coal and oil to carbon-light gas contribute to the gradual 'de-carbonization' of the world energy system.

Chernobyl, Three Mile Island, and other nuclear accidents have increased the fear of harnessing nuclear energy. Another issue with international and local implications is the storage and disposal of radioactive wastes: both from nuclear reactors making electricity and from the production of military weapons. Earlier disposal practices, such as dumping of nuclear waste at sea, have been completely stopped by formal treaty because of environmental concerns (and by cessation of furtive scuttling of nuclear submarines). Regimes for transport and temporary storage of civil and defence nuclear wastes now function, although sites and designs for permanent disposal have yet to be accepted.

People are concerned about both low- and high-level radioactive wastes; the latter, though smaller in volume is more technically problematic. With the rise of nuclear electrification, the volume of spent fuel and other wastes has risen substantially; but is still small. In some countries such as the US, the volume of high-level waste from commercial power plants has now reached hundreds of millions of tonnes.

1.20. NUCLEAR FISSION

Nuclear power stations use nuclear fission to generate energy by the reaction of uranium-235 inside a nuclear reactor. The reactor uses uranium rods, the atoms of which are split in the process of fission, releasing a large amount of energy. The process continues as a chain reaction with other nuclei. The energy heats water to create steam, which spins a turbine generator, producing electricity.

Depending on the type of fission fuel considered, estimates for existing supply at known usage rates varies from several decades for the currently popular Uranium-235 to thousands of years for uranium-238. At the present rate of use, there are (as of 2007) about 70 years left of known uranium-235 reserves economically recoverable at a uranium price of US$ 130/kg. The nuclear industry argue that the cost of fuel is a minor cost factor for fission power, more expensive, more difficult to extract sources of uranium could be used in the future, such as lower-grade ores, and if prices increased enough, from sources such as granite and seawater. Increasing the price of uranium would have little effect on the overall cost of nuclear power; a doubling in the cost of natural uranium would increase the total cost of nuclear power by 5 per cent. On the other hand, if the price of natural gas was doubled, the cost of gas-fired power would increase by about 60 per cent.

Opponents on the other hand argue that the correlation between price and production is not linear, but as the ores' concentration becomes smaller, the difficulty (energy and resource consumption are increasing, while the yields are decreasing) of extraction rises very fast, and that the assertion that a higher price will yield more uranium is overly optimistic; for example a rough estimate predicts that the extraction of uranium from granite will consume at least 70 times more energy than what it will produce in a reactor. As many as eleven countries have depleted their uranium resources, and only Canada has mines left which produce better than 1 per cent concentration ore. Seawater seems to be equally dubious as a source. As a consequence an eventual doubling in the price of uranium will give a marginal increase in the volumes that are being produced.

Another alternative would be to use thorium as fission fuel. Thorium is three times more abundant in Earth's crust than uranium, and much more of the thorium can be used (or, more precisely, bred into Uranium-233, reprocessed and then used as fuel). India has around 32 per cent of the world's reserves of thorium and intends on using it for itself because the country has run out of uranium.

Current light water reactors burn the nuclear fuel poorly, leading to energy waste. Nuclear reprocessing or burning the fuel better using different reactor designs would reduce the amount of waste material generated and allow better use of the available resources. As opposed to current light water reactors which use uranium-235 (0.7 per cent of all natural uranium), fast breeder reactors convert the more abundant uranium-238 (99.3 per cent of all natural uranium) into plutonium for fuel. It has been estimated that there is anywhere from 10,000 to five billion years worth of Uranium-238 for use in these power plants. Fast breeder technology has been used in several reactors. However, the fast breeder reactors at Dounreay in Scotland, Monju in Japan and the Superphénix at Creys-Malville in France, in particular, have all had difficulties and were not economically competitive and most have been decommissioned. The People's Republic of China intends to build breeders. India has run out of uranium and is building thermal breeders that can convert Th-232 into U-233 and burn it.

Some nuclear engineers think that pebble bed reactors, in which each nuclear fuel pellet is coated with a ceramic coating, are inherently safe and are the best solution for nuclear power. They can also be configured to produce hydrogen for hydrogen vehicles. China has plans to build pebble bed reactors configured to produce hydrogen.

The possibility of nuclear meltdowns and other reactor accidents, such as the Three Mile Island accident and the Chernobyl disaster, have caused much public fear. Research is being done to lessen the known problems of current reactor technology by developing automated and passively-safe reactors. Historically, however, coal and hydropower power generation have both been the cause of more deaths per energy unit produced than nuclear power generation. Various kinds of energy infrastructure might be attacked by terrorists, including nuclear power plants, hydropower plants, and liquified natural gas tankers. Nuclear proliferation is the spread from nation to nation of nuclear technology, including nuclear power plants but especially nuclear weapons. New technology like SSTAR ("small, sealed, transportable, autonomous reactor") may lessen this risk.

The long-term radioactive waste storage problems of nuclear power have not been fully solved. Several countries have considered using underground repositories. Nuclear waste takes up little space compared to wastes from the chemical industry which remains toxic indefinitely. Spent fuel rods are now stored in concrete casks close to the nuclear reactors. The amounts of waste could be reduced in several ways. Both nuclear reprocessing and fast breeder reactors could reduce the amounts of waste. Subcritical reactors or fusion reactors could greatly reduce the time the waste has to be stored. Subcritical reactors may also be able to do the same to already existing waste. The only long-term way of dealing with waste today is by geological storage.

The economics of nuclear power is not simple to evaluate, because of high capital costs for building and very low fuel costs. Comparison with other power generation methods is strongly dependent on assumptions about construction timescales and capital financing for nuclear plants.

Depending on the source different Energy Return on Energy Investment (EROI) are claimed. Advocates (using life cycle analysis) argue that it takes 4–5 months of energy production from the nuclear plant to fully pay back the initial energy investment. Opponents claim that it depends on the grades of the ores the fuel came from, so a full payback can vary from 10 to 18 years, and that the advocates' claim was based on the assumption of high grade ores (the yields are getting worse, as the ores are leaner; for less than 0.02 per cent ores, the yield is less than 50 per cent).

Advocates also claim that it is possible to increase the number of plants fairly rapidly. Typical new reactor designs have a construction time of three to four years. In 1983, 43 plants were being built, before an unexpected fall in fossil fuel prices stopped most new construction. Developing countries like India and China are rapidly increasing their nuclear energy use. However, a Council on Foreign Relations report on nuclear energy argues that a rapid expansion of nuclear power may create shortages in building materials such as reactor-quality concrete and steel, skilled workers and engineers, and safety controls by skilled inspectors. This would drive up current prices.

However, at present, nuclear energy is in decline, according to a report 'World Nuclear Industry Status Report 2007' presented by the Greens/EFA group in the European Parliament. The report outlines that the proportion of nuclear energy in power production has decreased in 21 out of 31 countries, with five fewer functioning nuclear reactors than five years ago. There are currently 32 nuclear power plants under construction or in the pipeline, 20 fewer than at the end of the 1990s.

Advantages

- The energy content of a kilogram of uranium or thorium, if spent nuclear fuel is reprocessed and fully utilized, is equivalent to about 3.5 million kilograms of coal.
- The cost of making nuclear power, with current legislation, is about the same as making coal power, which is considered very inexpensive. If a carbon tax is applied, nuclear does not have to pay anything because nuclear does not emit toxic gases such as CO_2, NO, CO, SO_2, arsenic, etc. that are emitted by coal power plants.
- Nuclear power does not produce any primary air pollution or release carbon dioxide and sulfur dioxide into the atmosphere. Therefore, it contributes only a small amount to global warming or acid rain.
- Raw material extration is much safer for nuclear power compared to coal. Coal mining is the second most dangerous occupation in the United States. Nuclear energy is much safer per capita than coal derived energy.
- For the same amount of electricity, the life cycle emissions of nuclear is about 4 per cent of coal power. Depending on the report, hydro, wind, and geothermal are sometimes ranked lower, while wind and hydro are sometimes ranked higher (by life cycle emissions).
- According to a Stanford study, fast breeder reactors have the potential to power humans on earth for billions of years, making it sustainable.

Disadvantages

- The improper operation of a badly designed nuclear reactor with no containment vessel near human settlements can be catastrophic in the event of an uncontrolled power increase in the reactor, as shown by the Chernobyl disaster in the Ukraine (former USSR), where large areas of Europe were affected by moderate radioactive contamination and the parts of the Ukraine and one fifth of Belarus continue today to be affected by radioactive fallout as of 2008.
- Transuranic waste produced from nuclear fission of uranium is poisonous and highly radioactive. Breeder reactors could burn this waste as fuel, fissioning transuranics into much faster-decaying fission products which stabilize at a relatively low level of radioactivity in 100-500 years, but recycling plutonium as MOX fuel in current light water reactors merely transmutes between isotopes of plutonium and offers little reduction in radioactivity.

Without nuclear reprocessing, whole spent fuel bundles containing transuranic waste must be stored in spent fuel pools, dry cask storage, or a geological repository.

- There can be connections between nuclear power and nuclear weapon proliferation, since many reactor designs require large-scale uranium enrichment facilities.
- Some claim that uranium ore is a limited resource and estimate that current supplies will fail to meet demand in 2026, provided no other deposits are discovered. This claim is strongly disputed; also, breeder reactors would extract about 100 times as much energy from the same amount of uranium.
- The limited liability for the owner of a nuclear power plant in case of a nuclear accident differs per nation while nuclear installations are sometimes built close to national borders.
- Since nuclear power plants are typically quite large power plants, and are, fundamentally, thermal engines, waste heat disposal becomes more difficult at higher ambient temperature. Thus, at a time of peak demand for power for air-conditioning, a power reactor may need to be shut down or operate at a reduced power level, as do large coal-fired plants, for the same reasons.

1.21. NUCLEAR FUSION

Fusion power could solve many of the problems of fission power (the technology mentioned above) but, despite research having started in the 1950s, no commercial fusion reactor is expected before 2050. Many technical problems remain unsolved. Proposed fusion reactors commonly use deuterium, an isotope of hydrogen, as fuel and in most current designs also lithium. Assuming a fusion energy output equal to the current global output and that this does not increase in the future, then the known current lithium reserves would last 3000 years, lithium from sea water would last 60 million years, and a more complicated fusion process using only deuterium from sea water would have fuel for 150 billion years.

1.21.1. Nuclear Fusion: Energy for the Future?

When it comes to energy sources, nuclear fusion used to be the wallflower. But now, scientists are working to see if it could be a safe and environmentally-friendly way of producing electricity.

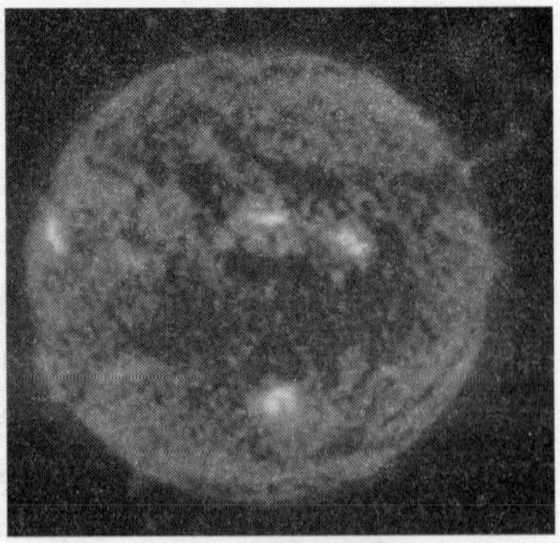

(Source: NASA)

Fig. 1.27. The Sun is a natural fusion reactor

The energy crisis has rocketed from a textbook concept into the most pressing political issue of our time. Future energy supplies are increasingly vulnerable and global consumption is expected to escalate dramatically, increasing by 71 per cent in 2030 and continuing to rise. Energy shortages would have a dramatic impact on every area of modern life: business, transport, food, health and communications. This looming crisis has drawn scientific minds and encouraged radical research into arcane technologies, such as the once neglected area of nuclear fusion.

1.21.2. Why Nuclear Fusion?

Our sun, and all the other stars in the universe, are powered by nuclear fusion. Similar to traditional nuclear power, or fission, it can produce huge amounts of carbon-neutral energy. But there is one vital difference: no dangerous, long-lasting radioactive waste. Waste from nuclear fusion is only radioactive for 50–70 years, compared to the thousands of years of radioactivity that result from fission. "This is a long-term supply of energy," says Professor Mike Dunne of the Rutherford Appleton Laboratory in Oxfordshire. "You can get a lot of energy from a small amount of fuel and the by-products are benign."

Raw materials for nuclear fusion – water and silicon – are plentiful and widespread on Earth. This should prevent the situations where energy supplies can be threatened by political instability; as demonstrated in January 2007 when Russia shut down a main oil pipeline to Europe after a political spat with Belarus.

Nuclear fusion could also help meet international climate change targets, such as those agreed by politicians in Washington last month. Current zero-carbon technologies are unlikely to meet our energy demands this century. Nuclear power is deeply unpopular while renewable energy sources – wind, solar and tidal – yield relatively little energy for their high cost. But nuclear fusion could render carbon dioxide-producing fossil fuels obsolete by 2100.

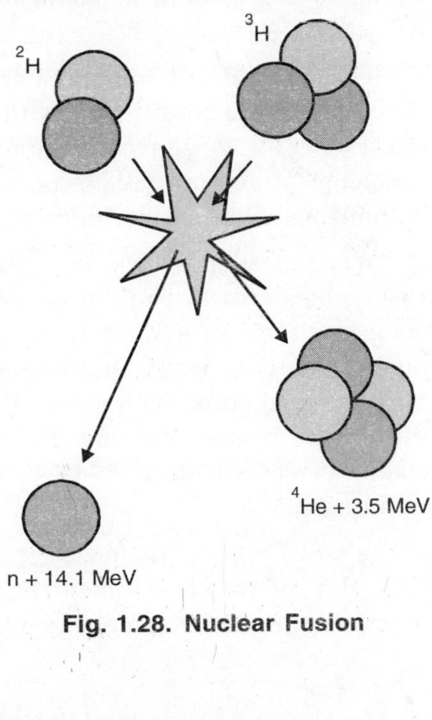

Fig. 1.28. Nuclear Fusion

1.21.3. The Challenge

If we have the potential for unlimited, clean energy, then why wait? Unfortunately, all previous attempts to produce large amounts of energy from nuclear fusion have failed. Secret tests during the atomic bomb programme in the 1950s discovered fusion was possible, but the continuous nuclear fusion reactions required to generate substantial amounts of energy have remained elusive.

DO YOU KNOW

The most promising nuclear reaction to be used for fusion power involves deuterium and tritium, two isotopes of hydrogen.

Huge amounts of energy are required for nuclear fusion. Atomic nuclei are forced to fuse together, in contrast to fission where nuclei are split apart. In the sun, temperatures of 15 million degrees Celsius and immense pressures force hydrogen nuclei to fuse and produce helium, thus releasing energy. Hydrogen exists as plasma and nuclear fusion reactions occur continually. "The trick is to get a self-sustaining reaction," says Dunne "It's like setting off an explosive, you have a little bit of energy – a detonator – and this sets off a chain reaction."

Energy production from nuclear fusion has proven an insurmountable challenge so far. Yet scientists are now saying that plans for larger and more sophisticated reactors around the world could finally make this possible in 50 years time. Is this more than just wishful thinking?

Nuclear Energy and Sustainable Development

The concept of sustainable development was described in the now famous Brundtland Report as "...development that meets the needs of the present without compromising the ability of future generations to meet their own needs." In a broad sense, sustainable development incorporates equity within and across countries as well as across generations, and integrates economic growth, environmental protection and social welfare. A key challenge for sustainable development policies is to address those three dimensions in a balanced way, taking advantage of their interactions and making relevant trade-offs whenever needed.

The analysis of nuclear energy characteristics within a sustainable development framework shows that the approach adopted within the nuclear energy sector is generally consistent with the fundamental sustainable development goal of passing on a range of assets to future generations while minimizing environmental impacts and burdens.

While existing nuclear power plants are economically competitive in most cases and perform well in deregulated electricity markets, the economic competitiveness of new nuclear power plants will remain an issue due to their high capital cost. However, nuclear power has already internalised a large part of its external costs, such as the decommissioning of the plant at the end of its life and the management and disposal of the radioactive waste produced. Such a claim cannot be made by fossil fuel technologies, all of which emit waste to the environment.

In OECD countries, nuclear energy in normal operation has a low impact on health and the environment, but in order to make a continuing contribution to sustainable development goals, nuclear energy will have to maintain its high standards of safety in spite of increasing competition in the electricity sector, ageing reactors, and the expansion of the industry to new countries and regions.

Geological disposal of long-lived radioactive waste has been identified as a technically safe solution that can be implemented without affecting the competitive position of nuclear energy. While there is no technical urgency to implement long-lived waste repositories, it is important to construct and commission such facilities to demonstrate that the goals of sustainable development can be fulfilled.

In order to meet sustainable development goals, nuclear energy will have to achieve a higher level of social acceptance than it enjoys in many countries today. The role of governments is to allow the public to put social, ethical and political issues related to nuclear energy into perspective with the issues raised by alternatives, in order to create the conditions for decision-making processes consistent with the goals of sustainable development.

Summary

- The word energy derives from Greek word energeia, which appears for the first time in the work Nicomachean Ethics of Aristotle in the 4th century BC.
- In physics, energy is a scalar physical quantity that describes the amount of work that can be performed by a force, an attribute of objects and systems that is subject to a conservation law.
- Different forms of energy include kinetic, potential, thermal, gravitational, sound, light, elastic, and electromagnetic energy.
- Energy is defined as the ability or the capacity to do work.
- Energy comes in different forms — heat (thermal), light (radiant), mechanical, electrical, chemical, and nuclear energy.
- Primary energy sources are those that are either found or stored in nature. Common primary energy sources are coal, oil, natural gas, and biomass (such as wood).
- Primary energy sources are mostly converted in industrial utilities into *secondary energy* sources; for example coal, oil or gas converted into steam and electricity.
- The energy sources that are available in the market for a definite price are known as commercial energy.
- The energy sources that are not available in the commercial market for a price are classified as non-commercial energy.
- Renewable energy is energy obtained from sources that are essentially inexhaustible. The five renewable sources used most often include hydropower (water), solar, wind, geothermal, and biomass.

- Non-renewable energy is the conventional fossil fuels such as coal, oil and gas, which are likely to deplete with time. Coal, petroleum, natural gas, propane, and uranium are nonrenewable energy sources.
- Solar energy is the most readily available source of energy. Solar power involves using solar cells to convert sunlight into electricity, using sunlight hitting solar thermal panels to convert sunlight to heat water or air, using sunlight hitting a parabolic mirror to heat water (producing steam), or using sunlight entering windows for passive solar heating of a building.
- Wind energy is the kinetic energy associated with the movement of atmospheric air.
- Wind turbines transform the energy in the wind into mechanical power, which can then be used directly for grinding etc. or further converting to electric power to generate electricity.
- The energy in the flowing water can be used to produce electricity. Waves result from the interaction of the wind with the surface of the sea and represent a transfer of energy from the wind to the sea.
- In hydro energy, the gravitational descent of a river is compressed from a long run to a single location with a dam or a flume.
- Geothermal energy harnesses the heat energy present underneath the Earth.
- Biomass is a renewable energy resource derived from the carbonaceous waste of various human and natural activities.
- Biomass is a biological, organic material cultivated from renewable sources such as agricultural waste and trees that can be utilized for fuel.
- Biodiesel is a renewable fuel that can be used instead of diesel fuel made from petroleum. Biodiesel can be made from vegetable oils, animal fats, or greases.
- Fuel cells are electrochemical devices that convert the chemical energy of a fuel directly and very efficiently into electricity (DC) and heat.
- A fuel cell consists of an electrolyte sandwiched between two electrodes.
- Fossil fuels are formed from organic material, both plant and animal remains, that have settled to the bottom of oceans.
- Fossil fuels can be burned to produce significant amounts of energy.
- Fossil fuels, in terms of energy, involve the burning of coal or hydrocarbon fuels, which are the remains of the decomposition of plants and animals.
- Coal is by far the most abundant fossil fuel on earth.
- Coal is the product of plants, mainly trees that died tens or hundreds of millions of years ago.
- There are three main types of coal: lignite, bituminous, and anthracite. Lignite and bituminous have a lesser percentage of carbon in them and therefore burn faster.

- Almost all oil and natural gas are found deep underground in tiny holes in rocks.
- Natural gas is the cleanest burning fossil fuel.
- LPG is the abbreviation or short form for liquefied petroleum gas. Like all fossil fuels, it is a non-renewable source of energy. It is extracted from crude oil and natural gas.
- The main composition of LPG is hydrocarbons containing three or four carbon atoms. The normal components of LPG thus, are propane (C_3H_8) and butane (C_4H_{10}).
- Nuclear energy is the energy that is trapped inside each atom.
- An atom's nucleus can be split apart. This is known as fission.
- Nuclear energy can also be harnessed by fusion. A fusion reaction occurs when two hydrogen atoms combine to produce one helium atom.

Descriptive Type Questions

1. Discuss the energy scenario in Indian context.
2. What are the alternative energy resources? Differentiate between renewable and non-renewable energy resources
3. Explain solar energy. Enumerate the application of solar energy in modern days.
4. Name and explain in brief, the various ways in which, energy from ocean can be obtain.
5. What is geothermal energy? Discuss its merits and limitations.
6. How you will get energy from biological degradable material?
7. How can we get energy from natural gas?
8. Enumerate the application of solar energy in modern days. Explain the use of solar energy for space heating of buildings.
9. Give a brief account of non-conventional resources of energy.
10. Name the various unconventional renewable energy resources and briefly discuss each of them.
11. Describe the various kinds of energy resources. Distinguish between renewable and non renewable energy resources.

Short Answer Type Questions

1. What do you mean by energy?
2. Write a short note on
   - Wind energy
   - Bio-energy
   - Hydro-energy
3. How many types of coal are there?
4. What are the uses of natural gas?

5. What are the effects of natural gas on environment?
6. How will you differentiate nuclear fusion and nuclear fission?

Objective Type Questions

Multiple Choice Questions

1. Which of the following is a non-renewable source?
   (*a*) Coal (*b*) Forest
   (*c*) Water (*d*) Wildlife
2. Which among the following is not a renewable source of energy?
   (*a*) Solar energy (*b*) Biomass energy
   (*c*) Hydro-power (*d*) Geothermal energy
3. Identify the non-renewable energy resource from the following :
   (*a*) Coal (*b*) Fuel cells
   (*c*) Wind power (*d*) Wave power
4. Which of the following is a disadvantage of most of the renewable energy sources :
   (*a*) Highly polluting (*b*) High waste disposal cost
   (*c*) Unreliable supply (*d*) High running cost
5. Photovoltaic energy is the conversion of sunlight into :
   (*a*) Chemical energy (*b*) Biomass
   (*c*) Electricity (*d*) Geothermal energy
6. Horizontal axis and vertical axis are the types of :
   (*a*) Nuclear reactor (*b*) Wind mills
   (*c*) Bio-gas reactor (*d*) Solar cell
7. A fuel cell in order to produce electricity, burns :
   (*a*) Helium (*b*) Nitrogen
   (*c*) Hydrogen (*d*) None of the above
8. Fuel cells are :
   (*a*) Carbon cells (*b*) Hydrogen battery
   (*c*) Nuclear cells (*d*) Chromium cells
9. Both power and manure is provided by :
   (*a*) Nuclear plants (*b*) Thermal plants
   (*c*) Biogas plants (*d*) Hydroelectric plant
10. Common energy source in Indian villages is :
    (*a*) Electricity (*b*) Coal
    (*c*) Sun (*d*) Wood and animal dung
11. The one thing that is common to all fossil fuels is that they :
    (*a*) Were originally formed in marine environment
    (*b*) Contain carbon
    (*c*) Have undergone the same set of geological process during their formation

(*d*) Represent the remains of one living organisms.

12. Lignite, bituminous and anthracite are different ranks of :

(*a*) Nuclear fuel (*b*) Coal

(*c*) Natural gas (*d*) Biogas

13. Crude oil is :

(*a*) Colourless

(*b*) Ordourless

(*c*) Smelly yellow to black liquid

(*d*) Ordourless yellow to black liquid

14. BTU is measurement of :

(*a*) volume (*b*) Area

(*c*) heat content (*d*) Temperature

Fill in the Blanks

1. The energy released by sun is by a mechanism which is similar to the energy released by a process, called nuclear ....
2. A closed chamber cover by transparent glass, which acts as a solar radiation collector to utilize solar energy for growing the plants, is called as solar ....
3. Instant of tungsten bulb, the use of florescent tube light and .... should be used to save energy.

True/False Questions

1. The secondary energy resources do exists abundantly in nature such as nuclear and geothermal energy.
2. Hydro-power energy and bio-mass energy are renewable energy sources.
3. Sunlight is composed of photons.
4. Solar cells are also known as fuel cells.
5. Wind is called a non-renewable energy source.

ANSWERS

Multiple Choice Questions

1. (*a*)	**2.** (*b*)	**3.** (*a*)	**4.** (*c*)	**5.** (*a*)
6. (*b*)	**7.** (*c*)	**8.** (*b*)	**9.** (*c*)	**10.** (*d*)
11. (*b*)	**12.** (*b*)	**13.** (*c*)	**14.** (*c*)	

Fill in the Blanks

1. Fusion **2.** Greenhouse **3.** CFL

True/False Questions

1. False **2.** True **3.** True **4.** False **5.** False

❑❑❑

CHAPTER 2

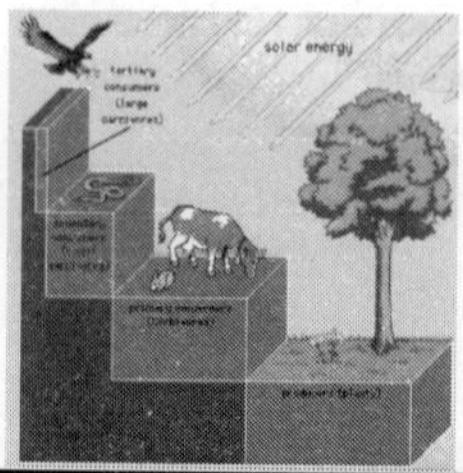

Ecosystem

INSIDE THIS CHAPTER

2.1. DEFINITION OF ENVIRONMENT

'Environment' is derived from the French word *Environner* which means to encircle or surround. All the biological and non-biological things surrounding an organism are thus included in environment. Thus according to Environment (Protection) Act, 1986, **environment is sum of total water, air and land, inter-relationship among themselves and also with the human beings, other living organism and property.** Thus we can say, Environment means our 'surroundings'.

Chemicals on earth are distributed among four major environmental compartments or conceptual spheres: atmosphere, hydrosphere, lithosphere and biosphere. While such a compartmentalization of nature is a somewhat arbitrary, it helps in organizing and extending our knowledge of the distribution and flow of chemicals. A schematic representation of the four environmental compartments and their interrelationships is shown in Fig. 2.1.

The circles represent the spheres and the curved arrows the flow pathways of matter. In the diagram, circles and curved arrows are used instead of boxes and straight line connections to emphasize the close, dynamic, inseparable, organic coupling among the environmental compartments; if one compartment or linkage changes, all other compartments respond.

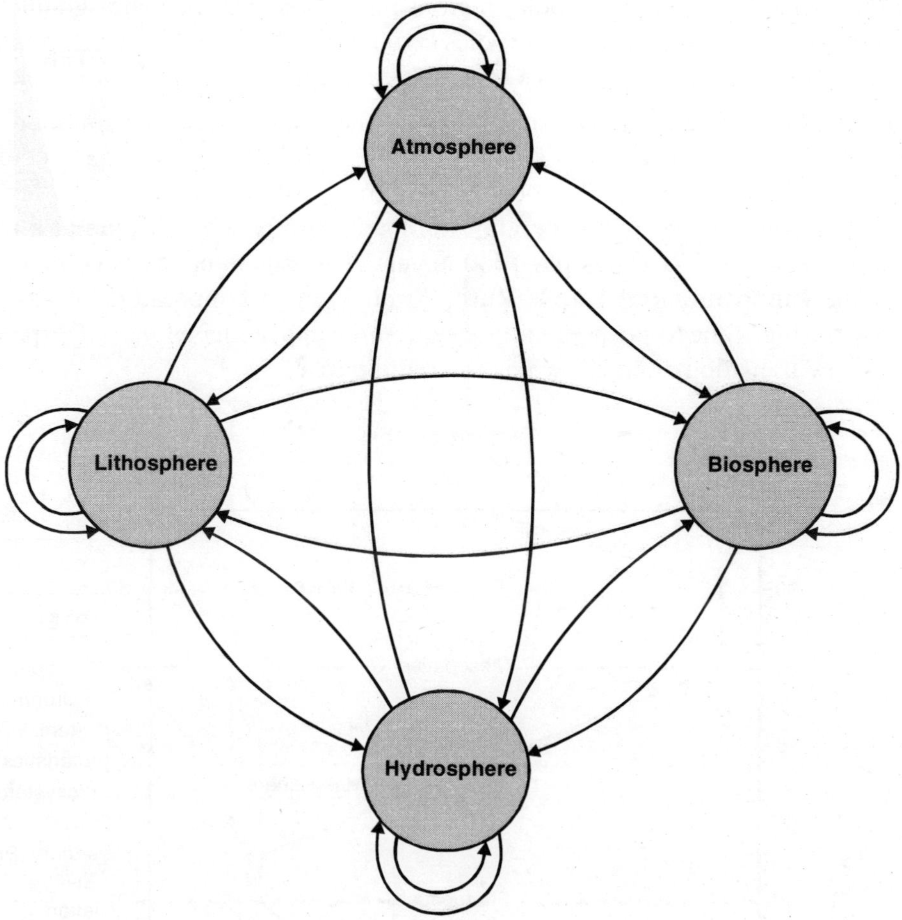

Fig. 2.1. The Four Environmental Spheres

In this conceptual frame, every sphere has a two-way linkage to every other sphere, including itself. The two-way linkage signifies that matter may flow from one compartment to another in both directions. Some arrows show the transfer within a given compartment from one location to another indicating movement of the substance from one physical location to another without leaving the sphere. Since matter cannot be created or destroyed, the question one seeks to answer is the location and chemical form of the substance at given time.

2.2. COMPONENTS OF ENVIRONMENT

2.2.1. Atmosphere

The Earth's atmosphere contains several different layers that can be defined according to air temperature. According to temperature, the atmosphere contains four different layers. The first layer is called the troposphere. The depth of this layer varies from about 8 to 16 kilometers. Greatest depths occur at the tropics where warm temperatures cause vertical expansion of the lower atmosphere.

From the tropics to the Earth's polar regions the troposphere becomes gradually thinner. The depth of this layer at the poles is roughly half as thick when compared to the tropics. Average depth of the troposphere is approximately 11 kilometers.

About 80 per cent of the total mass of the atmosphere is contained in troposphere. It is also the layer where the majority of our weather occurs. Maximum air temperature also occurs near the Earth's surface in this layer. With increasing height, air temperature drops uniformly with altitude at a rate of approximately 6.5° Celsius per 1000 meters. This phenomenon is commonly called the **Environmental Lapse Rate.** At an average temperature of –56.5° Celsius, the top of the troposphere is reached. At the upper edge of the troposphere is a narrow transition zone known as the **tropopause**.

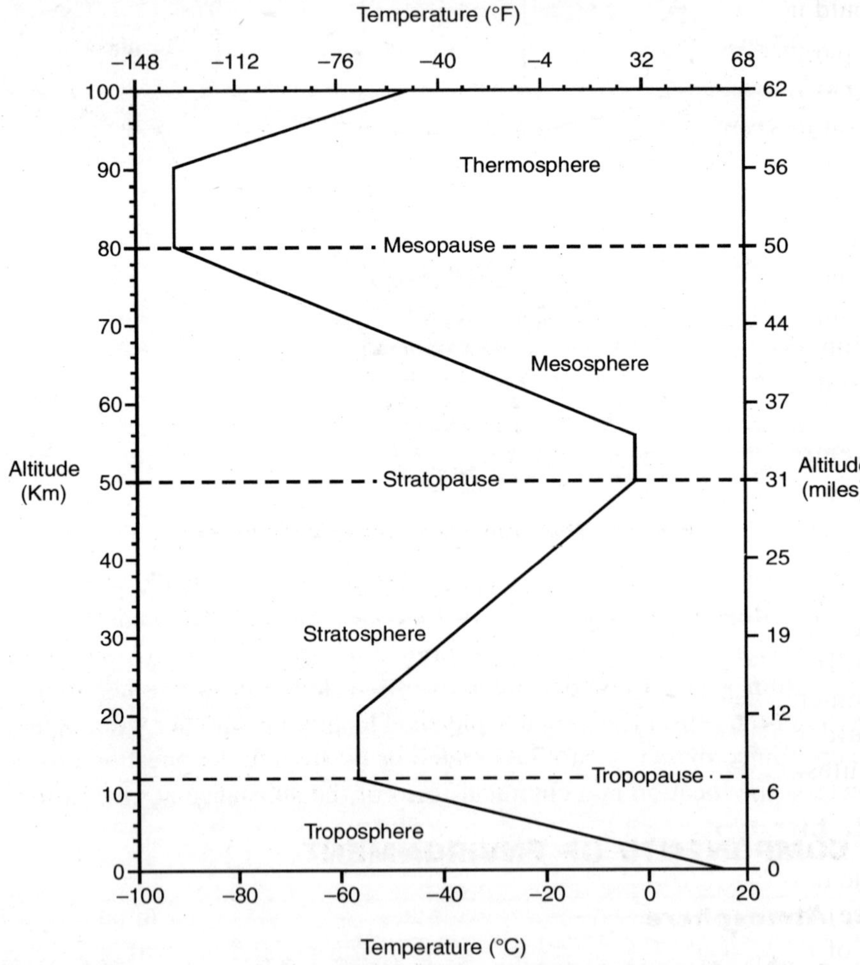

Fig. 2.2. Components of Environment

Above the tropopause is the **stratosphere**. This layer extends from an average altitude of 11 to 50 kilometers above the Earth's surface. This stratosphere contains about 19.9 per cent of the total mass found in the atmosphere. Very little weather occurs in the stratosphere. Occasionally, the top portions of thunderstorms breach this layer. In the first 9 kilometers of the stratosphere, temperature remains constant with height. A zone with constant temperature in the atmosphere is called an **isothermal layer**. From an altitude of 20 to 50 kilometers, temperature increases with an increase in altitude. The higher temperatures found in this region of the stratosphere occurs because of a localized concentration of **ozone** gas molecules. These molecules absorb ultraviolet sunlight creating heat energy that warms the stratosphere. Ozone is primarily found in the atmosphere at varying concentrations between the altitudes of 10 to 50 kilometers. This layer of ozone is also called the **ozone layer**. The ozone layer is important to organisms at the Earth's surface as it protects them from the harmful effects of the sun's ultraviolet radiation. Without the ozone layer life could not exist on the Earth's surface.

Separating the **mesosphere** from the stratosphere is transition zone called the **stratopause**. In the mesosphere, the atmosphere reaches its coldest temperatures (about –90° Celsius) at a height of approximately 80 kilometers. At the top of the mesosphere is another transition zone known as the **mesopause**.

The last atmospheric layer has an altitude greater than 80 kilometers and is called the **thermosphere**. Temperatures in this layer can be as high as 1200°C. These high temperatures are generated from the absorption of intense solar radiation by oxygen molecules (O_2). While these temperatures seem extreme, the amount of heat energy involved is very small. The amount of heat stored in a substance is controlled in part by its mass. The air in the thermosphere is extremely thin with individual gas molecules being separated from each other by large distances. Consequently, measuring the temperature of thermosphere with a thermometer is a very difficult process. Thermometers measure the temperature of bodies via the movement of heat energy. Normally, this process takes a few minutes for the conductive transfer of kinetic energy from countless molecules in the body of a substance to the expanding liquid inside the thermometer. In the thermosphere, our thermometer would lose more heat energy from radiative emission then what it would gain from making occasional contact with extremely hot gas molecules.

2.2.2. Lithosphere

The lithosphere contains all of the cold, hard solid land of the planet's crust (surface), the semi-solid land underneath the crust, and the liquid land near the center of the planet. It is composed of soil particles and the underlying rocks down to a depth of 50km. The soil layer is also referred to as the pedosphere, a mixture of inorganic and organic solid matter, air, water and microorganisms. Within the soil, biochemical reactions by microorganisms are responsible for most of the chemical changes of matter. However, soil and rock are mainly storage compartments for deposited matter.

The solid, semi-solid, and liquid land of the lithosphere form layers that are physically and chemically different. If someone were to cut through Earth to its center, these layers would be revealed like the layers of an onion. The outermost layer of the lithosphere consists of loose soil rich in nutrients, oxygen, and silicon. Beneath that layer lies a very thin, solid crust of oxygen and silicon. Next is a thick, semi-solid mantle of oxygen, silicon, iron, and magnesium. Below that is a liquid outer core of nickel and iron. At the center of Earth is a solid inner core of nickel and iron.

2.2.3. Hydrosphere

The hydrosphere is composed of all of the water on or near the earth. This includes the oceans, rivers, lakes, and even the moisture in the air. It ranges from 10 to 20 kilometers in thickness. The hydrosphere extends from Earth's surface downward several kilometers into the lithosphere and upward about 12 kilometers into the atmosphere.

The hydrosphere may be envisioned as two compartments: a conveyor, a river system that collects the substances within the water shed and delivers them to the second hydrologic compartment, oceans.

Ninety-seven per cent of Earth's water is salty. The salty water collects in deep valleys along Earth's surface. These large collections of salty water are referred to as oceans. The remaining three per cent is fresh water; three-quarters of the fresh water is solid and exists in ice sheets. This water flows as precipitation from the atmosphere down to Earth's surface, as rivers and streams along Earth's surface, and as groundwater beneath Earth's surface. Most of Earth's fresh water, however, is frozen.

2.2.4. Biosphere

The biosphere is composed of all living organisms. Plants, animals, and one-celled organisms are all part of the biosphere. Most of the planet's life is found from three meters below the ground to thirty meters above it and in the top 200 meters of the oceans and seas.

Within the biosphere, living things form ecological communities based on the physical surroundings of an area. These communities are referred to as **biomes**. Deserts, grasslands, and tropical rainforests are three of the many types of biomes that exist within the biosphere. It is impossible to detect from space each individual organism within the biosphere. However, biomes can be seen from space.

It occupies the least volume of all of the spheres but it is the heart, or the chemical pump, of much of the flow of matter through nature. Weathering through the hydrological cycle, blowing wind, and volcanic releases are the other mobilizing agents. The biosphere is responsible for the grand scale recycling of energy and matter on Earth. The mobilization of matter by biota is by no means restricted to small geographic regions. The periodic burning of forests and savannas, for example, not only change the chemical form of matter, but also result in long-range atmospheric transport and deposition. Some of the biologically released chemicals, including carbon, nitrogen, and sulfur have long atmospheric residence times, resulting in continental and global-scale redistribution.

2.3. CONCEPT OF ECOSYSTEM

The term Ecology was coined by Eaenst Haeckel in 1869. It is derived from the Greek words *Oikos*- home + *logos*- study. So ecology deals with the study of organism in their natural home interacting with their surroundings. An ecosystem includes a group of biotic communities of species interacting with one another in an area, plus their surroundings, plus all the ways in which they interact with each other and exchanging energy and matter. Now ecology is often defined as 'the study of ecosystem.'

An ecosystem was defined as a dynamic entity composed of a biological community and its associated abiotic environment. Often the dynamic interactions that occur within an ecosystem are numerous and complex. Ecosystems are also always undergoing alterations to their biotic and abiotic components.

2.4. ECOSYSTEM CHARACTERISTICS

Ecosystem show large variations in their size, structure, composition etc. however, all the ecosystems are characterized by certain basic structural and functional features which are common.

2.4.1. Structural Features

Ecosystems are composed of a variety of abiotic and biotic components that function in an interrelated fashion. Some of the more important components are:

2.4.2. Abiotic Structure

The physical and chemical components of an ecosystem constitute its abiotic structure. It includes climatic factors, soil factors, geographical factors, energy, nutrients and toxic substances.

Physical Factors and Climatic Regime

The sunlight and shade, intensity of solar flux, duration of sun hours, average temperature, annual rain fall, wind, latitude and altitude, soil type, water availability etc. are some of the important physical features which have a strong influence on the ecosystem.

Soils *are much more complex than simple sediments. They contain a mixture of weathered rock fragments, highly altered soil mineral particles, organic matter, and living organisms. Soils provide nutrients, water, a home,* *and a structural growing medium for organisms. The vegetation found growing on top of a soil is closely linked to this component of an ecosystem through nutrient cycling.*

The ***atmosphere*** provides organisms found within ecosystems with carbon dioxide for photosynthesis and oxygen for respiration. The processes of evaporation, transpiration, and precipitation cycle water between the atmosphere and the Earth's surface.

Solar radiation *is used in ecosystems to heat the atmosphere and to evaporate and transpire water into the atmosphere. Sunlight is also necessary for photosynthesis. Photosynthesis provides the energy for plant growth and metabolism, and the organic food for other forms of life.*

Most living tissue is composed of a very high percentage of *water*, up to and even exceeding 90 per cent. The protoplasm of a very few cells can survive if their water content drops below 10 per cent, and most are killed if it is less than 30-50 per cent. Water is the medium by which mineral nutrients enter and are translocated in plants. It is also necessary for the maintenance of leaf turgidity and is required for photosynthetic chemical reactions. Plants and animals receive their water from the Earth's surface and soil. The original source of this water is precipitation from the atmosphere.

Chemical Factors : Inorganic substances or the major essential nutrients like carbon, nitrogen, phosphorous, potassium, hydrogen, oxygen and sulphur which involve in the cycling of materials in the ecosystem and organic substances like protein, carbohydrates, lipids, humic substances etc. which largely form the living body and link the abiotic and biotic components. These chemical factors includes toxic substances, salt causing salinity and various organic substances present in the soil or water largely influence the functioning of the ecosystem.

2.4.3. Biotic Structure

Ecosystems are composed of a variety of *living organisms*. The plants, animal and microorganisms present in an ecosystem form the biotic components. These organisms have different nutritional behavior and status in the ecosystems and are accordingly known as Producers, consumers, based on how they get their food :

Producers or autotrophs are organisms that can manufacture the organic compounds they use as sources of energy and nutrients. Producers or energy transducers convert solar energy into chemical energy with the help of inorganic substances such as water and CO_2 and organic substances such as enzymes. These producers are *autotrophic* (auto-self, tropic-nourishing) organisms, largely green plants (trees, grasses, tiny phytoplankton etc.). Most producers are green plants that can manufacture their food through the process of photosynthesis. They possess a green pigment called chlorophyll, which transducts solar energy. These autotrophs are called *photoautotrophs* because they utilize light energy.

These are other type of autotrophs, called *chemoautotrophs*, which use energy generated in an oxidation-reduction process, but their importance in the ecosystem as producers is minimal.

Consumers or *heterotrophic* organism get their energy and nutrients by feeding directly or indirectly on producers. All the organisms that can not make their own food (and need producers) are called heterotrophs (hetro-other, tropic-nourishing). In an ecosystem heterotrophs are called consumers because they depend on others. They obtain food by eating other organisms. There are different levels of consumers. Those that feed directly from producers, *i.e.* organisms that eat plant or plant products are called primary consumers.

Organisms that feed on primary consumers are called secondary consumers. Those who feed on secondary consumers are tertiary consumers. Some organisms, like the squirrel are at different levels. When the squirrel eats acorns or fruits (which are plant product), it is a primary consumer; however, when it eats insects or nestling birds, is it is a tertiary consumer. Consumers are also classified depending on what they eat.

1. **Herbivores** are those that eat only plants or plant products. Examples are grasshoppers, mice, rabbits, deer, beavers, moose, cows, sheep, goats and groundhogs.
2. **Carnivores**, on the other hand, are those that eat only other animals. Examples of carnivores are foxes, frogs, snakes, hawks, and spiders.
3. **Omnivores** are the last type and eat both plants (acting primary consumers) and meat (acting as secondary or tertiary consumers). Examples of omnivores are :
   - ***Bears*** **:** They eat insects, fish, moose, elk, deer, sheep as well as honey, grass, and sedges.
   - ***Turtles*** **:** They eat snails, crayfish, crickets, earthworms, but also lettuce, small plants, and algae.
   - ***Squirrels*** **:** They eat insects, moths, bird eggs and nestling birds and also seeds, fruits, acorns, and nuts.
   - Human.

Plants and animals supply organic matter to the soil system through shed tissues and death. Consumer organisms that feed on this organic matter, or detritus, are known as **Detritivores** or **Decomposers**. Decomposers are also heterotrophic organisms but depend upon dead organic matter for their food and that organic matter is eventually converted back into inorganic nutrients in the soil. They are chiefly micro-organisms like bacteria, actinomycetes and fungi. Some invertebrate animals like protozoa, oligochaetes such as earth worms use the dead organic matter for their food. When producers, consumers and decomposers die, the decomposers recycle the nutrients tied up in dead bodies. Usually these nutrients end up back in the soil, and they are once again available to plants.

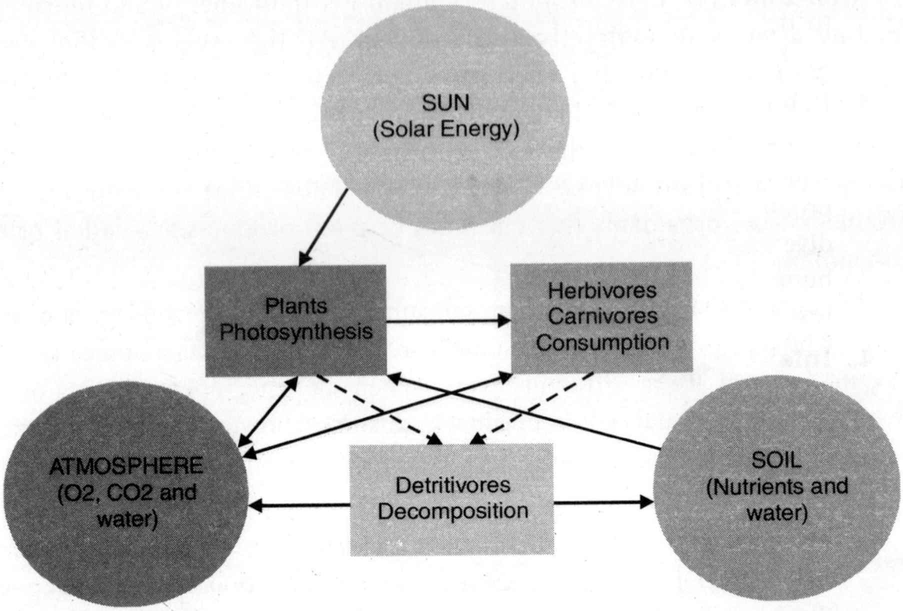

Fig. 2.3. The Major Ecosystem Components and Their Interrelationships

Abiotic Components	Biotic Components
Sunlight	Primary producers
Temperature	Herbivores
Precipitation	Carnivores
Water or moisture	Omnivores
Soil or water chemistry (e.g., P, NH_4+)	Detritivores
All of these vary over space/time	

Functional Attributes

Every ecosystem performs under natural conditions in a systematic way. From the operational point of view, the abiotic and biotic components of an ecosystem are so interwoven into the fabric of nature that their separation from each other becomes particularly very difficult.

Although a wide range of ecosystem functions and their associated goods and services are grouped into four primary categories (De Groot *et al*, 2000):

1. **Regulation functions :** This group of functions relates to the capacity of natural and semi natural ecosystems to regulate essential ecological processes and life support systems through bio-geochemical cycles and other biospheric processes. In addition to maintaining ecosystem (and biosphere) health, these regulation functions provide many services,

which have direct and indirect benefits to humans (such as clean air, water and soil, and biological control services).

2. **Habitat Functions :** Natural ecosystems provide refuge and reproduction-habitat to wild plants and animals and thereby contribute to the (in situ) conservation of biological and genetic diversity and evolutionary processes.
3. **Production Functions :** Photosynthesis and nutrient uptake by autotrophs converts energy, carbon dioxide, water and nutrients into a wide variety of carbohydrate structures which are then used by secondary producers to create an even larger variety of living biomass. This broad diversity in carbohydrate structures provides many ecosystem goods for human consumption, ranging from food and raw materials to energy resources and genetic material.
4. **Information Functions :** Because most of human evolution took place within the context of undomesticated habitat, natural ecosystems provide an essential 'reference function' and contribute to the maintenance of human health by providing opportunities for reflection, spiritual enrichment, cognitive development, re-creation and aesthetic experience.

   Although the rank-order of the function categories is somewhat arbitrary, there is an underlying logic in their ordering. The first two function-groups (regulation and habitat) are essential to the maintenance of natural processes and components, and are therefore conditional to the maintenance of the availability of the other two function-groups.

Table below provides an overview of the main functions, goods and services that can be attributed to natural ecosystems, and their associated ecological structures and processes.

S. No	Functions	Ecosystem Processes and Components	Goods and Services (examples)
Regulation Functions		*Maintenance of essential ecological processes and life support systems*	
1.	Gas regulation	Role of ecosystems in bio-geochemical Cycles (e.g. CO_2/O_2 balance, ozone layer, etc.)	UV-protection by O_3 (preventing disease); Maintenance of (good) air quality; Influence on climate ;
2.	Climate regulation	Influence of land cover and biol. mediated processes (e.g. DMS production)on climate	Maintenance of a favorable climate (temp., precipitation, etc.) for, for example, human habitation, health, cultivation;
3.	Disturbance prevention	Influence of ecosystem structure on dampening environmental disturbances	Storm protection (e.g. by coral reefs); Flood prevention (e.g. by wetlands and forests);

S. No	Functions	Ecosystem Processes and Components	Goods and Services (examples)
4.	Water regulation	Role of land cover in regulating runoff and river discharge	Drainage and natural irrigation; Medium for transport;
5.	Water supply	Filtering, retention and storage of fresh water (e.g. in aquifers)	Provision of water for consumptive use (e.g. drinking, irrigation and industrial use)
6.	Soil retention	Role of vegetation root matrix and soil biota in soil retention	Maintenance of arable land; Prevention of damage from erosion /siltation;
7.	Soil formation	Weathering of rock, accumulation of organic matter	Maintenance of productivity on arable land; Maintenance of natural productive soils;
8.	Nutrient regulation	Role of biota in storage and recycling of nutrients (e.g. N,P&S)	Maintenance of healthy soils and productive ecosystems
9.	Waste treatment	Role of vegetation and biota in removal or breakdown of xenic nutrients and compounds	Pollution control/detoxification; Filtering of dust particles; Abatement of noise pollution;
10.	Pollination	Role of biota in movement of floral gametes	Pollination of wild plant species; Pollination of crops;
11.	Biological control	Population control through Trophic – dynamic relations	Control of pests and diseases; Reduction of herbivory (crop damage);
Habitat Functions		*Providing habitat (suitable living space) for wild plant and animal species*	
12.	Refugium function	Suitable living space for wild plants and animals	Maintenance of biological and genetic diversity (and thus the basis for most other functions);
13.	Nursery Function	Suitable reproduction habitat	Maintenance of commercially harvested species
Production Functions		*Provision of natural resources*	
14.	Food	Conversion of solar energy into edible plants and animals	Hunting, gathering of fish, game, fruits, etc.; Small-scale subsistence farming and aquaculture;
15.	Raw materials	Conversion of solar energy into biomass for human construction and other uses	Building and Manufacturing (e.g. lumber, skins); Fuel and energy (e.g. fuel wood, organic matter); Fodder and fertilizer (e.g. krill, leaves, litter);

S. No	Functions	Ecosystem Processes and Components	Goods and Services (examples)
16.	Genetic resources	Genetic material and evolution in wild plants and animals	Improve crop resistance to pathogens and pests; Other applications (e.g. health care);
17.	Medicinal resources	Variety in (bio) chemical substances in, and other medicinal uses of, natural biota	Drugs and pharmaceuticals; Chemical models and tools; Test- and essay organisms;
18.	Ornamental resources	Variety of biota in natural ecosystems with (potential) ornamental use	Resources for fashion, handicraft, jewelry, pets, worship, decoration and souvenirs (e.g. furs, feathers, ivory, orchids, butterflies, aquarium fish, shells, etc.)
Information Functions		*Providing opportunities for cognitive development*	
19.	Aesthetic information	Attractive landscape features	Enjoyment of scenery (scenic roads, housing, etc.)
20.	Recreation	Variety in landscapes with (potential) recreational uses	Travel to natural ecosystems for eco-tourism, out door spots, etc.
21.	Cultural and artistic information	Variety in natural features with cultural and artistic value	Use of nature as motive in books, film, painting, folklore, national symbols, architect, advertising, etc.
22.	Spiritual and historic information	Variety in natural features with spiritual and historic value	Use of nature for religious or historic purposes (*i.e.* heritage value of natural ecosystems and features)
23.	Science and education	Variety in nature with scientific and educational value	Use of natural systems for school excursions, etc. Use of nature for scientific research

Adapted from Costanza *et al*, (1997), de Groot (1992), de Groot *et al*, (2000).

2.5. ENERGY FLOW IN ECOSYSTEM

Many of the most important relationships between living organisms and the environment are controlled ultimately by the amount of available incoming energy received at the Earth's surface from the sun. It is this energy which helps to drive biotic systems. The sun's energy allows plants to convert inorganic chemicals into organic compounds.

Only a very small proportion of the sunlight received at the Earth's surface is transformed into biochemical form. Several studies have been carried out to determine this amount. Most ecosystems fix less than 1 per cent of the sunlight available for photosynthesis.

Living organisms can use energy in basically two forms: radiant or fixed. *Radiant energy* exists in the form of electromagnetic energy, such as light. *Fixed energy* is the potential chemical energy found in organic substances. This energy can be released through respiration. Organisms that can take energy from inorganic sources and fix it into energy rich organic molecules are called autotrophs. If this energy comes from light then these organisms are called photosynthetic autotrophs. In most ecosystems plants are the dominant photosynthetic autotroph.

Organisms that require fixed energy found in organic molecules for their survival are called heterotrophs. Heterotrophs who obtain their energy from living organisms are called consumers.

Flow of energy in an ecosystem takes place through the food chain and it is this energy flow which keeps the ecosystem going. The most important feature of this energy flow is that it is unidirectional or one way flow. Unlike the nutrients which move in a cyclic manner and are reused by the producers after flowing through the food chain, energy is not reused in the food chain.

2.6. ENERGY FLOW MODELS

The flow of energy through various trophic levels in an ecosystem can be explained with the help of various energy flow models.

Universal energy flow model – energy flow through an ecosystem was explained by E. P. Odum as the universal energy flow model. As the flow of energy takes place, there is a gradual loss of energy at every level, thereby resulting in less energy available at next tropic level as indicated by narrower pipes (energy flow) and smaller boxes (stored energy in biomass). The loss of energy is mainly the energy not utilized (NU).

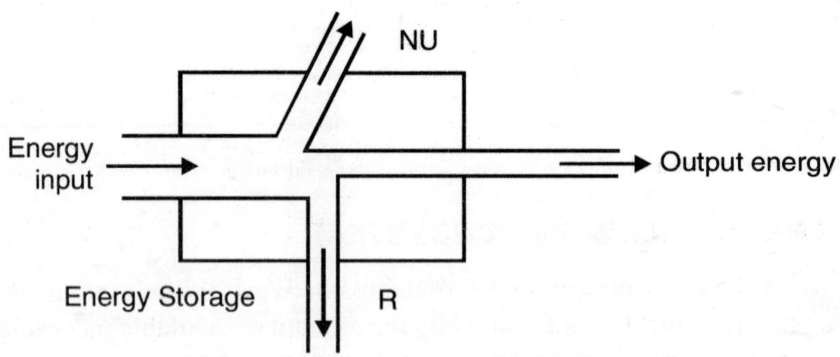

Fig. 2.4. Universal Energy Flow Model

Single channel energy flow model—the flow of energy takes place in a single channel of green plants or producers to herbivores and carnivores. Figure depicts such a model and illustrated the gradual decline the energy level due to loss of energy at each successive trophic level in a grazing food chain.

There are certain relationships to understand the above mentioned things :

1.	Solar energy assimilated by photoautrophs	=	NPP (chemical energy)	+	R (heat energy of respiration)
2.	NPP eaten by heterotrophs (herbivores)	=	Chemical energy of assimilated by herbivores	+	Chemical energy of faeces produced by herbivores
3.	Chemical energy of assimilated by herbivores	=	Chemical energy of secondary production of herbivores (growth and reproduction)	+	Metabolic energy loss by herbivores
4.	Chemical energy of secondary production of herbivores (growth and reproduction)	=	Chemical energy assimilated by carnivores	+	Chemical energy of faeces of carnivores
5.	Chemical energy assimilated by carnivores	=	Chemical energy of sec. production (growth and reproduction) of carnivores	+	Metabolic energy loss by carnivores
6.	Consumption	=	Assimilation	+	Egesta, faeces and urine
7.	Assimilation	=	Production	+	Respiration
8.	Production	=	Elimination (the biomass of individuals that have died or been killed)	+	Change in biomass (growth and reproduction)

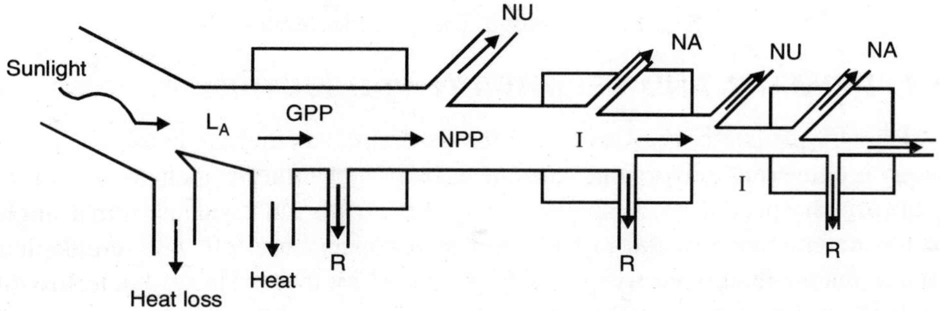

Fig. 2.5. Single Channel Energy Flow Model

The above figure shows a simplified energy flow diagram depicting three tropic levels in a linear food chain. I – total energy input; L_A – light absorbed by plant cover. GPP – Gross Primary Production; NPP – Net Primary Production; NU – energy not used (stored or exported); NA – energy not assimilated by consumers (egested); R – Respiration.

Double channel or Y-shaped energy flow model – in nature, both grazing food chain and detritus food chain operate in the same ecosystem. However, sometimes it is the grazing food chain which predominates. It happens in marine ecosystem where primary production in the open sea is limited and a major portion of it is eaten by herbivorous marine animals. On the other hand, in a forest ecosystem the huge quantity of biomass produced can not be all consumed by herbivores. Rather, a large proportion of the live biomass enters into detritus (dead) compartment in the form of litter. Hence the detritus food chain is more important there.

The two channels or Y shaped model of energy flow shows the passage of energy through both food chain i.e. grazing and detritus food chain, which separated in time and space.

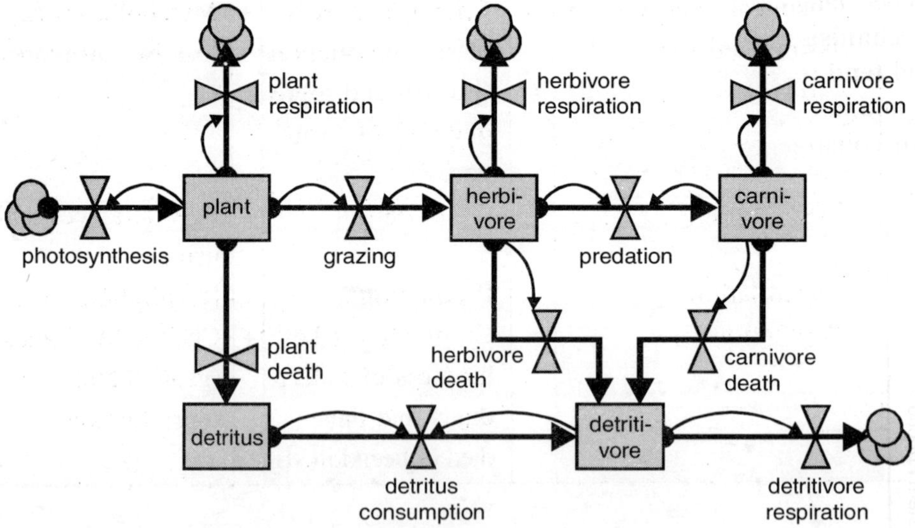

Fig. 2.6. Y-shaped Energy Flow Model

2.7. PRIMARY AND SECONDARY PRODUCTION

Primary productivity of an ecosystem is defined as the rate at which rediant energy is converted into organic substances by photosynthesis or chemo-synthesis by the primary producers. Respiratory loss of energy is must because it is required for the maintenance of the organism. The producers are left with a little less organic matter than what was actually produced by them. This is known as the Net Primary Production (NPP) and the respiratory loss (R) added to it gives the Gross Primary Production (GPP).

$$NPP = GPP - R$$

The energy stored at consumer level for use by next trophic level is thus defined as secondary production.

2.8. ECOSYSTEM REGULATION

All ecosystems regulate themselves and maintain themselves under a set of environmental conditions. Any environmental stress tries to disturb the normal ecosystem functions. However, the ecosystem, by itself, tries to resist the change and maintain itself in equilibrium with the environment due to a property known as **homeostasis**. Homeostasis is the inherent property of all living system to resist changes. However, the system can show this tolerance or resistance only within a maximum and a minimum range, which is its range of tolerance known as *homeostasis plateau.* Within this range, if stress tries to cause a deviation, then the system has its own mechanism to counteract these deviations which are known as **negative feedback mechanisms. So negative feedback mechanisms are deviation counteracting mechanisms which try to bring the system back to its ideal conditions.** But, if the stress is too high and beyond the range of homeostatic plateau, then another type of mechanism known as **positive feedback mechanisms. So the positive feedback mechanisms add to stress conditions and tend to take the system away form the optimal conditions.** Human being should not contribute to positive feedback mechanism otherwise the ecosystem will collapse.

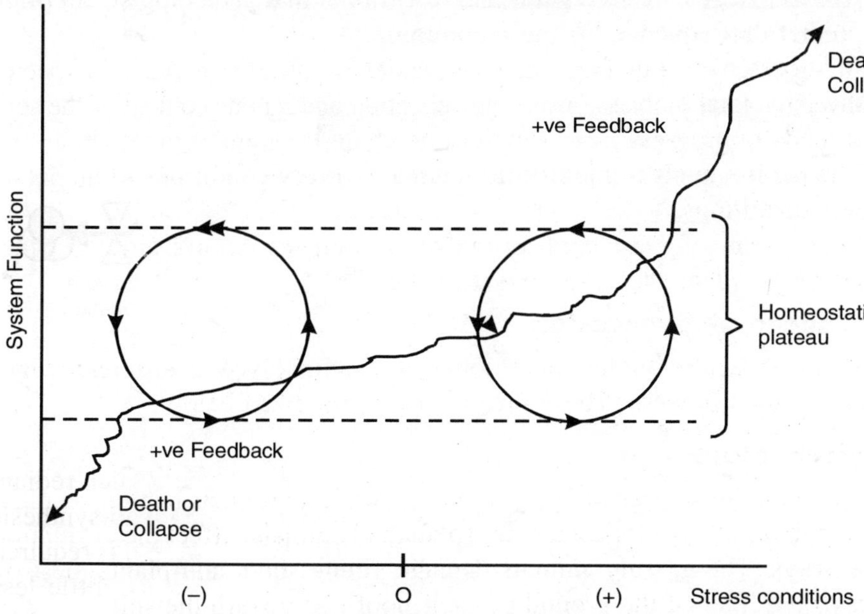

Fig. 2.7. Ecosystem Regulation

2.9. ECOLOGICAL SUCCESSION

A group of similar individuals able to breed among themselves constitute to species. A group of individuals of a particular species of a particular species occupying a definite space is referred as population. Many mutually adjusted populations of plants inhabiting a particular habitat make plant community. The community is not stable but a dynamic system. The gradual replacement of one type of plant community by the other is called plant succession. According to E. P. Odum (1971), "it is an orderly process of community change in a unit area". He prefers to call this orderly process as ecosystem development rather then the more often known as ecological succession.

Definition ***D. B. Sutton and N. P. Harmon (1973), define ecological succession as "the dynamic process by which ecosystem change in order to develop greater stability over the course of time."***

Ecological succession is defined as an orderly process of changes in the community structure and function with time mediated through modifications in the physical environment and ultimately culminating in a stabilized ecosystem known as climax.

2.9.1. Characteristics of Ecological Succession

1. It tends to progress from unstable biotic community to stable biotic community, i.e., complete adjustment with the environment.
2. Its seral stages are so regular and directional that an ecologist can often predict the sequence of future community.
3. In successive seral stages there is tendency towards increase in species diversity, total biomass, niche specialization and humus content of the soil.
4. It tends to progress from simple food chain to complex food chain.
5. The habitat tends to modify from aquatic or dry conditions to moderate wet conditions.
6. Succession of plant and animal communities occurs side by side. However, plant succession is easily visible.

2.9.2. Causes of Succession

A variety of factors, biotic and abiotic, may be involved at different stages of the successional process. These factors are summarized below:

(1) Biotic Factors

It include the influences of living organisms of a particular habitat on plant life of that habitat, many biotic agencies (plant and animals) affect the vegetation in several ways. The grazing animals damage young shoot and plants, destroy water absorbing litter of the ground by their hoof and disturb the soil.

Man is capable of destroying the whole of vegetation community of a given habitat within no time through his destructive activities such as by cutting, clearing, harvesting, cultivation and deforestation etc.

Vegetation may also be damaged by the attack of insects, pests and fungi. The competition among plants for space, light, nutrient, water also affect plant community in a specific habitat.

(2) Abiotic Factors

The development of plant community is affected and controlled by following abiotic factors :

Climatic factors : Temperature, sun-light, lightening, wind velocity, soil moisture, humidity, snow fall, hails etc.

Edaphic factors : Soil texture, soil structure, nutrient content of soils, soil reaction, soil temperature, soil erosion and soil organisms

Physio-graphic factors : It include nature of the ground, mountains, plains. Plateau, altitude, slope, depth etc.

Fire factors : Fire is very much involved in regulating the plant comities. The fire includes both natural forest fires through lightening and man caused forest fires due to deliberate actions.

Types of Succession

A Sere (the transitional series of communities) always starts on a bare area such as rock, sand, clay-gravel and water body. Depending upon the nature of bare area on which it develops, F.E. Clements (1916) has divided succession of plant community in two types:

Primary succession begins on extremely bare area or on the primitive substratum on which there was no previous existence of vegetation. Such areas or sites may be new2ly exposed rocks due to melting of ice from the glacial areas, newly formed sand dunes, flood plains formed by recent alluvia, heaps of debris, newly formed pond, newly formed sea floor, cooled and solidified basaltic surface due to recent lava flows, exposed lake bed due to drying of water etc where the conditions of existence may be favorable initially. The rate of succession is very slow in such type of newly formed bare habitat.

Secondary succession refers to community development on the secondary bare area which was once occupied by vegetation but now have been rendered nude or bare due to destruction of vegetation either by natural forces such as sudden change in climatic conditions, prolonged drought, glaciation, lava flow, severe storms, natural forest fires etc. or by human interference such as over grazing, deforestation, massive land use changes, intentional burning of vegetation, jhuming cultivation etc. it is important to note that such bear habitat still contain mature soil hence the secondary succession is comparatively more rapid.

Depending upon the nature of environment *i.e.* moisture present in substrata and staring on different types of areas or ecological successions is named differently as follows :

Hydrarch or Hydrosere : Starting in watery area like pond, swamp, bog.

Mesarch : Starting in an area of adequate moisture.

Xerarch or Xerosere : Starting in a dry area with little moisture. They can be the following types :

Lithosere : starting on a bare rock
Psammosere : starting on sand
Halosere : starting on saline soil

Process of Succession

The process of succession takes place in a systematic order of sequential steps as follows :

Nudation : It is the development of a bare area without any life form. The bare area may be caused due to topographic factor (land slide/ volcanic eruption), climatic factor (drought) or biotic factor (overgrazing).

Invasion : It is the successful establishment of one or more species on a bare through dispersal or migration, followed by ecesis or establishment. As growth and reproduction starts, these pioneer species increase in number and form groups or aggregations.

Competition and Coaction : As the number of individuals grows there is competition, both inter-specific and intra-specific, for space, water and nutrients. They influence each other in a number of ways, known as coaction

Reaction : The living organisms grow, use water and nutrients from the substratum, and in turn, they have a strong influence on the environment which is modified to a large extent and this is known as reaction.

Stabilization : The succession ultimately culminates in a more or less stable community called climax which is in equilibrium with the environment.

Importance of Succession

1. It provides information, which helps to have a controlled growth of one or more species in a forest by preventing the invasion of other species in the area.
2. Pastures can be maintained by not allowing biotic succession to proceed through grazing or fire.
3. It also helps in reforestation and forest management programmes.

Trends of changes in the ecosystem's characteristics during succession :

Ecosystem attribute/ characteristics	Seral stages	Climax stage
Community Structure		
(*i*) Size of individual	Small	Large
(*ii*) Ecological niches	Few, generalized	Many, specialized
(*iii*) Community organization	Simple	Complex
Community Functions		
(*i*) Food chains and food web	Simple	Complex
(*ii*) Efficiency of energy use	Low	High
(*iii*) Nutrient conservation	Low	High

2.10. NUTRIENT CYCLING

Recycling nutrients in the ecosystem is important because only plants can create new nutrients by combining molecules from the soil or air. Not all of the building blocks required by plants are readily available, and they must be carefully conserved within the ecosystem. Other members of the food chains re-use the nutrients assembled by plants.

Nutrients like carbon, nitrogen, sulphur, oxygen, hydrogen, phosphorus etc. Move in circular path through biotic and abiotic components are therefore known as biogeochemical cycles. Water also moves in a cycle, known as hydrological cycle.

2.10.1. Nitrogen Cycle

Nitrogen is the most important essential element required for the metabolism, growth, development and reproduction of all living organisms. The atmospheric air is the great reservoir of the gas. It constitutes about 78 per cent v/v of the atmosphere.

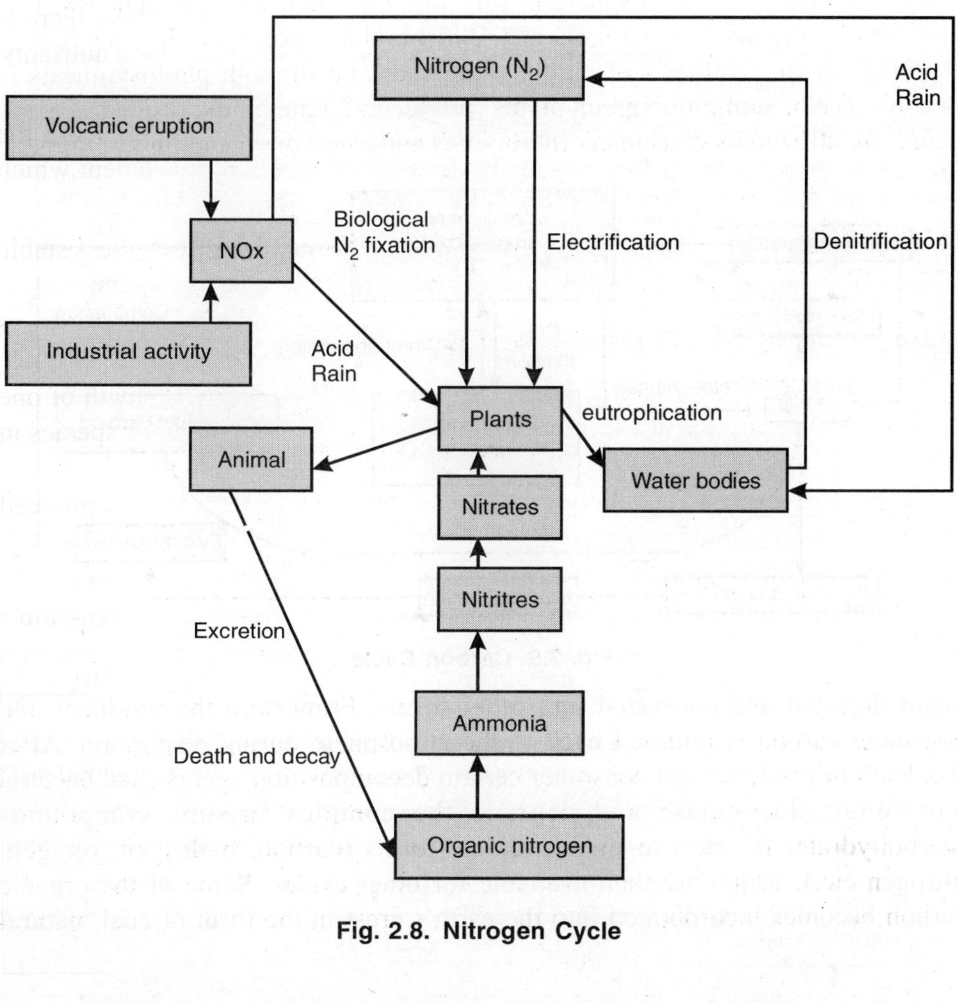

Fig. 2.8. Nitrogen Cycle

However, atmospheric nitrogen is chemically inert and can not directly utilize by most of the living organisms. They need it in the form of soluble nitrogen compounds such as nitrate (NO_3), nitrite (NO_2) and ammonia (NH_3) and amino acids. In fact the elemental nitrogen is converted into usable forms by the process of nitrogen fixation and added to the soil. Nitrogen fixation is possible by the activity of physical (lightening) and biological (micro-organism) agencies. Thus plants obtain their nitrogen supply from the soil and convert them into essential biomolecules needed for their healthy growth and development. On the other hand animal obtain their nitrogen requirements from plants. After the death of plants and animal nitrogen compounds again reach to the soil. The nitrogenous organic compounds in the soil undergo microbial decomposition by the process of ammonification and nitrification. These compounds are again absorbed by the plant root or get converted into free nitrogen through de-nitrification process. Thus a biogeochemical cycle exists and is called as nitrogen cycle.

2.10.2. Carbon Cycle

Carbon is the basic building element of all the living organisms. By volume, carbon dioxide in the atmosphere is very low *i.e.* about 0.03 per cent v/v. It is absorbed by the plant in the form of CO_2 from the atmosphere. It is absolutely essential for the synthesis of carbohydrates and fat through photosynthesis in the presence of sunlight in green plants (producers). Late on the formed in green plants are utilized by consumers (herbivores and carnivores) and thus compound

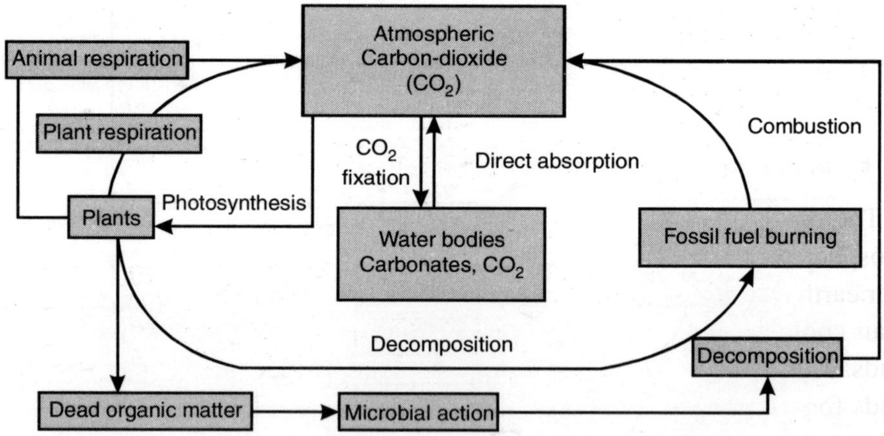

Fig. 2.9. Carbon Cycle

again digested and converted into other forms. From both the producer and consumer carbon is returned back to the atmosphere during respiration. After the death of producer and consumer certain decomposition agents (like bacteria and fungi) decompose and degrade the complex organic compounds (carbohydrate, fat etc.) in their simplest forms (carbon, hydrogen, oxygen, nitrogen etc.), which are then available for other cycles. Some of the organic carbon becomes incorporated into the earth's crust in the form of coal, natural gas, petroleum etc. these fossils (coal, natural gas, petroleum etc.) form the most important sources of energy. CO_2 is also released due to the combustion of these fossil fuels. Thus carbon cycle is mainly maintained by the process of photosynthesis, respiration, decomposition and combustion of fossil fuels.

Sea is the second major reservoir of carbon. Sea water contains 50 times mare CO_2 than air. This is in the form of carbonates and bicarbonates. The CO_2 dissolves in sea water to form carbonic acid which converts carbonate into bicarbonates. Thus the sea regulates the CO_2 content in the atmosphere.

CO_2 has the unique property of absorbing infrared rays of the sum, thus keeping the earth warm. But a huge amount of CO_2 has been released into atmosphere due to extensive industrialization and cutting of green plants. This has resulted in an excessive absorption of infrared radiations leading to an increase in atmospheric temperature. A green house effect has been experienced which had effected the functioning of organism adversely.

2.10.3. Water Cycle

The earth has a limited amount of water. That water keeps going around and around and around and around and in what we call the "Water Cycle".

This cycle is made up of a few main parts :

1. Condensation
2. Infiltration
3. Runoff
4. Evaporation (and transpiration)
5. Precipitation

The process begins with condensation, when water vapor condenses in the atmosphere to form clouds. Condensation occurs when the temperature of the air or earth changes. Water changes states when temperatures fluctuate. So when the air cools enough, water vapor has to condense on particles in the air to form clouds. This process is very noticeable on plants as they dew in the morning. As clouds form, winds move them across the globe, spreading out the water vapor. When eventually the clouds can't hold the moisture, they release it in the form of precipitation, which can be snow, rain, hail, etc.

The next three stages: infiltration, runoff, and evaporation occur simultaneously. Infiltration depends a lot on the permeability of the ground. Infiltration happens when water soaks into the soil from the ground level. It moves underground and moves between the soil and rocks. Some of the water will be soaked up by roots to help plants grow. The plant's leaves eventually release the water into the air through the plant's pores.

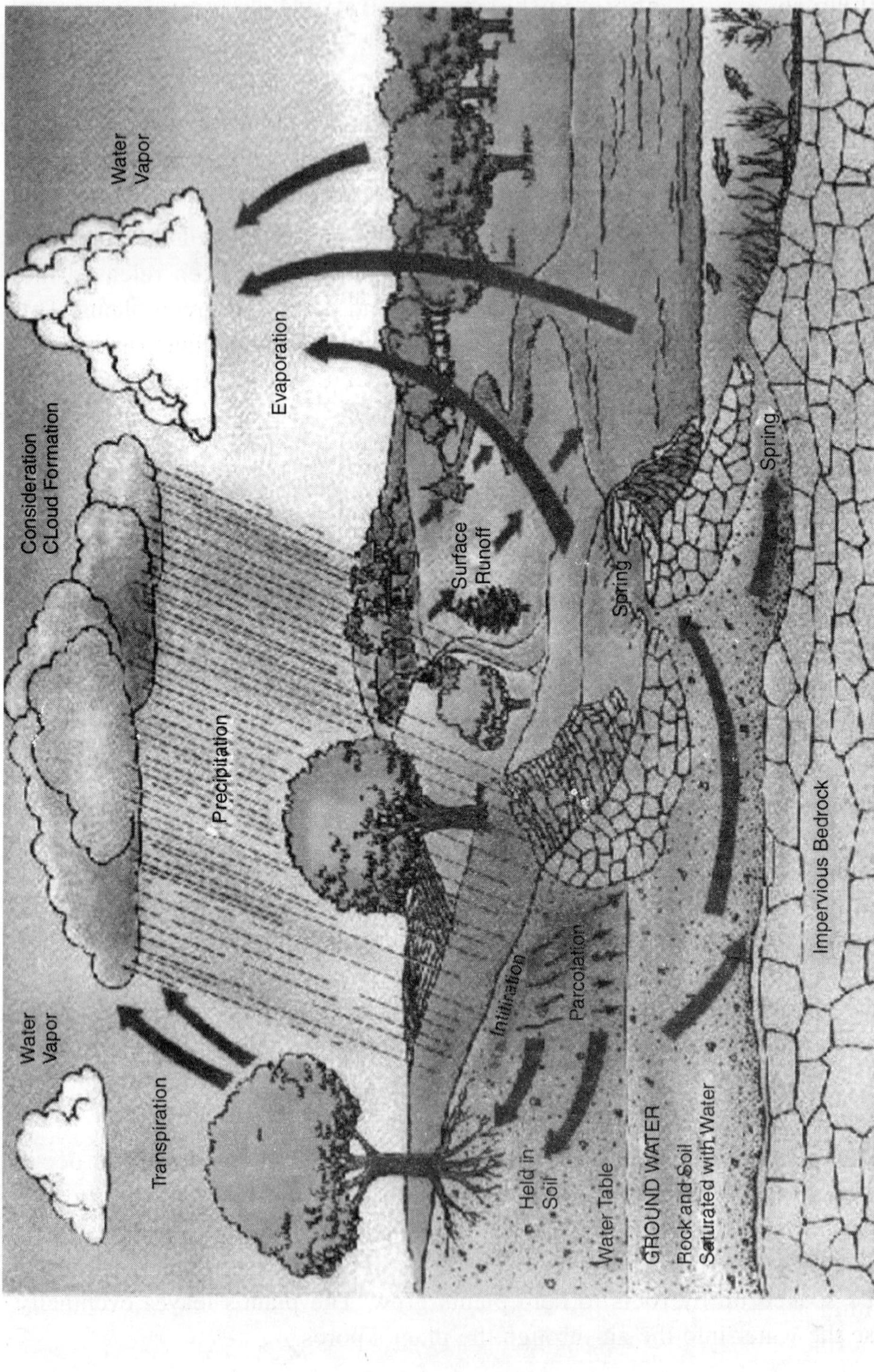

Fig. 2.10. Water Cycle

Some of the water keeps moving down into the soil to a level that is filled with water, called ground water. Ground water is simply water under the ground where the soil is completely filled or saturated with water. This water is also called an "aquifer." Ground water moves underground from areas where the elevation is high, like a hilltop, to places that are lowland areas. Water movement is slow and might move anywhere from less than a millimeter up to a mile in a day. The very top of this layer filled with ground water is called the water table. The Water Table is found underground where the rock and soil begin to be filled or "saturated" with water. It also marks the very top of the ground water layer.

Where the water table meets the land surface, a spring might bubble up or seep from the ground and flow into a lake, stream woodland, or the ocean. When ground water meets the land surface, it flows out and helps keep rivers, streams, lakes and wetlands filled with water. Permeability is the measure of how easily something flows through a substance. The more permeable, the more precipitation seeps into the ground.

If precipitation occurs faster than it can infiltrate the ground, it becomes **runoff.** Runoff remains on the surface and flows into streams, rivers, and eventually large bodies such as lakes or the ocean. Infiltrated groundwater moves similarly as it recharges rivers and heads towards large bodies of water.

As both of these processes are happening, warmth from the sun causes water from lakes, streams, ice, and soils to turn into water vapor in the air by causing **evaporation**. Evaporation is the change of liquid water to a vapor. Sunlight aids this process as it raises the temperature of liquid water in oceans and lakes. As the liquid heats, molecules are released and change into a gas. Almost all of the precipitated water (80 per cent) goes right back into the air because of evaporation. The rest runs off the land or soaks into the ground to become ground water. Transpiration happens when plants give off water vapor through tiny pores in their leaves. This is the plant's way of getting rid of waste, just like people and animals sweat when they're hot! This water vapor evaporates into the air and is stored in the atmosphere until it becomes clouds or precipitation or involved in condensation.

Precipitation is made up of any type of water that falls to the earth like snow, hail, mist, or rain. Precipitation occurs when water vapor in the atmosphere condenses into clouds and falls to the earth. About 300 cu km (about 70 cu mi) of precipitation falls on the land each day. Almost two-thirds of this precipitation re-evaporates into the atmosphere, while the rest flows down rivers to the oceans. Individual storms can produce enormous amounts of precipitation.

2.11. FOOD CHAIN

Within an ecosystem, nutrients are recycled as living things go about their daily business of eating food and excreting waste products. All living things exist as part of a nutrient recycling chain—an enormous chain defined by what eats what. This is called the **food chain**. Food chains are the pathways along which nutrients pass through an ecosystem. Food chains require constant supplies of new energy to make up for the continual losses

The transfer of food energy from the producers through a series of organism (herbivores to carnivores to decomposers) constitutes food chain. The sequence of eating and being eaten in an ecosystem is known as food chain. All organisms, living or dead, are potential food for some other organism and thus, there is essential no waste in the functioning of a natural ecosystem. Each organism in the ecosystem is assigned a feeding level or trophic level depending on its nutritional status.

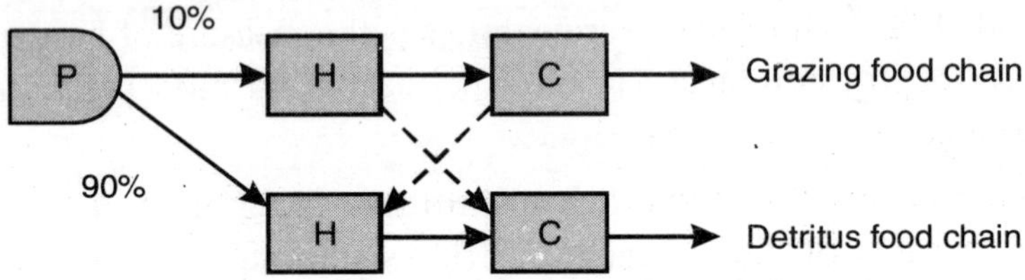

Fig. 2.11. Types of Food Chain

Three basic types of food chains are found in a typical ecosystem through which energy assimilated by plants can flow:

Grazing food chain- it starts with green plants (primary producers) and via herbivores culminates in carnivores. It may be terrestrial and aquatic type. For example:

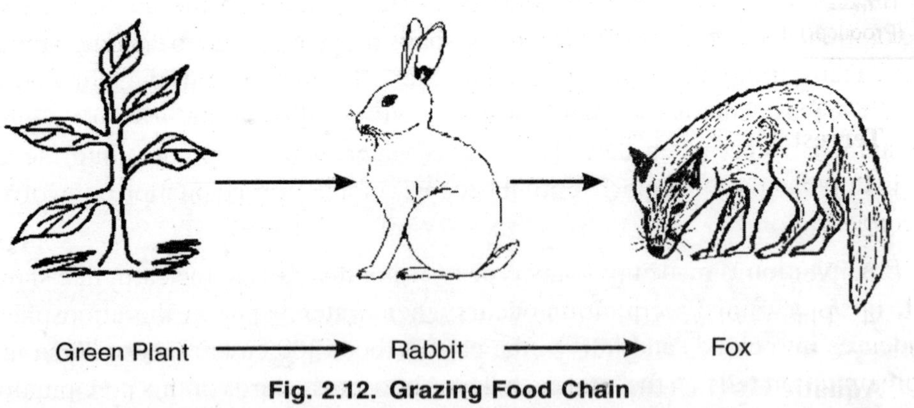

Fig. 2.12. Grazing Food Chain

Detritus Food Chain : Disintegrated dead bodies of plants is called **detritus**. A group minute organism those serves as a source of energy called **detritivores**. The food chain so formed is called detritus food chain. Millipedes, earthworms, fungi, moluscs etc. are the examples of detritivores. The detritus food chain is more common in terrestrial ecosystem than in aquatic ecosystem. It starts with dead organic matter and the detritivores and decomposers consume.

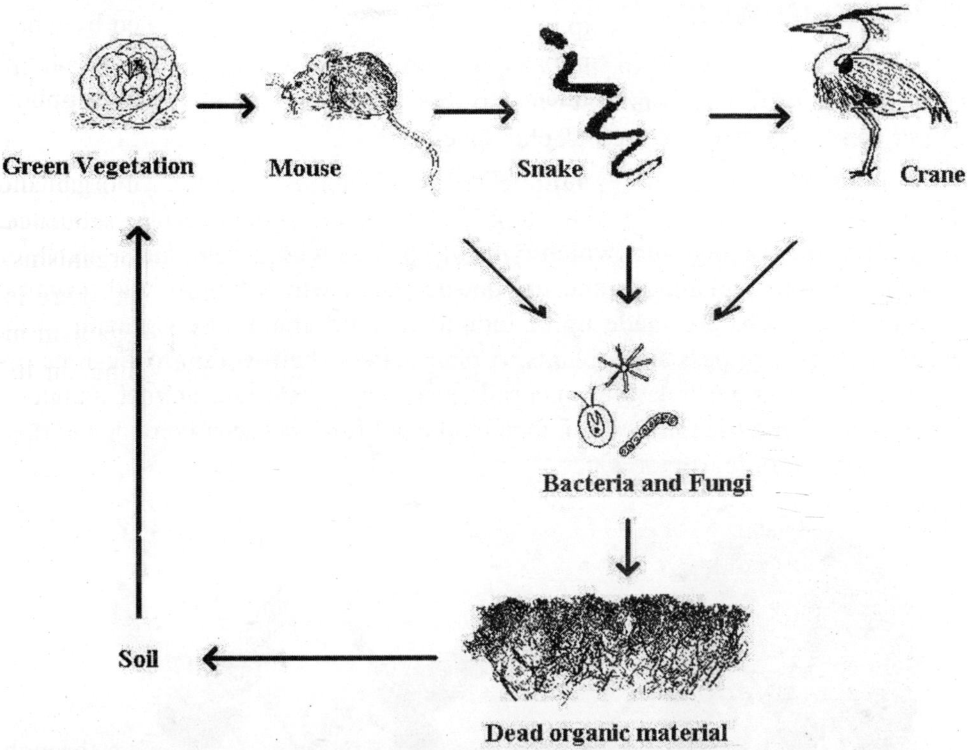

Fig. 2.13. Detritus Food Chain

Parasitic Food Chain : In parasitic food chain either the producer or the consumer is parasitized and therefore the food passes to smaller organism.

Tress (Producer) → Fruit eating birds (Herbivores) → Lice and Bugs (Parasite on Herbivores) → Bacteria and Fungi (Hyper parasite on lice and bugs)

Terrestrial Food Chains :

1. Vegetation → Grasshopper → Shrew → Hawk
2. Vegetation → Rabbit → Fox → Wolf → Tiger
3. Vegetation → Frog → Snake → Peacock
4. Plant → Butterfly → Frog → Snake → Hawk

Aquatic Food Chain :

1. Phytoplankton → Zooplankton → Small Crustaceans → Predator Insects → Small Fish → Stork/King Fisher → Crocodile
2. Phytoplanlton → Zooplankton → Small Fish → Large Fish → Shark
3. Phytoplankton → Zooplankton → Fish → Crane → Hawk

2.12. FOOD WEB

Feed web is a network of food chains where different types of organisms are connected at different trophic levels, so that there are a number of options of eating and being eaten at each trophic level.

Food Web is a set of interconnected food chains by which energy and materials circulate within an ecosystem. The food web is divided into two broad categories: the grazing web, which typically begins with green plants, algae, or photosynthesizing plankton, and the detrital web, which begins with organic debris. These webs are made up of individual food chains. In a grazing web, materials typically pass from plants to plant eaters (herbivores) to flesh eaters (carnivores). In a detrital web, materials pass from plant and animal matter to bacteria and fungi (decomposers), then to detrital feeders (detritivores), and then to their predators (carnivores).

Fig. 2.14. Terrestrial Food Web

Generally, many interconnections exist within food webs. For example, the fungi that decompose matter in a detrital web may sprout mushrooms that are consumed by squirrels, mice, and deer in a grazing web. Robins are omnivores, that is, consumers of both plants and animals, and thus are in both detrital and grazing webs. Robins typically feed on earthworms, which are detritivores that feed upon decaying leaves.

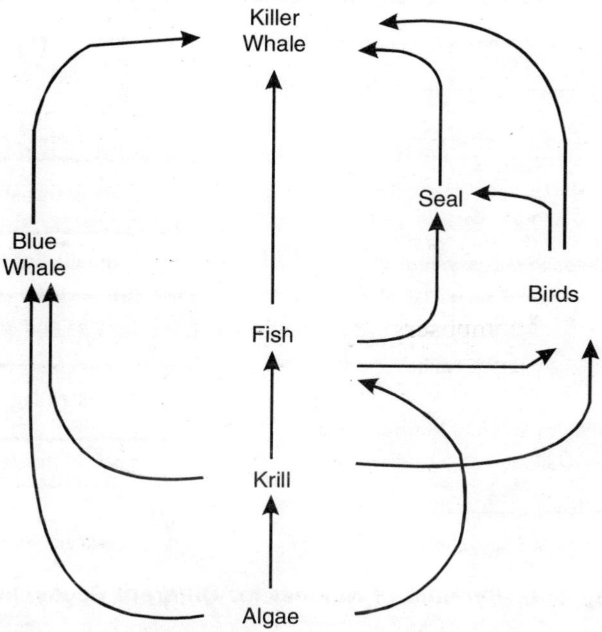

Fig. 2.15. Aquatic Food Web

Significance of Food Chains and Food Webs

- The energy flow and nutrient cycling take place through them.
- Help in maintaining the ecological balance
- Helps to understand the feeding relationships and the interaction between organisms in any ecosystem.
- Food chains show a unique property of biological magnification* of some chemicals. (* some metallic pollutants and pesticides are not biodegradable and are not easily excreted. These are deposited in higher concentration in the body of organisms of higher trophic levels through food chain).

2.13. ECOLOGICAL PYRAMIDS

Graphical representation of trophic structure and function of an ecosystem, starting with producers at the base and successive trophic levels forming the apex is knows as an ecological pyramid. Ecological pyramids are of three types:

2.13.1. Pyramid of Number

It shows the relationships among the number of producers (irrespective to taxonomic position), primary consumers (herbivores), secondary consumers (carnivores) and so on in any ecosystem. It also represents the number of individual organisms per unit area at each tropical level. We may have upright (grassland/pond ecosystem) or inverted pyramid (parasitic ecosystem) of number, depending upon the type of ecosystem and food chain.

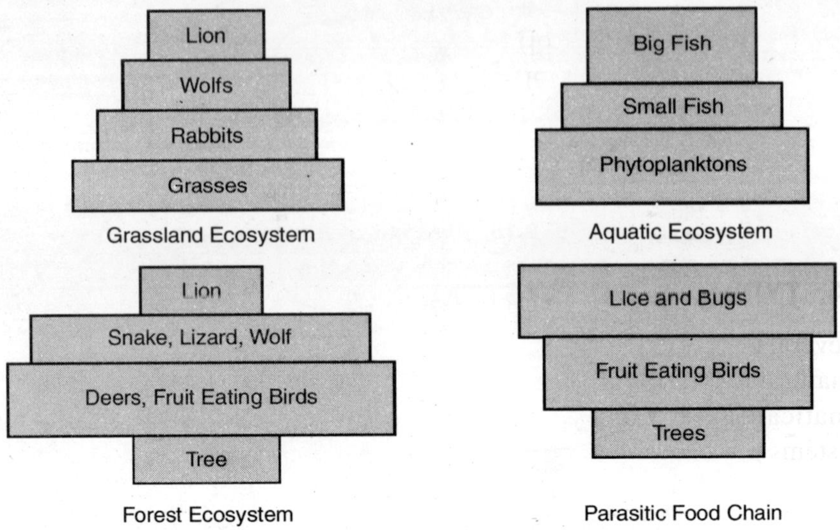

Fig. 2.16. Pyramid of Number for Different Ecosystems

2.13.2. Pyramid of Biomass

It is based upon the total biomass (dry matter) at each trophic level in a food chain. The pyramid of biomass can also be upright (grassland ecosystem) or inverted (pond ecosystem).

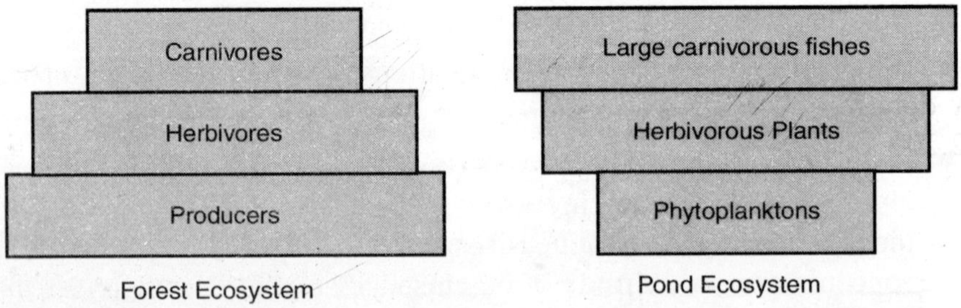

Fig. 2.17. Pyramid of Biomass for Different Ecosystems

2.13.3. Pyramid of Energy

The amount of energy present at each trophic level is considered for this type of pyramid. Pyramid of energy gives the best representation of the trophic relationship. According to Lindman (1942) only 10 per cent of energy is converted at each trophic level. For example in case of pond ecosystem each 1000 calory of energy produced in producers (algae etc.), 100 calories may be converted in primary consumers (zooplanktons), 10 calories by secondary consumer and 1 calory by tertiary consumers (top consumers). Thus the amount of energy gradually decreases at successive trophic levels and it is always upright.

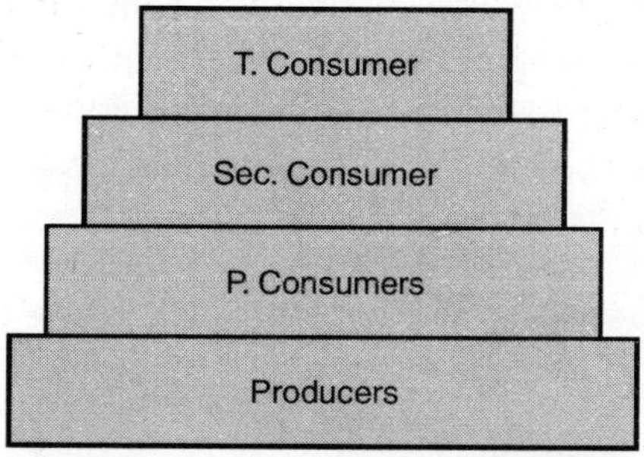

Fig. 2.18. Pyramid of Energy

2.14. TYPES OF ECOSYSTEM

Several kinds of ecosystems operate in biosphere. They are, broadly, grouped into natural and artificial type ecosystems. The natural ecosystem operates automatically under the activities of natural conditions, but the artificial ecosystems are controlled and manipulated by human beings.

Fig. 2.19. Two Types of Ecosystem

A detailed classification of different types of ecosystems is shown in Fig. 2.20.

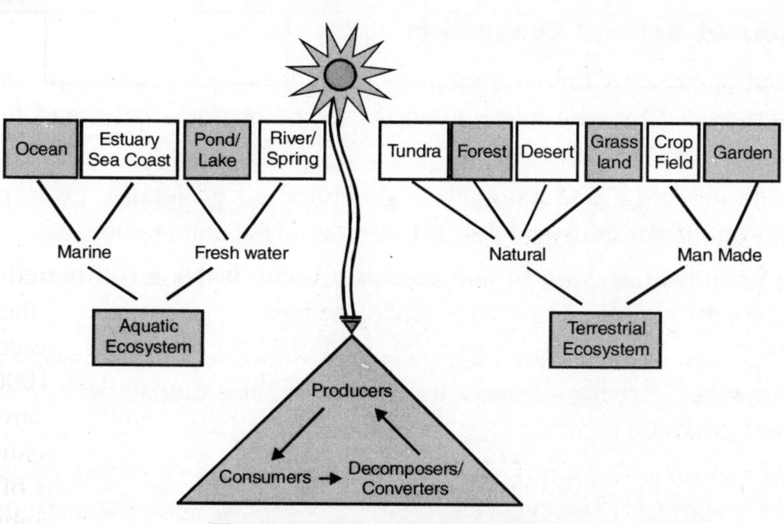

Fig. 2.20. Types of Ecosystem

Natural ecosystem of the world is broadly classified as terrestrial and aquatic ecosystem. Terrestrial ecosystems are known by the type of main vegetation in them. For example, if the main vegetation in an ecosystem is grass, then the terrestrial ecosystem will be known as grassland ecosystem. Aquatic ecosystems are another category on the basis of type of habitat. They may be fresh water or marine type water.

Fig. 2.21. Types of Natural Ecosystem

The fresh water ecosystems are lentic and lotic type. Pond and lakes etc. are the example of standing fresh water ecosystem (lentic ecosystem) while river springs etc. are examples of running fresh water (lotic ecosystem).

Functions of Natural Ecosystem

Natural ecosystem function themselves under natural conditions and they are not influenced by man-made activities. Some characteristics and functions of ecosystems are the following :

- The industrial and automobile generated air pollutants are trapped by leaves of the trees and convert into harmless compounds.
- The industrial, sewage and domestic waste water gets filtered through the natural soil and are thus made drinkable.
- The type of vegetation influences the climate and the soil structure. For example, if there is almost no vegetation. The climate will be very dry and of desert type.
- In natural ecosystems, there is a continued and rapid exchange of O_2, CO_2 and H_2O. The concentration of these compounds varies from season to season and also throughout the year.

Functions of Artificial Ecosystem

Such ecosystems are the creations of human beings. They do so for their convenience in order to fulfill certain needs. The kinds of species and the size of artificial ecosystems may vary widely. A few examples of artificial ecosystems may be as follows :

- A water-filtration tank/pond which purifies the water before supplying it for drinking.
- A cooling water spray pond in a thermal power station.
- Closed ecosystem created in scientific bottles with in the research laboratories.
- A home-acquairium or an exhibition acquairium such as Taraporewala acquairium, Marine Drive, Mumbai.

Forest Ecosystem

The 40 per cent land is covered by forest. In India 1/10th area of the total land is covered by forest. They are classified as :

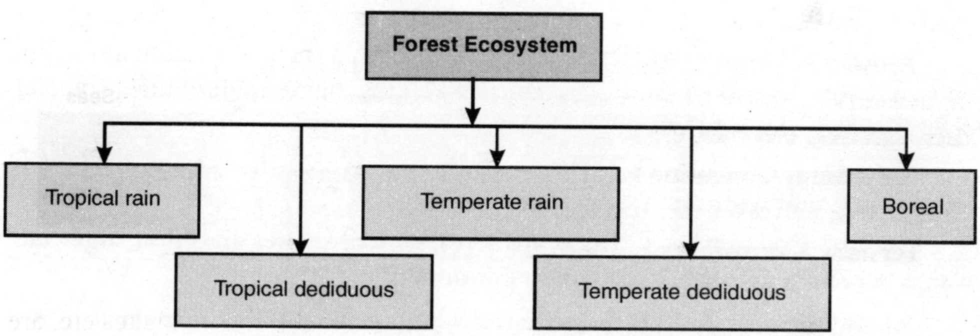

Fig. 2.22. Classification of Forest Ecosystem

Tropical Rain Forest : Such forests are located in equatorial regions of earth such as in Congo river basin of Africa, Central America where av. rain fall is high (<150 cm/year) and av. Annual temperature exceeds 18°C. these forest characteristic by warm and humid climate, high diversity of trees species, broad leaved and tall plants, high diverse animal life and abundance of insects and invertebrates.

Tropical deciduous forest of the continental regions of the tropics where av. rainfall range 100-120 cm/year, the rainfall is not evenly distributed but is restricted to a few months with prolonged dry season and are less luxuriant.

Temperate rain forest occurs in the region which are comparatively very cold and have winter rainfall, the summer are hot and dry.

Temperate deciduous forest occurs in very cold climate, the annual temperature is in between 7-15°C. the summer is hot and winter is very cold. These forest characteristic by tall deciduous trees, abundance of insects and birds and dominance of hard wood trees which are suitable for quality furniture and building purpose.

Boreal Forest : Such forests are found across east-west band of north Europe, north Asia and north America and in close vicinity below 60°N latitude to where the climate is cold. The annual rainfall varies from 100 mm to 350 mm and annual temperature ranges between 6° in winter to 20° in summer.

The major components of forest ecosystem are as follows :

Abiotic Components : It constitute the organic and inorganic substances present in the soil and the atmosphere. The different types of minerals are found in forests other.components are dead organic debris and eitter accumulation. The light conditions are different due to complex stratification of plant communities.

Biotic Components : The order of food chain in forest ecosystem among living organism is as under :

Producers : There are mainly trees with greater degree of stratification and much species diversity. The trees are of different kinds depending upon the kinds of forest formation developing in that climate. The shrubs and ground vegetation are also present in that area.

Consumers

Primary Consumers : These are herbivores that includethe animal feeding on tree leaves, shoots or fruit of producer like ants, bugs, spider, mice, squirrel, deer, giraffes, elephant etc.

Secondary Consumers : There are the carnivores like snake, birds, and lizards, fox, and wolf etc. feeding on herbivores.

Tertiary Consumers : There are top level carnivores like lion, tiger etc. which depend upon or eat secondary consumers.

Decomposers : A wide variety of decomposers are also present in forest ecosystem like fungi, bacteria and actinonycetes etc.

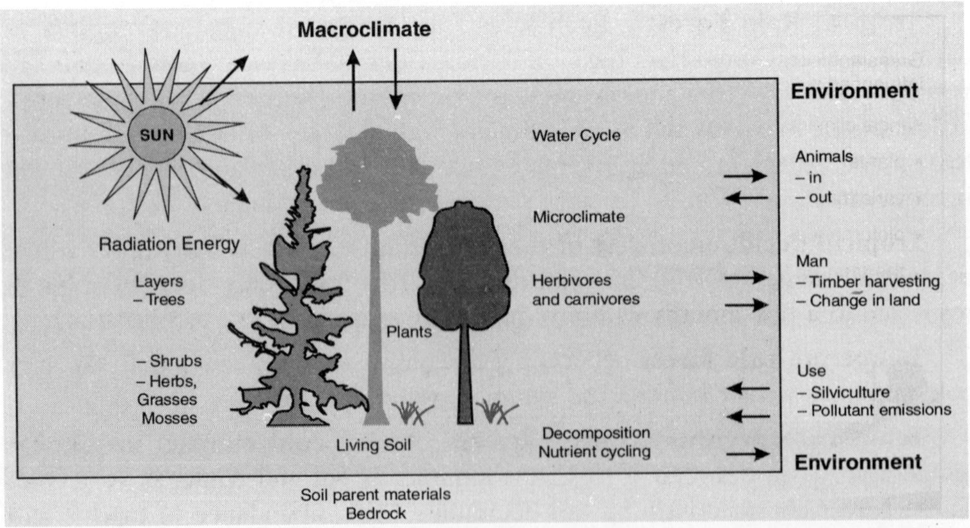

Fig. 2.23. The Forest Ecosystem and its Individual Components and the Environmental Factors Affecting the Ecosystem (Hannelius and Kuusela 1995).

Functions of Forest Ecosystem

Forest ecosystem enhances water resources in both quality and quantity, because of their capacity and ecological cleansing action. The hydrological cycle depends upon forest ecosystem and encourages absorption rather than runoff on precipitation. Forests give shelter to wildlife and fish.

Forest ecosystem considered as a pathway for the exchange and regulation of atmospheric gases, water and trace elements. The forest biomass above ground efficiently intercepts atmospheric particles and aerosols. The atmospheric pollutants move through leaves into the forest ecosystem, which are very important to the balance of life on earth.

Grassland Ecosystem

It is a type of terrestrial ecosystem. They may be of following types :

- Tropical grassland
- Temperate grassland
- Polar grassland

It occupies a comparatively small area and comprises following components:

Abiotic components : These are the nutrients present in soil and the aerial environment. The CO_2, H_2O, NO_3^-, PO_4^{4-} and SO_4^{2-} etc. supply C, H, O, N, P, S etc. from air and soil of the area. Some trace elements are also present in the soil.

Biotic components : These are categorized as under :

Producers : These are mainly grasses like species of Dichanthium, Cynodon, Desmodium, Digiteria, Setaria, sporobolus etc. some shrubs and fords also contributes to producers.

Fig. 2.24. Grassland Ecosystem

Consumers : These are mainly herbivores (Cows, Buffaloes, Deers, Sheep, Rabbit, Mouse and some insects and Termites and Millipeds) and carnivores (fox, jackals, snakes, frogs, lizards, birds, hawks)

Decomposers : These are microbes which decay the dead organic matter of different forms of higher life are fungi like Mucor, Aspergillus, Penicillum, Cladosporium etc. and some bacteria and actinomycetes. They bring about mineral back to the soil for use of producers.

Desert Ecosystem

About 17 per cent of land is covered by desert. The species composition of such ecosystem is much more varied and typical due to extremes of both, temperature and water factors. The various biotic components of a desert ecosystem are:

Producer : These are shrubs specially bushes, some grasses and a few trees. The shrubs have widespread, branched root system with their stem and branches variously modified. Sometimes a few succulents like cactus are also present. Some lower plants like Lichens and xerophytes mosses may also be present.

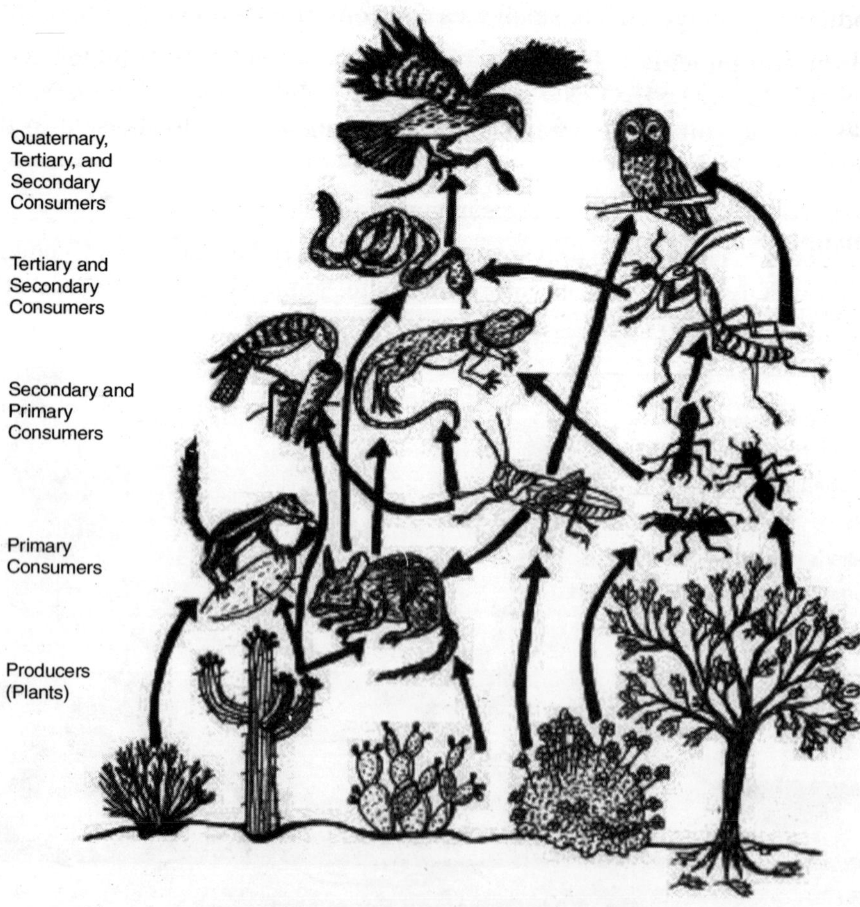

Fig. 2.25. Food-Web of Desert Ecosystem

Consumer : The animals like reptiles and insects are able to live under these desert conditions. Some nocturnal rodents and birds are also found. The camels 'ship of desert' feed on tender shoot of plants.

Decomposer : These are very few, as due to poor vegetation the amount of dead organic matter is correspondingly less. They are some fungi and bacteria. Most of them are thermophilic in nature.

Aquatic Ecosystem

Pond Ecosystem : The pond can be defined as a body of shallow standing water characterized by relatively quiet waters and abundant vegetation with thousands of microorganisms, large plant and animals. The important components of pond ecosystem are as follows:

Abiotic Components : These include basic inorganic and organic compounds such as water CO_2, O_2, Ca, N and P and their compounds as amino acid and humic acids etc. Amounts of various organic compounds like carbonates, proteins, lipids etc. are also estimated for biomass determination

Biotic Components **:**

Producers : These are auto-trophic, green plants and some photosynthetic bacteria. The producers fix radiant energy and with the help of minerals derived from the water and mud, they manufacture complex organic substances like carbonates, proteins, lipids etc. they are mainly Macrophytes (rooted hydrophytes, e.g. Trapa, Salvinia etc) and phytoplanktons (minute lower plants, e.g. Spirogyra etc.).

Consumers : These are nutritionally dependent or producers, the green plants. Most of the consumers are herbivores. They are mainly Zooplanktons-primary consumers (Euglene, Lecane etc.) secondary consumers (insects, small fishes and water beetles) and tertiary consumers (big fishes)

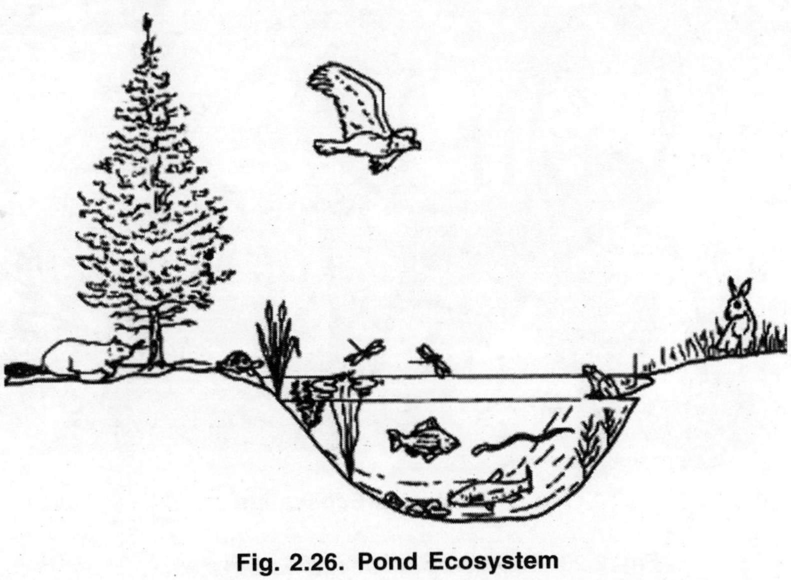

Fig. 2.26. Pond Ecosystem

Decomposers : These are also known as micro-consumers because they absorb only a fraction of the decomposed organic matter. They bring about the decomposition of complex dead organic matter of both producers as well as the macro-consumers to simple form. They play an important role in the return of mineral elements again to the medium of the pond.

Ocean (Marine) ecosystem – approximately 70 per cent of the earth's surface is covered by major oceans of the world. Each ocean represents a very large and stable ecosystem. The marine environment as compared with fresh water, appears to the more stable in their chemical composition due to being saeine and moreover others physico-chemical factors as dissolved oxygen content, light and temperature are also different. The *biotic components* of an ocean ecosystem as following :

Producers : These are auto-trophs, responsible for trapping the radiant energy of sum with the help of their pigments. Producers are mainly the phytoplanktons, such as diatoms, dinoflagellates and some microscopic algae and brown and red algae.

Consumers : These are dependent on primary producers for their nutrition are basically primary consumers (crustaceans, mollusks, fish etc.), secondary consumers (herring, shad, Mackeral etc.) and tertiary consumers (cod, haddock, halibut etc.)

Decomposers : The microbes active in the decay of dead organic matter of producers and macro-consumers are chiefly bacteria and some fungi.

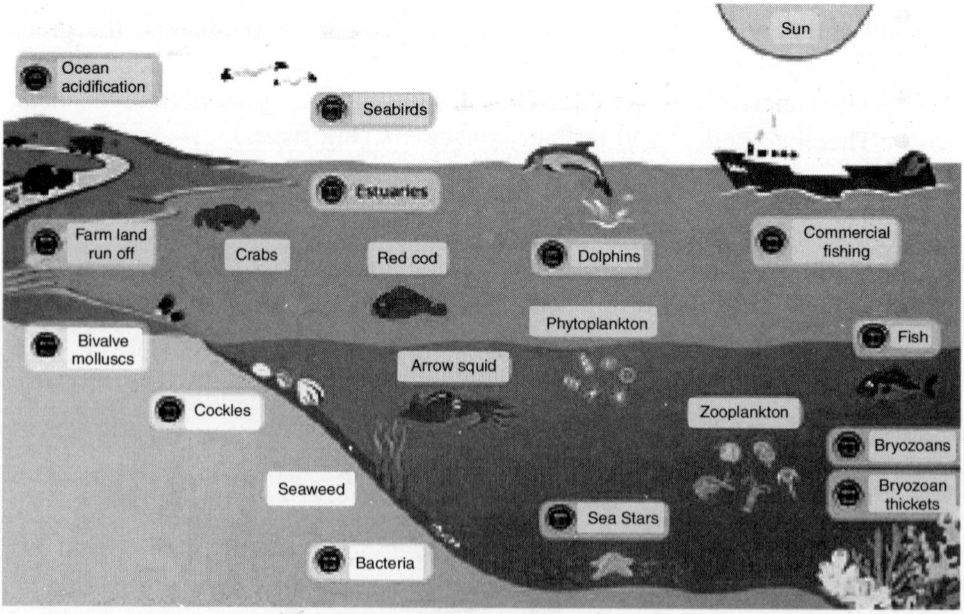

Fig. 2.27. Marine Ecosystem

Summary

- *Environment* is sum total of water, air, land, their inter-relationship among themselves, with the human beings, other living organism and property.
- Whole environment is divided in four major conceptual spheres or compartments on the basis of chemical distribution namely atmosphere, hydrosphere, lithosphere and biosphere.
- The Earth is surrounded by a blanket of air, which we call the *atmosphere*. It reaches over 560 kilometers (348 miles) from the surface of the Earth.
- The atmosphere absorbs the energy from the Sun, recycles water and other chemicals, and works with the electrical and magnetic forces to provide a moderate climate.
- The atmosphere also protects us from high-energy radiation and the frigid vacuum of space.
- Four distinct layers have been identified using thermal characteristics (temperature changes), chemical composition, movement, and density. These layers are naming the troposphere, the stratosphere, the mesosphere and the thermosphere.
- The *lithosphere* contains all of the cold, hard solid land of the planet's crust (surface), the semi-solid land underneath the crust, and the liquid land near the center of the planet.
- It is composed of soil particles and the underlying rocks down to a depth of 50km.
- The *hydrosphere* is composed of all of the water on or near the earth.
- This includes the oceans, rivers, lakes, and even the moisture in the air. It ranges from 10 to 20 kilometers in thickness.
- The *biosphere* is composed of all living organisms.
- Plants, animals, and one-celled organisms are all part of the biosphere.
- Most of the planet's life is found from three meters below the ground to thirty meters above it and in the top 200 meters of the oceans and seas.
- An ecosystem includes a group of biotic communities of species interacting with one another in an area, plus their surroundings, plus all the ways in which they interact with each other and exchanging energy and matter.
- Ecosystem show large variations in their size, structure, composition etc. however, all the ecosystems are characterized by certain basic structural and functional features which are common.
- Ecosystems are composed of a variety of abiotic and biotic components that function in an interrelated fashion.

- The physical and chemical components of an ecosystem constitute its abiotic structure.
- The plants, animal and microorganisms present in an ecosystem form the biotic components.
- These organisms have different nutritional behavior and status in the ecosystems and are accordingly known as Producers, Consumers and Decomposers.
- Consumers are also classified depending on what they eat as Herbivores, Carnivores and Omnivores.
- Every ecosystem performs under natural conditions in a systematic way. Although there are a wide range of ecosystem functions and are grouped into four primary categories as *Regulation functions, Habitat functions, Production functions* and *Information functions*.
- Flow of energy in an ecosystem takes place through the food chain and it is this energy flow which keeps the ecosystem going.
- The most important feature of this energy flow is that it is unidirectional or one way flow unlike the nutrients which move in a cyclic manner.
- Primary productivity of an ecosystem is defined as the rate at which radiant energy is converted into organic substances by photosynthesis or chemo-synthesis by the primary producers.
- The producers are left with a little less organic matter than what was actually produced by them, is known as the net primary production (NPP) and the respiratory loss (R) added to it gives the gross primary production (GPP).
- Any environmental stress tries to disturb the normal ecosystem functions. However, the ecosystem, by itself, tries to resist the change and maintain itself in equilibrium with the environment due to a property known as homeostasis.
- Nutrients like carbon, nitrogen, sulphur, oxygen, hydrogen, phosphorus etc. move in circular path through biotic and abiotic components are therefore known as biogeochemical cycles.
- Water also moves in a cycle, known as hydrological cycle.
- Food chains are the pathways along which nutrients pass through an ecosystem.
- Three basic types of food chains, grazing food chain, detritus food chain and parasitic food chain, are found in a typical ecosystem through which energy assimilated by plants can flow.
- Feed web is a network of food chains where different types of organisms are connected at different trophic levels, so that there are a number of options of eating and being eaten at each trophic level.

- Food Web is a set of interconnected food chains by which energy and materials circulate within an ecosystem.
- Graphical representation of trophic structure and function of an ecosystem, starting with producers at the base and successive trophic levels forming the apex is knows as an ecological pyramid.
- Ecological pyramids are of three types: Pyramid of number, Pyramid of Biomass and Pyramid of Energy.

Descriptive Type Questions

1. What is meant by environment? Enumerate and discuss its various components.

2. Classify the atmosphere on the basis of temperature variation. State main activities taking place in each of them.

3. With the help of neat sketch discuss the structure of atmosphere along with temperature profile of atmosphere and related phenomenon.

4. Explain the significance of biosphere, atmosphere, hydrosphere and lithosphere.

5. Name and describe the four major layers of the atmosphere. What are the important chemical species found in each layer?

6. Define the following :

(*a*) Tropopause (*b*) Stratopause (*c*) Mesopause (*d*) Thermopause

7. What is meant by structure of an ecosystem? Explain the various components of an ecosystem structure. Discuss the functions of an ecosystem also.

8. What is ecosystem? Discuss the biotic and abiotic components of an ecosystem.

9. Define ecology and ecosystem. Explain the role of producers, consumers and decomposers in an ecosystem.

10. Discuss the structure and function of an ecosystem in detail.

11. Explain in detail about producers, consumers and decomposers.

12. What is ecosystem? With the help of the suitable models explain the energy flow in the ecosystem.

13. Describe food chain. Name and explain various types of food chain with suitable example.

14. What is the significance of food chain and food web in any ecosystem?

15. What are the main factors, which influence the food chain?

16. Depict diagrammatically a terrestrial food web and oceanic food web.

17. Show diagrammatically a terrestrial food web in an ecosystem. Depict different food chains in this food web.

18. What is food chain and food web? Explain with the help of suitable examples.

19. What do you understand by ecological pyramids? Explain various types of pyramids with suitable example.

20. Discuss the models of energy flow in ecosystem

21. Explain the flow of energy through the various components of the ecosystem.

22. 'The flow of energy is unidirectional and continuous in an ecosystem.' Comment on this statement.

23. What is meant by 'flow of energy' in an ecosystem? What is its importance for ecosystem?

24. What are the biogeochemical cycles? How nitrogen circulation does occurs in nature.

25. Depict diagrammatically the carbon cycle.

26. With a neat sketch, explain how the elemental carbon is recycled in nature.

Short Answer Type Questions

1. List the four conceptual spheres in the earth's environment.

2. Differentiate between biosphere and lithosphere.

3. Define the term 'biomes'.

4. Define Ecology and Ecosystem.

5. Define biomagnification.

6. Differentiate between biome and ecosystem.

7. Explain food web with suitable example.

8. Why are decomposers called 'micro-consumers'?

9. What roles do autotrophs and heterotrophs play in the ecosystem?

10. "The pyramid of energy always takes a true upright shape." Why?

11. Write the short notes on

(*i*) Ecological pyramids

(*ii*) Food Chain and food web

12. Write short note on

(*a*) Trophic level

(*b*) Position of man on the food chain

(*c*) Bio-magmfication

13. Distinguish between the following:

(*a*) Biotic and abiotic components of ecosystem

(*b*) Food chain and food web

(*c*) Grazing food chain and detritus food chain

(*d*) Gross primary productivity and net primary productivity

(*e*) Autotrophs and heterotrophs

Objective Type Questions

Multiple Choice Questions

1. Which of the following conceptual components of the environment are effective transporters of matter?
   (*a*) Atmosphere and hydrosphere
   (*b*) Atmosphere and lithosphere
   (*c*) Hydrosphere and lithosphere
   (*d*) Lithosphere and biosphere
2. Biosphere is
   (*a*) The solid shell of inorganic materials on the surface of the Earth.
   (*b*) The thin shell of organic matter on the surface of earth comprising of all the living things.
   (*c*) The sphere which occupies the maximum volume of all the spheres.
   (*d*) All of the above
3. Which of the following conceptual sphere of the environment is having the least storage capacity for matter?
   (*a*) Atmosphere (*b*) Lithosphere
   (*c*) Hydrosphere (*d*) Biosphere
4. Which of the following is an example of impact of development activities on the Hydrosphere
   (*a*) Air pollution (*b*) Soil pollution
   (*c*) Soil erosion (*d*) Water pollution
5. The study of interactions between living and non-living organisms and environment is called
   (*a*) Ecosystem (*b*) Ecology
   (*c*) Phyto-geography (*d*) Phytosociology
6. Theterm ecology was introduced by
   (*a*) Haekel (*b*) Odum
   (*c*) Tansely (*d*) Ramdeo Mishra
7. An ecosystem consist of
   (*a*) Green plants and animals
   (*b*) Green plants and decomposers
   (*c*) Producers and consumers
   (*d*) Green plants, animals, decomposers and abiotic environment
8. In a food chain animals constitute the
   (*a*) First trophic level (*b*) Second trophic level
   (*c*) Intermediate trophic level (*d*) Ultimate trophic level
9. Which ecological pyramid is always upright?
   (*a*) Pyramid of number (*b*) Pyramid of biomass
   (*c*) Pyramid of energy (*d*) Pyramid of number and biomass

10. Which of the following statements is not true?

(*a*) Inorganic nutrients are recycled in an ecosystem
(*b*) Energy flows through the ecosystem in the form of C-C bonds
(*c*) Energy is recycled in an ecosystem
(*d*) Respiration process releases energy

11. Which one of the following is an abiotic component of the ecosystem?

(*a*) Bacteria (*b*) Plants
(*c*) Humus (*d*) Fungi

12. The primary producers in a forest ecosystem are

(*a*) Chlorophyll containing trees and plants
(*b*) Herbivores
(*c*) Carnivores
(*d*) Bacteria and other micro-organism

13. A trophic level refers to

(*a*) Area in the tropics
(*b*) An organism's portion in a food chain
(*c*) An organism's position in an ecosystem
(*d*) An organism's position in a biome

14. The food chain in which micro-organisms breakdown dead producers is called

(*a*) Consumer food chain. (*b*) Predator food chain.
(*c*) Parasitic food chain. (*d*) Detritus food chain.

15. A food web consists of

(*a*) Interlocking food chains (*b*) Producers, consumers and decomposers
(*c*) A portion of a food chain (*d*) A set of similar consumers

16. Tendency of pollutants to become concentrated in successive trophic levels is known as

(*a*) Biomagnification (*b*) Biorhythm
(*c*) Bioremediation (*d*) Biopiracy

17. Eutrophication is

(*a*) An improved water quality status of lakes.
(*b*) The result of accumulation of plant nutrients in water bodies.
(*c*) A process in the carbon cycle.
(*d*) A water purification technique.

18. Increase in fauna and decrease in flora would be harmful due to increase in

(*a*) Diseases (*b*) CO
(*c*) O_2 (*d*) Radioactive pollution

19. Which of the following is a possible producer in an ecosystem?

(*a*) Plants (*b*) Animals
(*c*) Human beings (*d*) Fish

20. The largest reservoir of nitrogen an our planet is

(*a*) Ocean (*b*) Atmosphere
(*c*) Biosphere (*d*) Fossil fuels

Fill in the Blanks

1. A zone with constant temperature in the atmosphere is called an ....
2. Temperatures in .... layer can be as high as 1200°C.
3. Separating the .... from the stratosphere is transition zone called the stratopause.
4. The term 'ccology' was coined by ....
5. The study of reciprocal relationship between organisms and their environment is called ....
6. Herbivores are .... consumers.
7. Every food chain starts with a ....
8. Each step in a food chain represent a ....
9. Autotrophic planktons are called ....
10. The pyramid of .... is always upright.
11. Dead plant parts and animal remains are called ....

ANSWERS

Multiple Choice Questions

1. (*a*)	**2.** (*b*)	**3.** (*a*)	**4.** (*d*)	**5.** (*b*)
6. (*a*)	**7.** (*d*)	**8.** (*c*)	**9.** (*c*)	**10.** (*c*)
11. (*c*)	**12.** (*a*)	**13.** (*c*)	**14.** (*d*)	**15.** (*a*)
16. (*a*)	**17.** (*b*)	**18.** (*b*)	**19.** (*a*)	**20.** (*b*)

Fill in the Blanks

1. Isothermal layer
2. Thermosphere
3. Mesosphere
4. Haekel
5. Ecology
6. Primary
7. Producer
8. Trophic level
9. Phytoplanktons
10. Energy
11. Detritus

❑❑❑

CHAPTER 3

Biodiversity

3.1. DEFINITION OF BIODIVERSITY

INSIDE THIS CHAPTER

Biological diversity or biodiversity refers to the variety of life forms: the different plants, animals and microorganisms, the genes they contain, and the ecosystems they form. This living wealth is the product of hundreds of millions of years of evolutionary history. The process of evolution means that the pool of living diversity is dynamic: it increases when new genetic variation is produced, a new species is created or a novel ecosystem formed; it decreases when the genetic variation within a species decreases, a species becomes extinct or an ecosystem complex is lost. The concept emphasises the interrelated nature of the living world and its processes.

In the convention of Biological diversity (1992) biodiversity has been defined as *the variability among living organisms from all sources including, inter-alia, terrestrial, marine and other aquatic ecosystems and the ecological complexes of which they are a part.* Thus, biodiversity can be defined as 'the totality of genes, species and ecosystems of a region'. Biological diversity is usually considered at three different levels: genetic diversity, species diversity and ecosystem diversity.

3.2. LEVELS OF BIODIVERSITY

Biodiversity is usually explored at three levels - genetic diversity, species diversity and ecosystem diversity. These three levels work together to create the complexity of life on Earth.

3.2.1. Genetic Diversity

Genetic diversity refers to the variety of genetic information contained in all of the individual plants, animals and microorganisms. Genetic diversity is the variety of genes within a species. Genetic diversity occurs within and between populations of species as well as between species. Each species is made up of individuals that have their own particular genetic composition. This means a species may have different populations, each having different genetic compositions. To conserve genetic diversity, different populations of a species must be conserved.

Genes are the basic units of all life on Earth. They are responsible for both the similarities and the differences between organisms. When the genes within the same species show different versions due to new combinations, it is called genetic variability or genetic diversity.

Genetic diversity refers to the variation of genes within species. This covers genetic variation between distinct populations of the same species, such as the four varieties of white-cheeked rosella, *Platycercus eximius*. It also covers genetic variation within a population, which tends to be relatively high in widespread eucalypts such as *Eucalyptus cloeziana, E. delegatensis,* and *E. saligna*. Genetic diversity can be measured using a variety of DNA-based and other techniques.

3.2.2. Species Diversity

Species diversity is the variety of species within a habitat or a region. Some habitats, such as rainforests and coral reefs, have many species. Others, such as salt flats or a polluted stream, have fewer. This is the variability found within the population of a species or between different species of a community. It represents broadly the species richness and their abundance in a community.

Species diversity refers to the variety of species. Aspects of species diversity can be measured in a number of ways. Most of these ways can be classified into three groups of measurement:species richness, species abundance and taxonomic or phylogenetic diversity.

Measures of species richness count the number of species in a defined area. Measures of species abundance sample the relative numbers among species. A typical sample may contain several very common species, a few less common species and numerous rare species. Measures of species diversity that simplify information on species richness and relative abundance into a single index are in extensive use. Another approach is to measure taxonomic or phylogenetic diversity, which considers the genetic relationships between different groups of species. These measures are based on analysis which results in a hierarchical classification usually represented by a 'tree' that depicts the branching pattern which is thought to best represent the phylogenetic evolution of the taxa concerned.

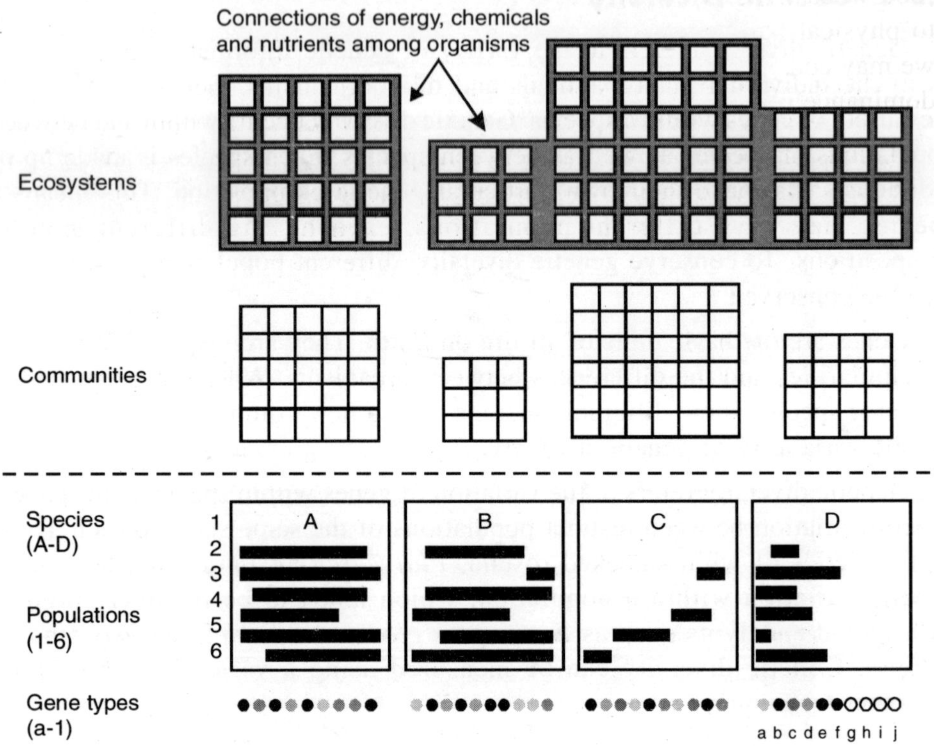

Fig. 3.1. Levels of Biodiversity

Different measures of taxonomic diversity emphasise various taxic characteristics and relationships. The species level is generally regarded to be the most appropriate to consider the diversity between organisms.

3.2.3. Ecosystem Diversity

Ecosystem diversity encompasses the broad differences between ecosystem types, and the diversity of habitats and ecological processes occurring within each ecosystem type. Generally these systems use a combination of a habitat type definition with a climatic descriptor; for example, tropical moist forest, or temperate grassland. Some systems also incorporate global biogeography to account for differences in biota between regions of the world which may have very similar climate and physical characteristics. An ecosystem is a community of organisms and their physical environment interacting together.An ecosystem can cover a large area, such as a whole forest, or a small area, such as a pond.

An ecosystem is a community of organisms and their physical environment interacting together. An ecosystem may be as large as the Great Barrier Reef or as small as the back of a spider crab's shell, which provides a home for plants and other animals, such as sponges, algae and worms. This is the diversity of ecological complexity showing variations in ecological niches, tropic structure, food-webs, nutrient cycle etc. The ecosystems also show variations with respect to physical parameters like moisture, temperature, altitude, precipitation etc. we may consider diversity in forest ecosystem which is suppose to have mainly dominance of trees.

Fig. 3.2. Levels of Biodiversity

3.3. BIODIVERSITY AT GLOBAL, NATIONAL AND LOCAL LEVEL

3.3.1. Global Biodiversity

Conservative estimates of the existing biodiversity is ten million species bet if estimates for insects are correct then it could be around 30 million species, we have till now estimated about 1.4 million species.

Globally, we have roughly 1, 70,000 flowering plants, 30,000 vertebrates and about 2, 50,000 other groups of species that have been described.

Terrestrial biodiversity of the earth is best describes as biomes, which are the largest ecological units present in different geographic areas and are named after the dominant vegetation e.g. the tropical rainforest, tall grass prairies etc. There is an estimated 1, 25,000 flowering plant species in tropical forests. Temperate forests have much less biodiversity, but there is much better documentation of the species.

It is interesting to know that marine diversity is even much higher than terrestrial biodiversity and ironically, they are still less known and described.

3.3.2. National Biodiversity

Every country is characterized by its own biodiversity depending mainly on its climate. India has a rich biodiversity depending mainly on its climate. India has a rich biological diversity of flora and fauna. Overall 6 per cent of global species are found in India. It is estimated that India ranks 10th among the plant rich countries of the world, 11th in terms of the number of endemic species of higher vertebrates and 6th among the centre of diversity and origin of agricultural crop.

India has a rich and varied heritage of biodiversity, encompassing a wide spectrum of habitats from tropical rainforests to alpine vegetation and from temperate forests to coastal wetlands. India figured with two hotspots - the Western Ghats and the Eastern Himalayas - in an identification of 18 biodiversity hotspots carried out in the eighties (Myers. 1988). Recently, Norman Myers and a team of scientists have brought out an updated list of 25 hotspots (Myers et. al. 2000). In the revised classification, the 2 hotspots that extend into India are The Western Ghats/Sri Lanka and the Indo-Burma region (covering the Eastern Himalayas); and they are included amongst the top eight most important hotspots. In addition, India has 26 recognised endemic centres that are home to nearly a third of all the flowering plants identified and described to date.

Of the estimated 5–50 million species of the world's biota, only 1.7 million have been described to date (Groombridge, and Jenkins. 2000), and the distribution is highly uneven. About seven per cent of the world's total land area is home to half of the world's species, with the tropics alone accounting for 5 million. India contributes significantly to this latitudinal biodiversity trend. With a mere 2.4 per cent of the world's area, India accounts for 7.31 per cent of the global faunal total with a faunal species count of 89,451 species (MoEF. 1999). Some salient features of India's biodiversity have been mentioned below.

India has two major realms called the Palaearctic and the Indo-Malayan, and three biomass, namely the tropical humid forests, the tropical dry/deciduous forests, and the warm desert/semi-deserts.

- India is one of the 12 centres of origin of cultivated plants.
- India has 5 world heritage sites, 12 biosphere reserves, and 6 Ramsar wetlands.
- Amongst the protected areas, India has 88 national parks and 490 sanctuaries covering an area of 1.53 lakh sq. km.

3.3.3. Regional or Local Biodiversity

R. H. Whittaker, (1972) described three terms for measuring biodiversity over geographic scales :

Alpha (α-) diversity refers to diversity within a particular area, community or ecosystem, and is measured by counting the number of taxa within the ecosystem (usually species). It refers to the diversity of organism sharing the same community or habitat.

Beta (β-) diversity is species diversity between community or ecosystems; this involves comparing the number of taxa that are unique to each of the ecosystems. There are differences in species composition of communities along anvironmental gradients, e.g., altitude, moisture etc.

Gamma (γ-) diversity is a measure of the overall diversity for different ecosystems within a region. It refers to the diversity of the habitats over the lotal land scape or geographical area.

3.4. VALUE OF BIODIVERSITY

The values of biodiversity in terms of its commercial utility, ecological services, social and aesthetic value is enormous. The multiple uses of biodiversity or biodiversity value have been classified by McNeely et al in 1990 as follows :

Consumptive Use Value

Goods that are consumed locally; do not appear in gross domestic product (GDP) because they are neither bought nor sold. Consumptive use value can be estimated for products by determining how much people would have to pay to buy equivalent products if local sources were not available.

Consumptive use values relate to the value of goods produced by the ecosystem that can be consumed and used by people Consumptive and productive use values are closely related to biodiversity because without biodiversity the productivity of the ecosystem responsible for the production of these goods would be significantly impaired. These are direct use values where the biodiversity product can be harvested and consumed directly e.g. fuel, food, drugs, fiber etc.

- 50 per cent of rainforest tree species are used by indigenous people for non-fuel products.
- 80 per cent of the world's population relies on traditional medicines from plants and animals.
- Protein from wild game, fish, and insects is vital to people in many countries, including the U.S.
- Fuel wood and animal dung provides energy for people in many countries.

Productive Use Value : Products that are harvested and sold in markets; these are the commercially usable values where the product is marked and sold. Standard economic methods value these products as the price paid at the first point of sale minus the costs incurred up to that point (rather than the final retail price). Approximately 4.5 per cent ($87 billion) of the U.S. GDP depends on wild species; likely to be higher in developing nations. These may include the animal products like tusk of elephant, musk from musk deer, silk from silk worm, wool from sheep, fur from many animals, lac from lac insects etc, all of which are traded in the market.

Forests—wood products, game, fruits, gum, resins, medicinal plants.

New medicines—AIDS drugs, cancer drugs...

Social Value

These are the values associated with the social life, customs, religion and psycho-spiritual aspects of the people. The cultural value of biological diversity conservation for present and future generations is an important reason for conserving it today. Human cultures coevolve with their environment, and the conservation of biological diversity can be important for cultural identity. Many of the plants and animals are considered holy and sacred in our country like Tulsi, Peepal, Mango, Lotus, Bael, Cow, Snake, Bull, Peacock, Owl etc.

Ethical Value

It is also sometimes known as existence value. People love live, a phenomenon called "biophilia" by E. O. Wilson (1984). Sometimes, our feeling for biodiversity is revealed in the way we spend our money; sometimes they are not. Economists are trying to devise methods for estimating the monetary value of the blue whales and harp seals for people whose only relationship with them knows that they exist.

Aesthetic Value

The aesthetic values of our natural ecosystems and landscapes contribute to the emotional and spiritual well-being of a highly urbanised population. Natural and wild landscapes are aesthetically pleasing and provide opportunities to get away from human-dominated landscapes. They also provide opportunities for recreational activities such as hiking, canoeing, bird watching and nature photography In North America, protected wild areas where indigenous organisms live undisturbed give people a sense of satisfaction in knowing that there are bears and wolves and rare plants and insects that still exist on their continent. Ecotourism (enjoyment of aesthetical value) is estimated to generate about 12 billion dollars of revenue annually that roughly gives the aesthetic value of biodiversity.

Option Values

A species' **potential to produce economic benefits to society in the future**; our need for pieces of nature we once ignored is often important and unpredictable. Time after time we have rushed back to nature's cupboard for cures to illnesses or for infusions of tough genes from wild plants to save our crops from pest outbreaks.

When we assess the instrumental value of species, we generally focus on their usefulness hear and now, but this is a shortsighted viewpoint as revealed in our discussion of medicinal research and biodiversity. Our rudimentary understanding of biology and ecology has leaved an enormous gap between the currently realized values of a species and its potential future value. This gap is particularly wide because we have vague idea of what our future lives will be like- technologically, culturally and ecologically. It may be difficult to guess at the potential ecological role that a species might assume in the future. It would certainly have taken a very prescient biologists to guess that the shrewlike mammals tat shared the earth with dinosaurs would lead to the earth-dominating *Homo sapiens.* The evolution of our scientific knowledge has already taught us that: keep options alive. We must take this approach because we know so little. We can never say of species, a part of biodiversity that lacks of values. The problem is **we don't know which species have the benefits**; so, it is **important to preserve them all.**

Ecosystem Service Value

A non-consumptive use value related to self maintenance of the ecosystem and various important ecosystem services has been recognized. "*Goods and Services" provided by ecosystems include:*

- Provision of food, fuel and fibber.
- Provision of shelter and building materials.
- Purification of air and water.
- Detoxification and decomposition of wastes.
- Stabilization and moderation of the Earth's climate.
- Moderation of floods, droughts, temperature extremes and the forces of wind.

- Generation and renewal of soil fertility, including nutrient cycling.
- Pollination of plants, including many crops.
- Control of pests and diseases.
- Maintenance of genetic resources as key inputs to crop varieties and
- Livestock breeds, medicines, and other products.
- Ability to adapt to change.

3.5. BIO-GEOGRAPHICAL CLASSIFICATION OF INDIA

India, being a vast country, shows a great diversity in climate, topography and geology and hence the country is very rich in terms of biological diversity. India is the seventh largest country in the world and Asia's second largest nation with an area of 3,287,263 sq. km encompassing a varied landscape rich in natural resources. India is shielded by the world's highest mountains, the Himalayas, in the north. The southern part of India takes the shape of a peninsula and divides the Indian Ocean into the Bay of Bengal to the southeast and the Arabian Sea to the southwest. The southern tip of Kanyakumari is washed by the Indian Ocean. The Andaman and Nicobar Islands in the Bay of Bengal and the Lakshadweep group of islands in the Arabian sea are also a part of India.

Zone Name	Per centage of India	Existing Protected Areas		Biotic provinces
		Number of National Parks and Wildlife Sanctuaries	% of Zone	
Trans-Himalaya	5.62	7	9.20	Ladakh mountains, Tibetan plateau
Himalaya	6.41	67	9.92	Northwest, West, Central and East Himalayas
Desert	6.56	6	7.45	Thar, Kutch
Semi-Arid	16.60	91	2.80	Punjab plains, Gujarat Rajputana
Western Ghats	4.03	56	10.12	Malabar plains, Western Ghats
Deccan Peninsula	41.99	137	3.71	Central highlands, Chotta Nagpur, Eastern highlands, Central Plateau, Deccan South
Gangetic Plain	10.79	36	2.16	Upper and Lower Gangetic plains
Coasts	2.52	26	6.79	West and East coast, Lakshadweep
North-East	5.21	37	2.54	Brahmaputra valley, Northeast hills
Islands	0.25	103	18.54	Andaman and Nicobar

Source: Wildlife Protected Area Network in India: A Review, Wildlife Institute of India, 2000

India has a great diversity of natural ecosystems from the cold and high Himalayan ranges to the sea coasts, from the wet northeastern green forests to the dry northwestern arid deserts, different types of forests, wetlands, islands and the oceans. The climate of India is determined by the southwest monsoon between June and October, the northeast monsoon between October and November and dry winds from the north between December and February. From March to May the climate is dry and hot. **The country has 10 different biogeographic zones and 26 biotic provinces.** *A biographic zone is a large distinctive unit of similar ecology, biome representation, community and species,* e.g. Trans Himalaya, The Deccan Peninsula etc. *A biotic province is secondary unit with a biogeographic zone, giving weight to particular community separated by dispersal barrier or gradual change in environmental factors.*

1. The Trans-himalayan Region

This area is very cold and arid (4,500 - 6,000 m above mean sea level). The only vegetation is a sparse alpine steppe. Extensive areas consist of bare rock and glaciers.

The faunal groups best represented here are wild sheep and goats (chief ancestral stock), ibex, snow leopard, marbled cat, marmots and black-necked crane.

2. The Himalayan Region

The fantastic altitude gradient results in the tremendous biodiversity of the Himalayan region. Flora and fauna vary according to both altitude and climatic conditions: tropical rainforests in the Eastern Himalayas and dense subtropical and alpine forests in the Central and Western Himalayas. The lower levels of the mountain range support many types of orchids. On the eastern slopes, rhododendrons grow to tree height.

Animals of Himalayas show several behavioural and physiological adaptations. Sambar and muntjac are found in the subtropical foothills; serow, goral and the Himalayan thar are found in the temperate and subalpine regions; snow leopard and brown bear inhabit the alpine region. Carnivores are the most elusive of all mammals in the Himalayas. There are a variety of carnivores in the higher mountains, some of which are rare and threatened with extinction.

3. The Indian Desert

The natural vegetation consists of tropical thorn forests and tropical dry deciduous forests, sandy deserts with seasonal salt marshes and mangroves are found in the main estuaries. Typical shrubs are growing on sand dunes. Sewan grass covers extensive areas called pali.

Thar Desert possesses most of the major insect species. 43 reptile species and moderate bird endemism are found here. No niche of the Thar is devoid of birds. The black buck was once the dominant mammal of the desert region now confined only to certain pockets. The gazelle is the only species of the Indian antelope of which the females have horns. Nilgai the largest antelope of India and the wild ass, a distinct subspecies, is now confined to the Rann of Kutch which is also the only breeding site in the Indian subcontinent for the flamingoes. Other species like desert fox, great Indian bustard, chinkara and desert cat are also found.

4. The Semi-arid Region

The semi-arid region in the west of India includes the arid desert areas of Thar and Rajasthan extending to the Gulf of Kutch and Cambay and the whole Kathiawar peninsula.

The natural vegetation consists of tropical thorn forests and tropical dry deciduous forests, moisture forests (extreme north) and mangroves. The sandy plains have a few scattered trees of Acacia and Prosopis. The gravelly plains have Calotropis, Gymnosporia, etc. The rocky habitats are covered by bushes of Euphorbia while species of Salvadora and Tamarix occur mainly near saline depressions. The lion of Gir is the endemic species in this zone.

5. The Western Ghats

They cover only 5 per cent of India's land surface but are home to more than about 4,000 of the country's plant species of which 1800 are endemic. The monsoon forests occur both on the western margins of the ghats and on the eastern side where there is less rainfall. This zone displays diversity of forests from evergreen to dry deciduous.

The Nilgiri langur, lion tailed macaque, Nilgiri tahr, Malabar grey hornbill and most amphibian species are endemic to the Western Ghats.

6. The Deccan Peninsula

The Deccan Peninsula is a large area of raised land covering about 43 per cent of India's total land surface. It is bound by the Sathpura range on the north, Western Ghats on the west and Eastern Ghats on the east. The elevation of the plateau varies from 900 m in the west to 300 m in the east. There are four major rivers that support the wetlands of this region which have fertile black and red soil. Large parts are covered by tropical forests. Tropical dry deciduous forests occur in the northern, central and southern part of the plateau. The eastern part of the plateau in Andhra Pradesh, Madhya Pradesh and Orissa has moist deciduous forests.

Fauna like tiger, sloth bear, wild boar, gaur, sambar and chital are found throughout the zone along with small relict populations of wild buffaloes, elephants and barasingha.

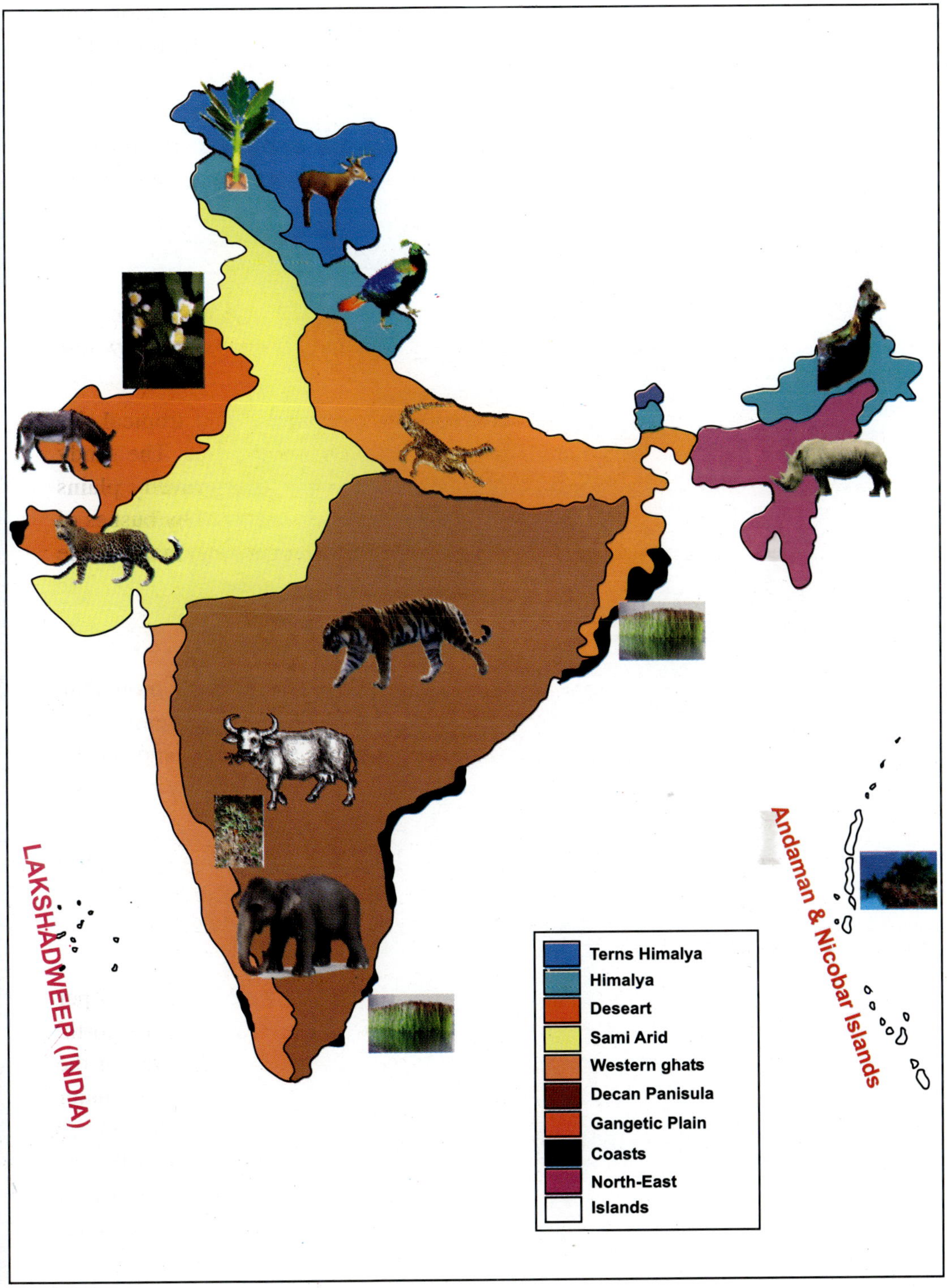

We are not Responsible for any mistake in the Map

7. The Gangetic Plain

The Gangetic plain is one of India's most fertile regions. The soil of this region is formed by the alluvial deposits of the Ganges and its tributaries. The five important surface differences recognized in the geomorphology of the plains are

- **Bhabar :** Pebble studded zone with porous beds.
- **Terai :** Marshy tract.
- **Bhangar :** Older alluvium of the flood plain.
- **Terai :** Marshy tract.
- **Khadar :** Newer alluvium.

The Gangetic plains stretching from eastern Rajasthan through Uttar Pradesh to Bihar and West Bengal are mostly under agriculture. The large forest area is under tropical dry deciduous forest and the southeastern end of the Gangetic plain merges with the littoral and mangroves regions of the Sunderbans.

The fauna includes elephants, black buck, gazelle, rhinoceros, Bengal florican, crocodile, freshwater turtle and a dense waterfowl community.

8. The Coastal Region

The natural vegetation consists of mangroves. Animal species include dugong, dolphins, crocodiles and avifauna. There are 26 species of fresh water turtles and tortoises in India and 5 species of marine turtles, which inhabit and feed in coastal waters and lay their eggs on suitable beaches. Tortoise lives and breed mainly on the land.

Over 200,000 Olive Ridley turtles come to Orissa to nest in the space of three or four nights. The highest tiger population is found in the Sunderbans along the east coast adjoining the Bay of Bengal.

Lakshadweep consists of 36 major islands - 12 atolls, 3 reefs and 5 submerged coral banks - make up this group of islands more than three hundred kilometers to the west of the Kerala coast. The geographical area is 32 sq. km. and the usable land area is 26.32 sq. km. The fauna consists mainly of four species of turtles, 36 species of crabs, 12 bivalves, 41 species of sponges including typical coral, ornamental fishes and dugongs. A total of 104 scleractinian corals belonging to 37 genera are reported.

9. The North-East

Biological resources are rich in this zone. The tropical vegetation of northeast India is rich in evergreen and semi-evergreen rain forests, moist deciduous monsoon forests, swamps and grasslands.

Mammalian fauna includes 390 species of which 63 per cent are found in Assam. The area is rich in smaller carnivores. The country's highest population of elephants is found here.

10. The Indian Islands

It is a group of 325 islands: Andaman to the north and Nicobar to the south. The two are separated by about 160 km by the Ten Degree Channel of the sea. The rainfall is heavy, with both Northeast and Southwest monsoons. At present, 21 of the 325 islands in the Andaman and Nicobar Islands are inhabited. Many unique plants and animals are found here. About 2,200 species of higher plants are found here of which 200 are endemic. The Andaman & Nicobar Islands have tropical evergreen forests and tropical semi-evergreen forests as well as moist deciduous forests, littoral and mangrove forests.

112 endemic species of avifauna, the Andaman water monitor, giant robber crab, 4 species of turtles, wild boar, Andaman day gecko and the harmless Andaman water snake are found only in these islands. The Narcondam hornbill found only in Narcondam is a large forest bird with a big beak. Coral reefs are stretched over an area of 11,000 sq. km in the Andamans and 2,700 sq. km in Nicobar.

3.6. INDIA AS A MEGA-DIVERSE NATION

India, being a vast country, shows a great diversity in climate, topography and geology and hence the country is very rich in terms of biological diversity. India's biological diversity is one of the most significant in the world, since India has only 2 per cent of the total landmass of the world containing about 6 per cent of the world's known wildlife.

India is one of the 12 mega-diverse countries in the world. The MOEF, GOI (2000) records 47,000 species of plants and 81,000 species of animals which are about 7 per cent and 6.5 per cent respectively of global flora and fauna.

3.6.1. Endemism

Species which are restricted only to a particular area are known as endemic. The endemism of Indian biodiversity is high. About 33 per cent of the country's recorded flora are endemic to the country and are concentrated mainly in the North-East, Western Ghats, North-West Himalaya and the Andaman and Nicobar islands.

Of the 49,219 plant species, 5150 are endemic and distributed into 141 genera under 47 families corresponding to about 30 per cent of the world's recorded flora, which means 30 per cent of the world's recorded flora is endemic to India. Of these endemic species, 3,500 are found in the Himalayas and adjoining regions and 1600 in the Western Ghats alone. About 62 per cent of the known amphibian species are endemic with the majority occurring in the Western Ghats.

Nearly 50 per cent of the lizards of India are endemic with a high degree of endemicity in the Western Ghats. India's record in agro-biodiversity is equally impressive. There are 167 crop species and wild relatives. India is considered to

Earth's 25 Terrestrial hot spots of biodiversity

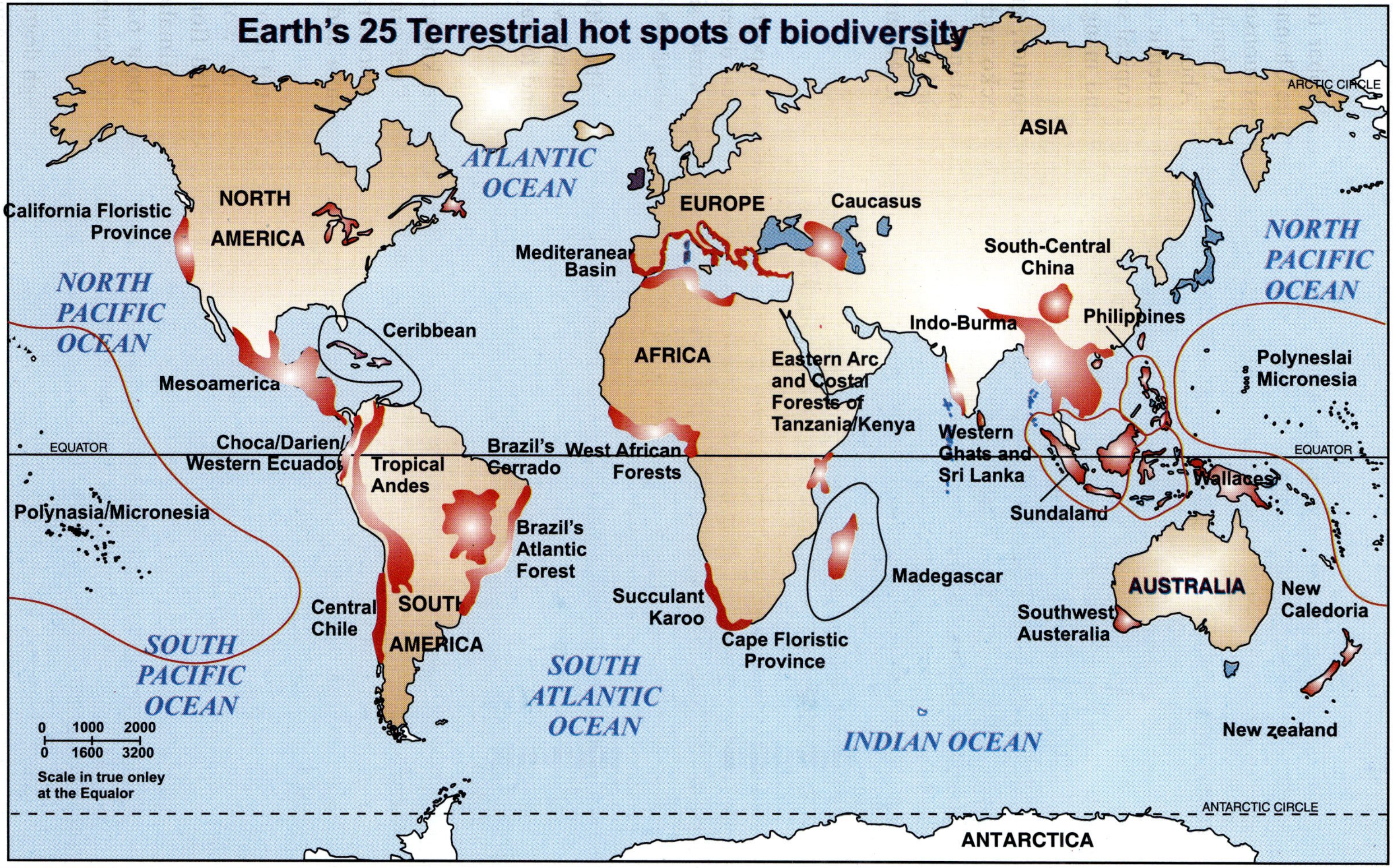

We are not Responsible for any mistake in the Map

be the centre of origin of 30,000-50,000 varieties of rice, pigeon-pea, mango, turmeric, ginger, sugarcane, gooseberries etc and ranks seventh in terms of contribution to world agriculture.

3.6.2. Centre of Origin

A large number of species are known to have originated in India. Nearly 5000 species of flowering plants had their origin in India. India is a centre of crop diversity—the homeland of 167 cultivated species and 320 wild relatives of crop plants.

3.6.3. Marine Diversity

Along 7500 km long coastline of our country in the mangroves, estuaries, coral reefs, back water etc are rich biodiversity. More than 340 species of corals of the world are found here.

A large proportion of the Indian biodiversity is still unexplored. There are about 93 major wet lands, coral reefs and mangroves which need to be studied in detail.

Coral reefs are undersea living communities with self-contained steady-state ecosystems and well-defined tropic levels of producers and consumers. Reefs are home to thousands of marine creatures and make available all basic necessities like food, shelter and breeding areas to their refugee floras and faunas. Human expansion and development is one the prime threats to coral reefs; various anthropogenic activities raise pollutants and cause eutrophication in water that boosts the growth of such organisms like sponges, which out-compete the corals for space on crowded reefs.

Runoff, industrial outflows and huge sewage disposals are ruthlessly degrading the delicate and frail coral reef frameworks. Excess fishing has considerably reduced the population of reef fishes in many parts of the world leading to imbalances in coral reef ecology and allowing more competitive organisms to prevail over the reefs. Usage of explosives or poisons by fishermen to get more yields from sea in some regions is severely damaging the corals. Natural disturbances like hurricanes, typhoons, powerful stroms, etc. also cause irreparable damage to coral reefs. Ever minor changes in the susceptible reef environment can bring detrimental impact on the health of entire coral colonies.

3.7. HOT SPOTS OF BIODIVERSITY

Hot spots are areas that are extremely rich in species, have high endemism, and are under constant threat. The term was introduced by Mayer's (1988). The hot spots covering less than 2 per cent of the world's land area are found to have 50 per cent of terrestrial biodiversity. According to

Mayer's et al (2000) an area is designated as a hot spot when it contains at least 0.5 per cent of the plant species as endemic.

About 40 per cent of terrestrial plants and 25 per cent of vertebrates' species are endemic and found in these hot spots. After the tropical rain forest, the second highest numbers of endemic plant species are found in the Mediterranean (Mittermeier). Broadly, these hot spots are :

1. Tropical Andes
2. Mesoamerica
3. Caribbean
4. Brazil's Atlantic Forest
5. Choc. Darien/ Waetern Ecuador
6. Brazil's Cerrado
7. Central Chile
8. California Floristic Province
9. Madagascar
10. Western African Forest
11. Cape Floristic Province
12. Succulent Karoo
13. Mediterranean Basin
14. Caucasus
15. Sundaland
16. Wallacea
17. Philippines
18. Indo-Burma
19. South Central China
20. Western Ghats/Sri Lanka
21. South West Australia
22. New Caledonia
23. New Zealand
24. Polynesia/Micronesia
25. Eastern Arc and Coastal Forest of Kenya and Tanzania

These are the areas of high diversity, endemism and are also threatened by human activities. More than $1/6^{th}$ of world's population, most of whom are desperately poor people, live in there areas. Among the 25 hot spots of the world, two are found in India extending into neighbouring countries - the Western Ghats/ Sri Lanka and the Indo-Burma region (covering the Eastern Himalayas). These areas are particularly rich in floral wealth and endemism, not only in flowering plants but also in reptiles, amphibians, swallow-tailed butterflies, and some mammals.

Eastern Himalayas

Phytogeographically, the Eastern Himalayas form a distinct floral region. The area comprises Nepal, Bhutan, and neighbouring states of northern India, along with a continuous sector of the Yunnan province in southwest China. Although all Himalayan forests lie well north of the Tropic of Cancer, and some of them are at altitudes of 1780 to 3500 metres (rather akin to temperate forests in physiognomy and structure), they can be considered tropical forests since they occur largely within the climatic tropics.

It is believed that forest cover in the Eastern Himalayas has dwindled from 340,000 km^2, to 110,000 km^2, with a mere 53,000 km^2 of primary forests. Despite this loss, the north-eastern region is home to some botanical rarities. One of these is the *Sapria* himalayana, a parasitic angiosperm that has been sighted only twice since 1836.

This region is the meeting ground of the Indo-Malayan and Indo-Chinese biogeographical realms as well as the Himalayan and Peninsular Indian elements, formed when the peninsular plate struck against the Asian landmass, after it broke off from Gondwana land.

- The Eastern Himalayas display an ultra-varied topography, a factor that fosters species diversity and endemism.
- Many deep and semi-isolated valleys are exceptionally rich in endemic plant species. In Sikkim, in an area of 7298 km^2, of the 4250 plant species, 2550 (60 per cent) are endemic (Myers. 1988). In India's sector of the area, there are about 5800 plant species, of which roughly 2000 (36 per cent) are endemic.
- In Nepal, there are around 7000 plant species, many of which overlap with those of India, Bhutan, and even Yunnan. Of these species, at least 500 (almost 8 per cent) are believed to be endemic to Nepal.
- Bhutan possesses an estimated 5000 species, of which as many as 750 (15 per cent) are considered to be endemic to the Eastern Himalayas.
- The numerous primitive angiosperm families found in this region include Magnoliaceae, Degeneriaceae, Himantandraceae, Eupomatiaceae, Winteraceae, Trochodendraceae, Tetracentraceae, and Lardizalbaleaceae. The primitive genera are *Alnus, Aspidocarya Betula, Decaisnea, Euptelea, Exbucklandia, Haematocarpus, Holboellia, Houttuynia, Magnolia, Mangelietia, Pycnarrhena,* and *Tetracentrol* (Malhotra and Hajra. 1977).
- Studies have shown that north-east India along with the contiguous region of the Chinese provinces of Yunnan and Schezwan is an active centre of organic evolution and is the cradle of flowering plants.

- This has now been confirmed by studies in south-east China where an altogether new large mammal (*Muntiacus gongshanensis*) and four new genera of flowering plants (*Xizangia, Sinoleontopodium, Sindoxa,* and *Tetradoxa*) have been discovered (Myers. 1988).

Western Ghats

Of India's 49,219 plant species, 1600 endemics (40 per cent of the total number of endemics) are found in a 17,000 km^2 strip of forest along the seaward side of the Western Ghats in Maharashtra, Karnataka, Tamil Nadu, and Kerala (WCMC. 1992). Forest tracts up to 500 metres in elevation, comprising one-fifth of the entire forest expanse, are mostly evergreen, while those in the 500–1500 metres range are semi-evergreen. There are two main centres of diversity, the Agastyamalai Hills and the Silent Valley/New Amambalam Reserve basin (Myers. 1988).

Very little has been documented recently about the status of the forest cover, except that it seems to have declined between 1972 and 1985 at a rate paralleling that for India as a whole, which implies a loss of over 2.4 per cent annually. If we extrapolate from 1986 to 1989, this means a total loss of almost 34 per cent for 1972 to 1989. According to Myers et al. 2000, only 6.8 per cent of the original extent of vegetation exists today. Still worse is the decline of the primary forest: the amount remaining seems to be no more than 8000 km^2. All but isolated pockets of original forest have been opened up by shifting cultivation, allowing a take-over by deciduous species and bamboo among other forms of 'degenerate' vegetation.

3.8. THREATS TO BIODIVERSITY

Extinction or elimination of a species is a natural process of evolution. In the geologic period the earth has experienced mass extinction. During evolution, species have died out and have been replaced by others. However, the rate of loss of species in geologic past has been a slow process, the process of extinction has become particularly fast in the recent years of human civilization. In this century, the human impact has been so severe that thousands of species and varieties are becoming extinct annually. One of the estimates by the noted ecologist, E.O.Wilson puts the figure of extinction at 10,000 species per year or 27 per day! This startling figure raises an alarm regarding the serious threat to biodiversity.

India's biodiversity is one of the most significant in the world. However, this rich biodiversity of India is under severe threat owing to habitat destruction and over exploitation. The country has several problems such as overpopulation, large number of cattle heads, growing demand for land, energy and water supply. Unplanned developmental works and over exploitation of resources have made its living resources most vulnerable. The major causes and issues related to threat to biodiversity are discussed on next page :

(i) Loss of Habitat

Destruction and loss of natural habitat is the single Largent cause of biodiversity loss. Billions of hectares of forests and grasslands have been cleared over the past 10,000 years for conversion into agriculture land, pastures, and settlement area and development projects. These natural forests and grasslands were the natural homes of thousands of species which perished due to loss of their natural habitat. The wetlands are destroyed due to draining, filling and pollution thereby causing huge biodiversity loss. In India, several rare butterfly species are facing extinction with the uncannily swift habitat destruction of the Western Ghats.

(ii) Poaching

Illegal trade of wildlife products by killing prohibited endangered animal *i.e.* poaching is another threat to wildlife. Man has hunted for food. Commercially wild animals are hunted for their products such as skin, tusk, antlers, fur, meat, pharmaceutical perfumes, cosmetics and decoration purposes. Despite international ban on trade in products from endangered species of wildlife items like furs, hides, horns, tusk, live specimen and herbal products worth millions of dollars per year. The developing nations in Asia, Latin America and Africa are richest sources of biodiversity and have enormous wealth of wild life. The rich countries in Europe and North America and some affluent countries in Asia like Japan, Taiwan and Hong Kong are the major importers of the wildlife products and wildlife itself. We can certainly help in preserving biodiversity by not purchasing fur coats, purses, bags or items made of crocodile skin or python skin.

(iii) Man-Wildlife Conflicts

Sometimes we come across conflicting situation when wildlife starts immense damage and danger to man and under such situation it becomes very difficult for the forest department to pacify the affected villagers and gain local support for wild-life conservation. the root causes of these conflicts are discussed below:

(*a*) Loss of habitats or shirking of forest cover.

(*b*) Ill, wear and injured animals have a tendency to attack man.

(*c*) Due to lack of food.

(*d*) Due to electric wiring in fields, animal get injured, suffer in pain and turn violent.

(*e*) Due to settlement in wildlife corridors.

(*f*) Due to low compensation, the agonized farmers get revengeful and kill the wild animals.

Remedial measures to curb the conflict :

(*a*) By dealing tactfully with any imminent danger

(*b*) By adequate crop, cattle and human life compensation

(*c*) By providing solar powered fencing with electric wiring

(*d*) By changing crop patter near forest borders

(*e*) By adequate fooder, fruit and water supply for wild animal in forest zone.

(iv) Over Exploitation of Resources

This is one of the main causes of the loss of not only economic species but also biological curiosities like the insectivorous and primitive species and other texa needed for teaching of laboratory work. Commercial exploitation of wild plants has invariably meant its overuse and eventual destruction. This has been as true in case of Indian wild mango trees, which were turned into playwood as of the whales that were hunted for tallow, in the oceans.

(v) Collection for Zoo and Research

Animals and plants are collected throughout the world for zoos and biological laboratories for study and research in science and medicine.

(vi) Introduction of Exotic Species

Native species are subjected to competition for food and space due to introduction of exotic species. For example, introduction of goats and rabbits in the Pacific and Indian region has resulted in destruction of habitats of several plants, birds and reptiles.

(vii) Control of Pests and Predators

Predator and pest control measures, generally kill predators that are a component of balanced ecosystem and may also indiscriminately poison non-target species.

(viii) Pollution

Pollution alters the natural habitat. Water pollution especially injurious to the biotic components of estuary and coastal ecosystem. Toxic wastes entering the water bodies disturb the food chain, and so to the aquatic ecosystem. Insecticides, pesticides, SO_2, NO_x, acid rain, ozone depletion and global warming too, affect adversely the plant and animal species. The impact of coastal pollution is also very important, it is seen that coral refs are being threatened by pollution from industrialization along the coast, oil transportation and offshore mining. Noise pollution is also the cause of wildlife extinction.

(ix) Deforestation

One of the main causes of wildlife is population and resultant deforestation. Deforestation mainly results from population settlement, shifting cultivation, development projects, demand for fuel wood, demand of wood as material for many industries such as paper and pulp, match, plywood, furniture etc.

3.9. ENDANGERED SPECIES OF INDIA

A species is said to be ***extinct*** *when it is not seen in the wild for 50 years at a stretch e.g. dodo, passenger pigeon. A species is said to be in* ***vulnerable*** *category if its population is facing continuous decline due to over exploitation or habitat destruction.* Such species is still abundant but under a serious threat of becoming endangered if casual factors are not checked e.g., plants as Cupressus cashmeriana (Kashmir cypress) and animals as Antelope cervicapra. Species those are not endangered or vulnerable at present, but are at a risk, are categorized as rare species. These species are usually localized with in restricted area *i.e.* they are usually endemic. Some times they are thinly scattered over a more extensive area.

A species is said to be ***endangered*** *when its number has been reduced to a certain level or whose habitats have been drastically reduced and if such species is not protected and conserved, it is in immediate danger of extinction.* The International Union for conservation of Nature and Natural Resources publishes the **Red Data Book** which includes the list of endangered species of plant and animals. The red data symbolizes the warning signal for those species which are endangered and if not protected are likely to become extinct in near future. On the global level, the IUCN published Red Data Book in two volumes. **Red Data Book** is the name given to the book dealing with threatened plants or animals of any region. Many countries have prepared their own Red Data Book.

Its opposite is the **Green Book** published by Botanical Survey of India (BSI) which lists rare plants growing in protected area like botanic gardens. Green Book consists of 100 rare plant species. The United Nation Environmental Programme (UNEP) has complied endangered species of the world under the title, 'Blue Book'.

In India, nearly 450 plant species, 150 mammals and 150 species of birds is estimated to be threatened and identified in the categories of endangered or rare while an unknown number of species of insects are endangered.

Few of the endangered reptiles, birds, mammals and plants are given below:

Reptiles	Gharial, green sea turtle, tortoise, python
Birds	Great India bustard, peacock, pelican, great Indian hornbill, Siberian white crane
Animals	Indian wolf, red fox, sloth bear, red panda, tiger, leopard, striped hyena, Indian lion, golden cat, desert cat, dugong
Plants	Sapria Himalayan, Uvaria Iurida, Nepenthes Khasiana, Pedicularis perroter, Piper longum, Sesamum indicum, Butea, a large number of species of orchids, medicinal plants and the sandal wood tree etc.

Dodo

Passenger Pigeon

Fig. 3.3. Extinct Species of India

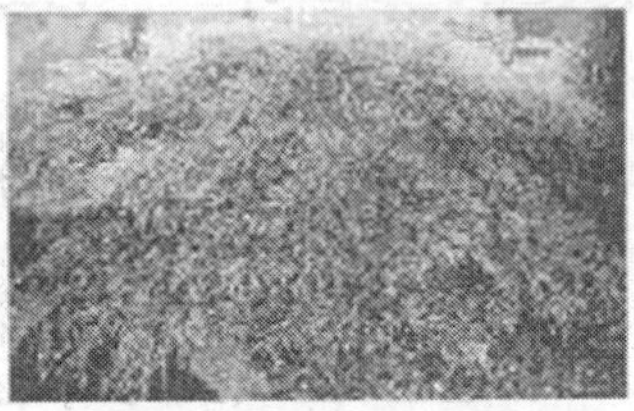

Cupressus Cashmeriana

Antelope Cervicapra

Fig. 3.4. Vulnerable Species of India

Green Sea Turtle

Indian Python

Green Indian Burtard

Fig. 3.5. Endangered Reptiles and Animal Species of India

Fig. 3.6. Endangered Animal and Plant Species of India

3.10. ENDEMIC SPECIES OF INDIA

Total number of plant species recorded in the world 2, 50,000 species. In India 45,000 species 33 per cent of the above are native. There are 15,000 flowering plant species which is 6 per cent of the world's total. Areas rich in endemism are the Northeast, the Western Ghats and the Northwestern and Eastern Himalayas. Andaman and Nicobar Islands contribute at least 200 endemic species to the endemic flora.

Endemic Species of Animals

Group	No. of species
Mollusca	
Land	878
Freshwater	89
Insects	16,214
Amphibians	110
Reptiles	214
Aves	69
Mammals	38

Source: MoEF 1999.

Endemic Species of Plants

Group	No. of species
Pteridophyta	200
Angiosperms	4950

Source: MoEF 1999.

3.11. CONSERVATION OF BIODIVERSITY

The enormous value of biodiversity due to their genetic, commercial, medical, aesthetical, ecological and optional importance emphasized the need to conserve biodiversity and aspiration of future generation is known as conservation of biodiversity. Conservation process performs the following functions:

(*i*) Maintenance of essential ecological processes and life-supporting system

(*ii*) Preservation of diversity of species

(*iii*) Suitable utilization of species and ecosystem.

Gradually we are coming to realize that wildlife is not just 'a game to be hunted' rather it is a 'gift of nature' to be nurtured and enjoyed. A number of measures are now being taken the world over to conserve biodiversity including plants and wildlife.

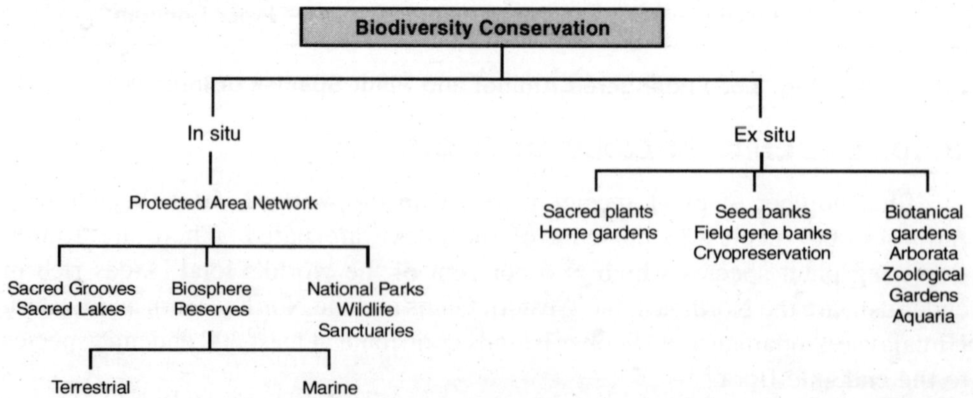

There are two approach of biodiversity conservation:

In-situ Conservation (Within Habitat) : Conservation of species in its natural residing place is known as in-situ conservation. It may be the natural ecosystem or the artificial one. This type of conservation is applicable only wild fauna and flora and not to domestic animals. This is achieved by protection of wild flora and fauna in nature itself e.g. biosphere reserves, national park, wildlife sanctuaries, reserve forest etc.

At present we have 7 major biosphere reserves, 80 national parks and 420 wildlife sanctuaries in our country covering 4 per cent of the geographic area.

The concept of biosphere reserve was started in 1975 under the 'Man and Biosphere' Pragramme of UNESCO. In this type, the multiple land use is permitted and the biosphere is not restricted to any one, two or more species but to the whole ecosystem. Thus, biosphere is ecosystem oriented. The biosphere reserves conserve some representative ecosystems as a whole for long term in site conservation. Within a biosphere reserves we may have one or more national park. For example Nanda Devi (UP), Sunderbans (WB) and Nilgiri (which have Bandipur and Nagarhole National Park in it).

Biosphere Reserves of India

Name of the site	Date of notification	Area in Sq. km	Location (State)
Nilgiri	01.08.86	5,520	Parr of Wynad , Nagarhole, Bandipur and Madumalai, Nilambur, Silent Valley and Siruvani hills (Tamil Nadu, Kerala and Karnataka)
Nanda Devi	18.01.88	5,860.69	Par of Chamoli, Pithoragarh, Almora Districts (Uttaranchal)
Nokrerk	01.09.88	820	Part of Gora Hills (Meghalaya)
Manas Wildlife Sanctuary	14.03.89	2,837	Part of Kokrajhar, Bongaigaon, Barpeta, Nalbari, Kamprup and Darang district (Assam)
Sunderbans	29.03.89	9,630	Part of delta of Ganga & Brahamaputra river system (West Bengal)
Gulf of Mannar	18.02.89	10,500	Indan part of Gulf of Mannar between India and Sri Lanka (Tamil Nadu)
Great Nicobar	06.01.89	885	Southern most islands of Andaman and Nicobar (A&N islands)
Similpal	21.06.94	4,374	Part of Mayurbhanj district (Orissa)
Dibru-Saikhowa	28.07.97	765	Part of Dibrugarh and Tinsukia district (Assam)
Dehang Debang	02.09.98	5,112	Part of Siang and Debang velley (Arunachal Pradesh)
Pachmarhi	03.03.99	4,926.28	Parts of Betul, Hoshangabad and Chindwara districts (Madhya Pradesh)
Kanchanjanga	07.02.00	2,619.92	Part of Kanchanjanga Hills (Sikkim)

Source: MoEF 2000

A national park is an area dedicated for the conservation of wildlife along with its environment. National parks are meant for the betterment of wild life. It is also meant for enjoyment tourism but without impairing the environment. Grazing of domestic animals, all private rights and forestry activities are prohibited within national park. Corbett and Dudwa are the two examples of national park in Utter Pradesh.

India's World Heritage Sites

Site	Location
Kaziranga National Park	Assam
Keoladeo Ghana National Park	Rajasthan
Nanda Devi National Park	Uttar Pradesh
Sundarban National Park	West Bengal

Source: MoEF 1999.

Wildlife sanctuaries are also protected area for wild lives where killing, hunting, shooting or capturing of wildlife is prohibited except under the control of highest authority. However private ownership rights are permissible and forestry operations are also permitted to an extent that they do not affect the wildlife adversely. Ghana Bird Sanctuary- Rajasthan (300 species of birds).

Ex-situ Conservation (Outside Habitat) : Conservation of species away from their natural habitat is called ex-situ conservation. This is done by establishment of gene banks, seed banks, zoos, botanical garden, culture collection etc.

This type of conservation is mainly done for conservation of crop varieties with the main objective of conserving the total genetic variability of crop species for future improvement or afforestation programmes. Steps involved in ex-situ conservation :

(*i*) Establishment of minimum target population goals to provide for maintenance of captive genetic diversity.

(*ii*) Distribution of founders through various captive breeding programmes

(*iii*) Circulation of compiled programmes to all breeding facilities.

(*iv*) Implementation of an overall plan contributing to the objectives of maintaining viable populations across the globe.

In India, we have the following important gene bank/seed bank facilities:

(*a*) National Bureau of Plant Genetic Resource (NBPGR), New Delhi.

(*b*) National Bureau of Animal Genetic Resource (NBAGR), Karnal, Haryana.

(*c*) National Facility for Plant Tissue Culture Repository (NFPTCR), New Delhi.

Advantages of Ex-situ Conservation

1. Assurance of food, shelter and security to the organism.

2. Increase in life span and breeding activity.
3. Increase of chance of survival.
4. Possibility of use of genetic techniques to improve concerned species.

Disadvantages of ex-situ conservation

1. It can be adopted only for a few species due to limitation of space, finance and facilities in the institution undertaking captive breeding
2. Deprival of opportunity by organism to adopt to be ever-changing natural environment.

Case Study

Project Tiger

Tiger sightings have become quite rare these days in India, reason being the Tiger killings because of its multitude of medicinal or magical properties that is why tiger trade is very profitable. Genuinely the tiger skin is not fashionable but the smuggling of Tiger fur coats and rugs are not difficult for the impoverished hunters.

In the beginning of the 1970s, once tiger hunting had officially been banned in India, a tiger count was done across the entire country. This lead to the shocking discovery that only 1800 specimens of this magnificent animal were left. This jolted the concerned authorities and some serious thought went into devising plans to save the tiger. The result was the launch of "Project Tiger" in 1972 on the 1st April at the Dhikala Forest Rest House in Corbett National Park. The main idea behind the project was to provide safe havens for tigers where they could flourish as a species and hopefully reverse the startling decline in their population. The project initially had 9 parks that were chosen for it's implementation. This number has slowly risen and a total of 19 parks are now attached to the project. The project was begun in association with and still receives its main funding from the WWF (World Wildlife Fund).

Some of the best examples of this programmes success can be seen in the national parks situated in the high Himalayan region, to the mangrove swamps of the Sundarbans and the thorny scrubs of Rajasthan. But more wildlife conservation laws and awareness among people is still required to make Indian sanctuaries a safe haven for tigers.

Major Regions	Madhya Pradesh, Uttaranchal, West Bengal, Rajasthan, Kerala
Famous Tiger Reserves	Kanha National Park, Bandhavgarh National Park, Corbett National Park, Sundarbans National Park.
Other Wildlife Attractions	Leopards, Antelopes, Deers, Sambar, Gaurs.

Although the experts say that the project has its shortcomings, the increase in the populations of the tiger is clearly evident to even the common man.

Many experts had predicted that the tiger would be extinct by the turn of the century, but, whoever may be responsible, the tiger has proudly proved them wrong. Tiger population may not still be in thrilling numbers and poaching still may be quite rampant but a lot more effort is being put into saving this beautiful animal. This is good news for the entire natural treasure of the country because if the tiger flourishes, so will the jungle and vice-versa.

It's aims were :

(*i*) To ensure maintenance of a viable population of tigers in India for scientific, economic, aesthetic, cultural and ecological values.

(*ii*) To preserve, for all times, areas of such biological importance as a national heritage for the benefit, education and enjoyment of the people.

Project Elephant

Project Elephant was launched in 1991-92 to assist states having ranging populations of wild elephants to ensure long term survival of identified viable population of elephants in their natural habitats. States are being given financial as well as technical and scientific assistance in achieving the objectives of the project. Main activities of Project Elephants are as follows :

(*i*) Ecological restoration of existing natural habitats and migratory routes of elephants;

(*ii*) Development of scientific and planned management for conservation of elephant habitats and viable population of Wild Asiatic elephants in India;

(*iii*) Promotion of measures for mitigation of man elephant conflict in crucial habitats and moderating pressures of human and domestic stock activities in crucial elephant habitats;

(*iv*) Strengthening of measures for protection of Wild elephants form poachers and unnatural causes of death;

(*v*) Research on Project Elephant management related issues;

(*vi*) Public education and awareness programmes;

(*vii*) Eco-development

(*viii*) Veterinary care

During the year an amount of Rs. 3.05 crores (approx.) was allocated to the elephant range states to enable them to provide financial assistance for habitat management of man-elephant conflict, payment of ex-gratia relief for loss of life etc., strengthening of anti-poaching measures, capture and translocation of problem elephant populations etc. Permission to capture 33 problem elephants in West Bengal and Assam was given in order to reduce man-elephant conflict. An amount of Rs. 142.30 lakhs was released to the elephant range states to strengthen their anti-poaching and anti-depredation activities.

Summary

- Biodiversity is the verity and variability among living organisms from all sources and the ecological complexes in which they occur.
- Biodiversity is usually explored at three levels - genetic diversity, species diversity and ecosystem diversity.
- Genetic diversity refers to the variety of genetic information contained in all of the individual plants, animals and microorganisms.
- Species diversity is the variety of species within a habitat or a region.
- Ecosystem diversity encompasses the broad differences between ecosystem types, the diversity of habitats and ecological processes occurring within each ecosystem type.
- India, being a vast country, shows a great diversity in climate, topography and geology and hence the country is very rich in terms of biological diversity.
- The values of biodiversity in terms of its commercial utility, ecological services, social and aesthetic value is enormous.
- The multiple uses of biodiversity or biodiversity value have been classified as Consumptive use value, Productive use value, Social value, Ethical value, Aesthetic value, Option values and Ecosystem service value.
- Hot spots are areas that are extremely rich in species, have high endemism, and are under constant threat.
- Among the 25 hot spots of the world, two are found in India the Western Ghats/Sri Lanka and the Indo-Burma region.
- The country has several problems such as overpopulation, large number of cattle heads, growing demand for land, energy and water supply, unplanned developmental works and over exploitation of resources causes the loss of biodiversity.
- The major causes and issues related to threat to biodiversity are loss of habitat, poaching, man-wildlife conflicts, over exploitation of resources, collection of species for zoo and research, introduction of exotic species, control of pests and predators, pollution and deforestation.
- Species which are restricted only to a particular area are known as endemic species.
- A species is said to be extinct when it is not seen in the wild for 50 years at a stretch e.g. dodo, passenger pigeon.

- A species is said to be in vulnerable category if its population is facing continuous decline due to over exploitation or habitat destruction.
- Species those are not endangered or vulnerable at present, but are at a risk, are categorized as rare species.
- A species is said to be endangered when its number has been reduced to a certain level or whose habitats have been drastically reduced and if such species is not protected and conserved, it is in immediate danger of extinction.
- The International Union for Conservation of Nature and Natural Resources (IUCN) publishes the Red Data Book which includes the list of endangered species of plant and animals.
- The red data symbolizes the warning signal for those species which are endangered and if not protected are likely to become extinct in near future.
- In general, conservation of biodiversity refers to human efforts to correct, reverse, prevent, or discontinue activities that are causing decline in biodiversity, like pollution, deforestation and chemical waste.
- In-situ conservation (within habitat) is the conservation of species in its natural residing place.
- In-situ conservation is achieved by protection of wild flora and fauna in nature itself e.g. biosphere reserves, national park, wildlife sanctuaries, reserve forest etc.
- Ex-situ conservation (outside habitat) is the conservation of species away from their natural habitat is called ex-situ conservation.
- Ex-situ conservation is done by establishment of gene banks, seed banks, zoos, botanical garden, culture collection etc.

Descriptive Type Questions

1. What is meant by biodiversity? What is its concept? Explain the three hierarchical levels of biodiversity.
2. What is biodiversity? Explain genetic, species and ecosystem diversity. Explain the consumptive and productive value of biodiversity. Explain the levels of biodiversity.
3. What is meant by 'Biodiversity'? What is its concept; describe the uses and importance of biodiversity.
4. What do you mean by consumptive use value, productive use value, social value, ethical value and option value of biodiversity?
5. What do you mean by value of biodiversity? Explain the consumptive use or productive use of biodiversity.

6. Comment upon Indian biodiversity with special reference as a mega diverse nation.

7. What do you understand by 'Hotspots of Biodiversity'? Name and briefly describe, along with vital signs, the two hotspots of biodiversity that extend into India.

8. What are hotspots of biodiversity? Name the hotspots found in India? Discuss it.

9. Write an explanatory note on threats to biodiversity.

10. Is biotechnology a threat to biodiversity? Substantiate your answer with examples.

11. With the suitable example, explain how poaching and human-wildlife conflict are responsible for biodiversity depletion.

12. What are the endemic, extinct, endangered, vulnerable and rare species? Give example for each.

13. What is 'Red Data Book'? How do we declare species threatened or endangered? Name two species each of endangered reptiles, birds, mammals and plants of our country.

14. What is 'Red Data Book'? How 'Red Data Book' is different from 'Green Book' and 'Blue Book'?

15. Describe the international efforts that have been made to protect the biodiversity of world. In this context, explain any one international treaty.

16. Explain 'In-Site' and Ex-Situ' approaches of conservation of biodiversity. Compare their advantages and limitations.

17. Write the short notes on

(*a*) Habitat loss

(*b*) Endangered and endemic species of India

(*c*) Man-wildlife conflict

(*d*) Biodiversity treaty.

Short Answer Type Questions

1. Define biodiversity.

2. Enumerate the Biosphere Reserves of India.

3. Expand IUCN, BSI and WWF.

4. What is the main objective use of Red list?

5. What is meant by man-animal conflicts?

6. What are the roles of Biosphere reserves?

7. What are the ecosystem services which are provided by the biological diversity?

8. Name the plant which yield drug for (*i*) Malaria (*ii*) Cancer.

9. Define the following :

(*a*) Exotic species

(*b*) Rear species

(*c*) Hot spots

10. Differentiate between the
(*a*) National park and sanctuary
(*b*) In-situ and ex-situ conservation
(*c*) Species and genetic biodiversity
(*d*) Endangered and extinct species

11. Write short notes on :
(*i*) Rare species and Threatened species
(*ii*) Poaching of wild-life
(*iii*) In-situ conservation of biodiversity.

Objective Type Questions

Multiple Choice Questions

1. Genetic variation between distinct populations of the same species is known as :
(*a*) Species diversity (*b*) Ecosystem diversity
(*c*) Genetic diversity (*d*) Biodiversity

2. Which of the following is an endemic species of India?
(*a*) Asian elephant (*b*) Lion-tailed macaque
(*c*) Whales (*d*) Panda

3. Which of the following is a biodiversity hotspot in India?
(*a*) Gulf of Mannar (*b*) Western Ghats
(*c*) Pachmarhi (*d*) Sunderbans

4. Which of the following is an in-situ conservation measure taken by India?
(*a*) Project Elephant (*b*) Project Lion
(*c*) Project Rhino (*d*) All of the above

5. Who introduced the concept of biodiversity hotspot?
(*a*) Christopher Columbus (*b*) Norman Myers
(*c*) WWF (*d*) Charles Darwin

6. Lion-tailed macaque is found in :
(*a*) Western Ghats (*b*) Eastern Ghats
(*c*) Caucasus (*d*) Western Himalaya

7. Which of the following is a cause of loss of biodiversity?
(*a*) Habitat degradation and loss (*b*) Invasion of non-native species
(*c*) Pollution (*d*) All of the above

8. The animal that has recently become extinct is :
(*a*) Dinosaur (*b*) Dodo
(*c*) Wolly Mammonth (*d*) Archaeopteryx

9. The Dodo was extinct due to :
(*a*) Pollution
(*b*) Invasion of non-native species

(*c*) Over-exploitation of resources

(*d*) Global environmental change

10. Indian bustard has become endangered species due to :

(*a*) Deforestation in Rajasthan (*b*) Spread of desert

(*c*) Hunting (*d*) Change in climate

11. Which group of vertebrates comprises the highest number of endangered species?

(*a*) Birds (*b*) Mammals

(*c*) Fishes (*d*) Reptiles

12. Spot the site, which is not a biodiversity hotspot :

(*a*) Brazil's Cerrado (*b*) Central Chile

(*c*) California Floristic Province (*d*) Pakistan

13. The area reserved for the welfare of wildlife is called :

(*a*) National Park (*b*) Forest

(*c*) Botanical garden (*d*) Sanctuary

14. Kaziranga National Park is famous for :

(*a*) Rhinoceros (*b*) Tiger

(*c*) Elephant (*d*) Musk deer

15. Animals and Plants are best protected :

(*a*) Zoos (*b*) Botanical garden

(*c*) National Park (*d*) Sanctuaries

16. New approach to conservation is the establishment of :

(*a*) Biosphere Reserve (*b*) Reserve forest

(*c*) National Park (*d*) Sanctuaries

Fill in the Blanks

1. Biodiversity refers to the totality of ...., ... and .... of a region.

2. Exotic species have large impact especially in .... ecosystem.

3. In India .... and .... are .... among the 25 global biodiversity hot spot.

4. As of September 2002, India had .... protected areas.

True/False Questions

1. National parks are reserved for the conservation of animals only.

2. Extinction of species refers to its disappearance from the earth.

3. There are 13 biosphere reserves.

Match the Following Terms

Match the following terms of column I with appropriate terms of column II

Column I	Column II
(i) Doda	(a) In-situ conservation
(ii) Red data book	(b) Threat to biodiversity
(iii) National park	(c) Extinct species
(iv) Seed bank	(d) Endangered species
(v) Poaching	(e) Ex-situ conservation

ANSWERS

Multiple Choice Questions

1. (c)	**2.** (b)	**3.** (b)	**4.** (a)	**5.** (b)
6. (a)	**7.** (d)	**8.** (b)	**9.** (b)	**10.** (c)
11. (b)	**12.** (d)	**13.** (a)	**14.** (a)	**15.** (d)
16. (a)				

Fill in the Blanks

1. Genes, species, ecosystem **2.** Island

3. Western ghats, eastern himalaya, two **4.** 581

True/False Questions

1. False **2.** True **3.** True

Match the Following Terms

(i) (c) (ii) (d) (iii) (a) (iv) (e) (v) (b)

❑❑❑

CHAPTER

4

Air and Noise Pollution

4.1. INTRODUCTION

INSIDE THIS CHAPTER

The environment is made up of air, water and land, technically known as atmosphere, hydrosphere and lithosphere respectively which together constitute the biosphere. In the biosphere, apart from human beings, plants, animals, birds, fishes, insects and microorganisms (algae, bacteria and virus) also exist. The atmosphere provides oxygen, while the hydrosphere and lithosphere provide food, water and space. Whenever a change, physical or chemical, occurs in the atmosphere, hydrosphere or lithosphere, all living beings are affected. This change is termed as pollution and the agents that institute these changes are called pollutants.

Pollution can be defined as an undesirable change in the physical, chemical or biological characteristics of our air, water or land that may or will harmfully affect the life or create a potential health hazard and threaten the survival or activities of humans or other living organisms.

Pollution is thus direct or indirect change in any environmental components that is causing harmful effect adversely the industrial progress, culture or natural assets or general environment.

It is difficult to estimate the desirable and undesirable effects of any activity which alters the environment. Sometimes short-term gains can cause immeasurable damage in the future, as seen in the case of use of nuclear energy, motorcars, air-conditioners and refrigerators, etc. In an age of fast material change, pollution is an unavoidable result. History has shown that societies pollute first and pay later. As the decline of the biosphere continues unchecked, people must find the will to force governments and industries to change existing conditions.

4.2. TYPES OF POLLUTANTS

Environmental pollution may be defined as the inorganic, organic, radioactive or biological substances, noise odor etc. which tends to damage the quality of the environment so as to constitute a hazard or impair the usefulness of the environment.

On the basis of **decomposition nature,** pollutants are of following two types :

4.2.1. Non-degradable Pollutants

They include many radioactive materials, heavy metals and some plastics which cannot be degraded by natural or artificial means. They are not cycled in ecosystem naturally. They do not only accumulate but are often biologically magnified with their subsequent movement in food chains and biogeochemical cycles. They must be controlled or prevented from reaching the environment.

4.2.2. Degradable Pollutants

Degradable pollutants are those that can be decomposed, removed or consumed or reduced to acceptable levels either by natural or artificial means. However, pollutants such as human sewage and animal and crop wastes can decompose only if the system is not overloaded. Certain chemicals decompose slowly, and can persist at harmful levels for decades like detergents and pesticides.

On the basis of nature, the pollutants are of following four types :

Physical pollutants : Heat, noise, radioactive substances

Chemical pollutants : Oxides of carbon, oxides of sulphur, oxides of nitrogen, halogen gases etc.

Biological pollutants : Micro-organisms such as bacteria, virus and fungi etc.

On the basis of their **source of origin,** these are divided in following three types :

Point source pollutants : When the emission or origin source is point, i.e. the area are very less as compare to the environment. It includes emissions from stack.

Line source pollutants : When the emission source is in the form of line or curve e.g. transportation of material by conveyor belts, road rollers and run-off seepage etc.

Area source pollutants : When the emission or origin source is an area, i.e. the area are large e.g. crushing, grinding and blasting in mines etc.

4.3. AIR POLLUTION

Clean air, which is essential for the survival of all living organisms, is rapidly becoming scarce. At mean sea levels air contains 20.94 per cent oxygen and 78.09 per cent nitrogen. Other elements present comprise less than one percent of its composition.

Air pollution may be defines as the presence of one or more contaminants in the atmosphere in such a quantity and of such duration as is or tends to be injurious to human health of welfare, animal or plants or properties or would unreasonable interfere with enjoyment of life and property.

4.3.1. Air Pollutants

Primary pollutants are those which are emitted directly into the atmosphere, like sulphur dioxide, nitric oxides and carbon monoxide. They are categorized as follows :

1. Particulate Matter (Aerosol)

Any kind of small particle found floating in the air is referred to as aerosols or particulates. Particulates in air are the result of both natural and human activities. Various types of particulate present in air are given below:

(*a*) **Dust :** The particulates having the size 1mm to 200mm in solid forms is referred as dust. This includes metallic and non-metallic dust. Copper, lead, steel, aluminium zinc are examples of metallic dust while non-metallic industrial operations produce cotton dust, cement dust, asbestos dust, ceramic dust etc.

(*b*) **Smoke :** It mostly consists of soot, fly ash and other liquid or solid particles 0.01 to 1mm in diameter. Smoke arises from several sources such as factories, automobiles, oil refineries, steel plant etc.

(*c*) **Fog :** It is special type of thin cloud consisting of microscopically small water droplets which are kept in suspension in the air near the ground surface.

(*d*) **Smog (smoke + fog) :** Smog is two types- one is found in place having coal as the principal fuel. The other one is photo-chemical smog.

2. Oxides of Carbon

CO_2 and CO are the two important pollutants.

(*a*) **Carbon monoxide :** CO, a product of incomplete combustion, is a deadly poisonous gas in high concentration. The chief sources are automobile, furnaces, domestic fires, forest fires, burning of coal in factories, houses and power plants.

(*b*) **Carbon dioxide :** A large amount of CO_2 is released in the atmosphere from burning of fossil fuels. Domestic and thermal power plant furnaces, industries are the main sources of CO_2 emission.

3. Oxides of Nitrogen (NOx)

Nitrous Oxide (N_2O), nitrogen oxide (NO), nitrogen dioxide (NO_2), di-nitro trioxide (N_2O_3), di-nitro tetraoxide (N_2O_4), di-nitro pentaoxide (N_2O_5) and nitrogen trioxide (NO_3) collectively called as oxides of nitrogen. Of these seven oxides NO and NO_2 are the most important air pollutants. NO formed during lighting discharges and stratospheric oxidation of ammonia. A small amount of NO and N_2O are also emitted out of the soil through denitrification process by bacterial activity.

4. Oxides of Sulphur (SOx)

Sulphur dioxide, hydrogen sulphide and sulphate particles are the three forms of sulphur found in the atmosphere. The chief source of SO_2 emission are burning of fossil fuels as coal in thermal power plant, smelting of sulphur containing ore and other chemical industries related to fertilizer and sulphuric acid. The chief source of H_2S is decomposition of vegetation and animal waste especially in aquatic habitats. Coal pits, volcanic eruption, sulphur springs and sewage system are also give rise to not a little but considerable amount of hydrogen sulphide.

5. Hydrocarbons

There are many hydrocarbons known to be present in atmosphere as primary pollutants. They basically include methane, ethane, toluene, n-butane, iso-bentane, acetylene etc. coming out from automobile industrial exhaust and decomposition process.

6. Halogens Gases

Halogen remains in the atmosphere in two forms, namely chlorine and fluorine. The chief source of Cl_2 is caustic soda industries. Motor vehicle exhaust contains lead halid aerosols and their photochemical decomposition releases chlorine atoms. Fluorine which is an essential element of protoplasm becomes pollutant at high concentration. It occurs mainly in smoke form brick kiln and iron and steel industries.

Secondary pollutants are pollutants formed by the photochemical reaction of primary pollutants. For example, "smog", is a combination of smoke and fog. Smoke consists of carbon particles and fog is an emulsion of water vapour in air. Smog has become very common in large cities, especially during winter. Similarly acid rain is formed by the combination of sulphur dioxide and water vapour present in the air.

4.3.2. Sources of Air Pollution

Air pollution can be due to natural or man-made causes. The former is beyond our control as natural disasters like dust storms, earthquakes and volcanic eruptions throw up large quantities of dust and gases into the atmosphere. Man-made causes, however, should be prevented or controlled as they pose a greater danger by way of toxic emissions from factories, power plants, vehicular traffic, etc. Industries such as mining, thermal plants, brick kilns, etc. also pollute the air. These emissions are particularly intense in urban conglomerations where the density of human habitation is very high.

Natural sources : Volcanic eruption, forest fires, bio-pollutants etc.

Man-made Sources

(*i*) Burning of fossil fuels

(*ii*) Automobiles

(*iii*) Steel plants

(*iv*) Thermal power plants

(*v*) Textile industries

(*vi*) Fertilizer industries

(*vii*) Chemical and Pesticides Plants

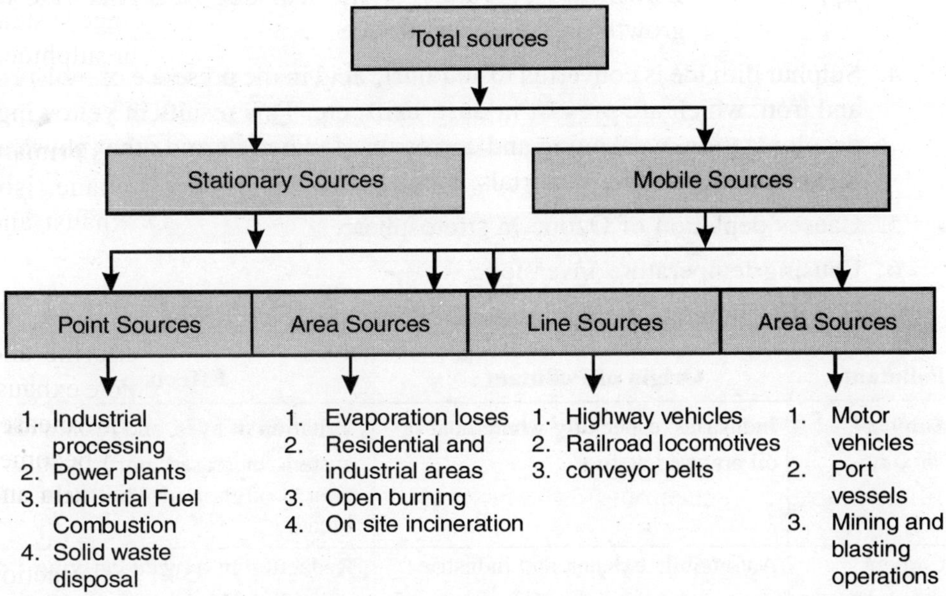

Fig. 4.1. Classification of Air Pollution Sources

4.3.3. Effects of Major Air Pollutants on Environment and Human Health

Most air pollutants do not lead to specific diseases. Depending on the pollutant, the concentration and the duration of exposure, some organs are more affected than others. The most frequent disorders are those caused by irritant gases and particulates on the mucous membranes and respiratory organs. The consequences are eye, nose and throat inflammations, diminished lung function, increased susceptibility to respiratory infection and a higher incidence of chronic bronchitis. These disorders and diseases are, of course, influenced by other factors as well, such as immune deficiency, allergies, occupational exposure to pollutants, and particularly smoking.

Other Effects of Air Pollution

1. Ozone causes reduced pollination and yellow spot formation on leaves, thus affecting the rate of photosynthesis and decreases in plant productivity
2. Causes global warming (green house effect), acid rain and helps in the formation of photo-chemical smog
3. Causes chlorsis, nacrosis and epinesty in plants

   Nacrosis : Killing or collapse of tissues

   Chlorosis : Reduction in the chlorophyll

   Abcission : Dropping of leaves

   Epinasty : Downward curvature of the leaf due to higher rate of growth on the upper surface
4. Sulphur dioxide is converted to sulphuric acid in the presence of moisture and iron, which are present in dust, nails, etc. This results in yellowing, decolonization, weakening and corrosion of materials and other physical distortion of building materials, monuments, status etc.
5. Causes depletion of Ozone in stratosphere.
6. Causing temperature inversion.

Major Air Pollutants, their origin and effects are given below :

Pollutant	Origin of Pollutant	Effects
Sulphur dioxide	Industries, especially where coal or oil are used as fuel	Irritation of eyes, and respiratory system, increased mucus production, cough and shortness of breath
Carbon monoxide	Automobile exhaust and industries	Reduction in oxygen-carrying capacity of blood
Oxides of nitrogen	Automobile exhaust	Irritation of pulmonary tract affecting functioning of lungs
Hydrocarbons Chlorine	Automobile exhaust Chloralkali industry	Lung cancer Irritation of mucous membrane
Ammonia	Fertiliser, industry, agriculture and in poultry farming	Irritation of mucous membrane
Hydrogen	Manufacture of coke, viscose rayon, sulphide	Distillation of tar and petroleum excessive inhalation leads to death

Acids and aldehydes	Chemical industries	Eyes, nose and throat irritation
Suspended particulate matter (SPM)	industries, automobile exhaust	respiratory diseases
Dust	Industries and automobile exhaust	Silicosis
Asbestos	Roofings, brake linings	Asbestosis
Lead	Automobile exhaust	Cumulative poison, impairment of central nervous system
Manganese	Mining operations	Damages nerves and reproductive system
Benzene	Automobile exhaust and manufacture of chemicals	Leukemia, chromosomal damage
Pesticides	Manufacture and application of pesticides	Depression; leads to death if inhaled in excess

4.3.4. Lapse Rate and Temperature Inversion

The degree of stability of the atmosphere depends on the rate of change of ambient temperature with altitude. **The decrease in temperature with altitude is known as lapse rate.**

Based on meteorological data in the troposphere up to 10 km where the temperature decreases linearly with altitude, the environmental lapse rate is found to be about – 6.5°C/km.

The adiabatic decrease in temperature with altitude is especially important in the vertical movement of air pollutants. As the air rises in the atmosphere, it passes through a region of decreasing pressure and expands to equalize the decreasing pressure. As it expands, it does work on the surroundings. Since the process is usually rapid, there is no heat transfer between air parcel and the surrounding air, the gas undergo adiabatic cooling. The temperature lapse rate for dry air parcel moving upward adiabatically and is known as dry adiabatic lapse rate and has been calculated to be nearly – 10° C/km for dry air and for humid air that is – 6° to –7° C/km.

A comparison of the adiabatic lapse rate with the environmental lapse rate gives an idea of the stability of the atmosphere. We now consider the following situations from the point of view of air pollution:

When the environmental lapse rate and the dry adiabatic lapse rate are exactly same and air would experience no buoyant force, such an atmosphere is said to be **neutrally stable.**

When the environmental lapse rate is greater than the dry adiabatic lapse rate, the atmosphere is said to be *super-adiabatic*. Then the atmosphere is said to be in an **unstable equilibrium**. Under these conditions the atmospheric pollutants are quickly dispersed due to considerable vertical mixing of air.

When the environmental lapse rate is less than the dry adiabatic lapse rate, such a condition is called **stable** and the lapse rate is said to be *sub-adiabatic*. The extreme case of a stable atmosphere occurs when the temperature increases with altitude, called an **inversion**. This occurs in the atmosphere due to the following reasons :

1. **Subsidence inversion :** This is usually associated with subtropical anticyclone where the air is warmed by compression as it descends in a higher pressure system and achieves temperature higher than that of the air underneath. If the temperature increase is sufficient, an inversion will result.
2. **Radiation inversion :** This result from the normal diurnal cooling cycle. After sunset, the ground cools quickly by radiational heat transfer, and the lowest layer of air in contact with surface loses sensible heat through conduction and small scale mixing. Consequently, a temperature inversion is set up between the cool low-level air and the warmer air above, in the first few hundred meters above the surface.
3. **Advection inversion :** This is formed when warm air moves over a cold surface or cold air. An example of an elevated advection inversion occurs when a hill range forces a warm land breeze to flow at high levels and a cool sea breeze flows at low levels in the opposite direction.

4.3.5. Control of Air Pollution

Air pollution can be controlled by two fundamental approaches: prevention techniques and emission control. They are as following:

1. The particulate matter can be removed by gravitational settling chamber, cyclone separator, fabric filter, electrostatic precipitator and wet scrubber.
2. The gaseous pollutants are controlled through absorption in liquid, adsorption on a solid surface and chemical alteration of the pollutants through combustion or catalytic treatment.
3. By using desulphurized fuel
4. By absorbing the toxic or pollutants gases in different solutions
5. By plantation and foresting
6. NOx can be control by using excess amount of air
7. Building higher smoke stacks facilities the discharge of pollutants as far away from the ground as possible.

4.4. AIR POLLUTION CONTROL EQUIPMENT

The most common control technologies employed for particulates include the following :

- Gravitational settling chamber,
- Cyclone separator,
- Fabric filter,
- Electrostatic precipitator
- Wet scrubber.

The first two are used for coarser particulates while last three are for finer.

4.4.1. Gravitational Settling Chamber

They are generally used to remove large, abrasive particles (usually > 50 mm) from gas streams. Since most of the troublesome particles have much smaller size than 50mm, there devices are usually used as per cleaners prior to passing the gas stream through high efficiency collection device.

Settling chambers, which rely on gravitational settling as a collection mechanism, are the simplest and oldest mechanical collectors. Settling chambers are generally built in the form of long, horizontal, rectangular chambers with an inlet at one end and an exit at the side or top of the opposite end. Flow within the chamber must be uniform and without any macroscopic mixing. Hoppers are used to collect the settled-out material, though drag scrapers and screw conveyers have also been employed. The dust removal system must be sealed to prevent air from leaking into the chamber which increases turbulence, causes dust re-entrainment, and prevents dust from being properly discharged from the device.

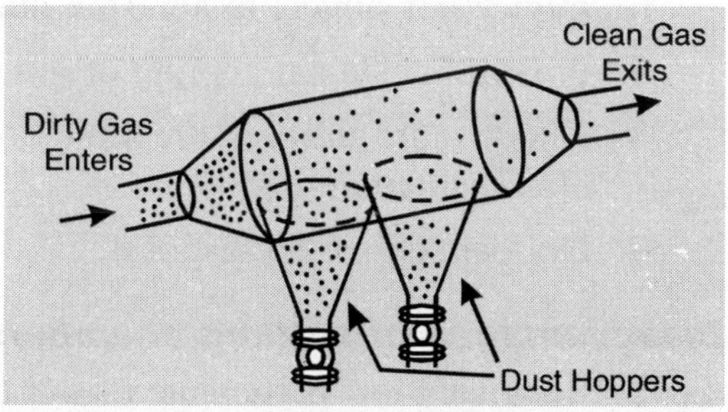

Fig. 4.2. Gravitational Settling Chamber

There are two primary types of settling chambers: the expansion chamber and the multiple-tray chamber. In the expansion chamber, the velocity of the gas stream is significantly reduced as the gas expands in a large chamber. The reduction in velocity allows larger particles to settle out of the gas stream.

A multiple-tray settling chamber is an expansion chamber with a number of thin trays closely spaced within the chamber, which causes the gas to flow horizontally between them. While the gas velocity is increased slightly in a multiple-tray chamber, when compared to a simple expansion chamber, the collection efficiency generally improves because the particles have a much shorter distance to fall before they are collected. Multiple-tray settling chambers have lower volume requirements than expansion-type settling chambers for the collection of small particles.

The efficiency of settling chambers increases with residence time of the waste gas in the chamber. Because of this, settling chambers are often operated at the lowest possible gas velocities. In reality, the gas velocity must be low enough to prevent dust form becoming re-entrained, but not so low that the chamber becomes unreasonably large. The size of the unit is generally driven by the desired gas velocity within the unit, which should be less than 3 m/s (10 ft/sec), and preferably less than 0.3 m/s (1 ft/sec).

Advantages of settling chambers include

1. Low capital cost;
2. Very low energy cost;
3. No moving parts, therefore, few maintenance requirements and low operating costs;
4. Excellent reliability;
5. Low pressure drop through device;
6. Device not subject to abrasion due to low gas velocity;
7. Provide incidental cooling of gas stream;
8. Temperature and pressure limitations are only dependent on the materials of construction; and
9. Dry collection and disposal.

Disadvantages of settling chambers include

1. Relatively low particulate matter collection efficiencies, particularly for particulate matter less than 50 μm in size;
2. Unable to handle sticky or tacky materials;
3. Large physical size; and
4. Trays in multiple-tray settling chamber may warp during high-temperature operations.

4.4.2. Cyclone Separator

A cyclone separator is a very useful piece of equipment for the removal from air streams of particles above 10 micrometer in diameter. The equipment is a settling chamber in the form of a vertical cylinder, so arranged that the particle laden air spirals round the cylinder to create centrifugal forces which throw the particles to the outside walls.

The gas or liquid stream is injected at high velocities through the inlet pipe, which is positioned tangentially to the body of the cyclone (Seinfeld, 1975). The shape of the cone induces the stream to spin, creating a vortex. Larger or denser particles are forced outward to the walls of the cyclone where the drag of the spinning air as well as the force of gravity causes them to fall down the sides of the cone into an outlet (Seinfeld, 1975; Svarovsky, 1984). Meanwhile the lighter and/or less dense particles as well as the gas or liquid medium exit through the top of the cyclone via the low pressure center.

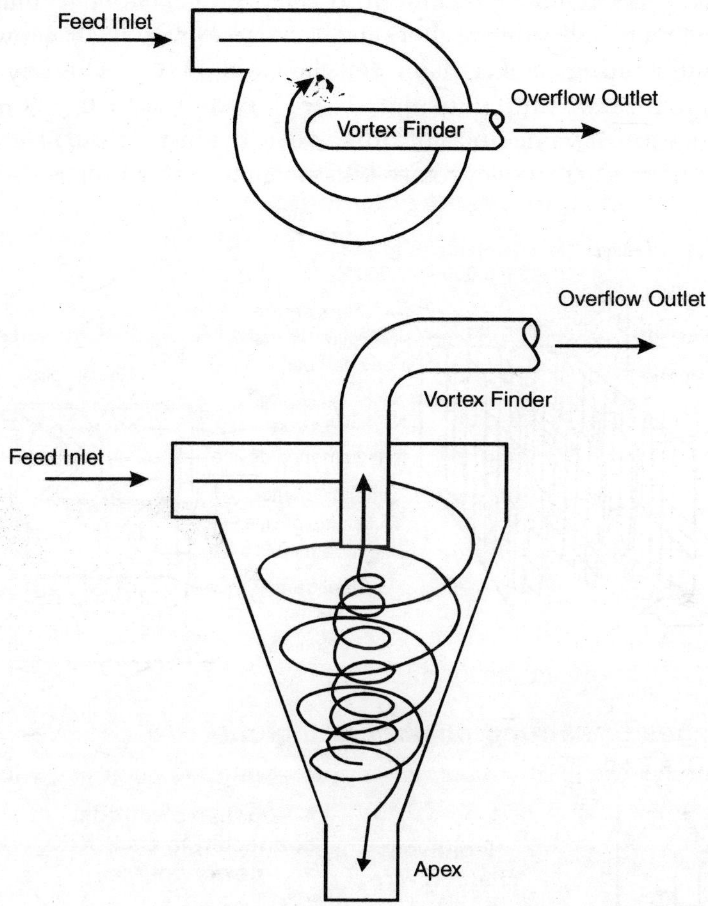

Fig. 4.3. Cyclone Separator

The separation process in cyclones requires a steady flow, free of fluctuations or short term variations in the flow rate. Cyclone separators are customarily operated with the top and bottom open to the atmosphere so that there is no pressure difference between the two (Singh and Eckhoff, 1995). According to Singh and Eckhoff (1995), if one of the outlets has to be operated.

4.4.3. Electrostatic Precipitator

The electrostatic precipitator has two electrodes : discharge electrode and collecting electrode. When smoke containing particulates is ionized while passing between the discharge electrode (which induces corona discharges) and the collecting electrode, the particulates are negatively charged. Attracted by positive electricity, the particulates then adhere to the collecting electrode. Accumulated particulates, periodically hammered, fall to the bottom of the precipitator and are removed.

The main applications of ESP are in textile, pulp and paper, coke ovens in steel industry, hog fuel boilers, veneer and particle board dryers, glass furnaces, stannic chloride collection, SO_2 control, fly ash control, pharmaceutical detergent dryers, cogeneration, distilling liquors and beers, phosphorous furnace emissions, silicon manufacturing, power plant emissions, ammonia removal, phosphate fertilizer manufacturing, phosphoric acid manufacturing, liquid waste incinerators, solid waste incinerators, corn drying, sulfuric acid plants, incineration of sewage sludge, rotary kiln cleaning, cement plants.

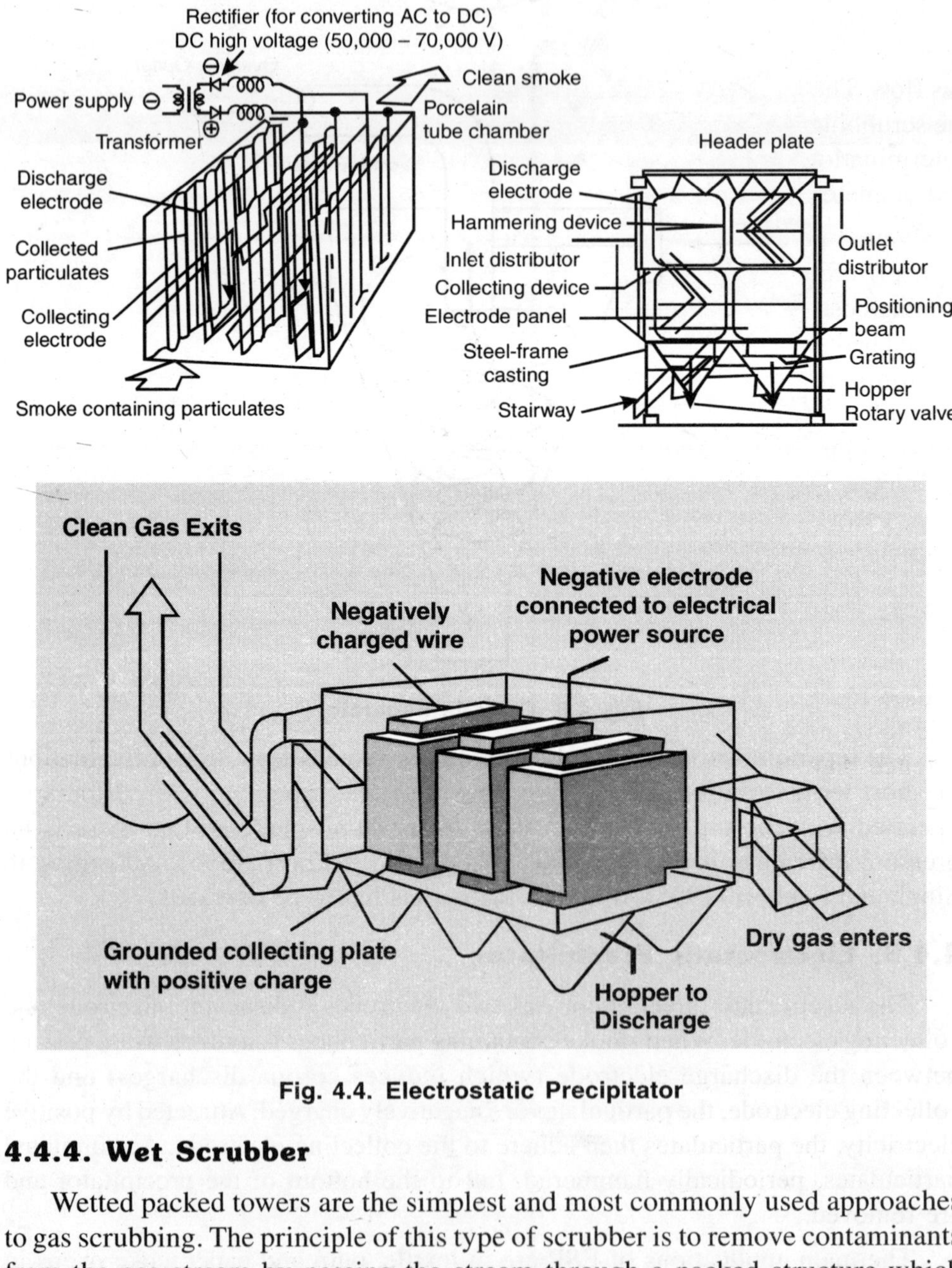

Fig. 4.4. Electrostatic Precipitator

4.4.4. Wet Scrubber

Wetted packed towers are the simplest and most commonly used approaches to gas scrubbing. The principle of this type of scrubber is to remove contaminants from the gas stream by passing the stream through a packed structure which provides a large wetted surface area to induce intimate contact between the gas and the scrubbing liquor. The contaminant is absorbed into or reacted with the scrubbing liquor.

The packing of the tower is normally a proprietary loose fill random packing designed to encourage dispersion of the liquid flow without tracking, to provide maximum contact area for the 'mass transfer' interaction and to offer minimal back pressure to the gas flow. The reactivity between the contaminant and the scrubbing liquor influences the system designer's determination of gas and liquor flow and the height and diameter of the packed bed.

Wetted packed towers can be designed for very high efficiencies with relatively low capital and running costs. The low pressure drop associated with packed bed scrubbers permits the use of smaller more economical fans. Although efficiency may be affected, a packed tower will usually function when gas or liquor flows vary from its original design parameters. The major types are: spray tower (5 - 10 mm), centrifugal scrubber, packed bed and plate column, venture scrubber (> 3 mm)

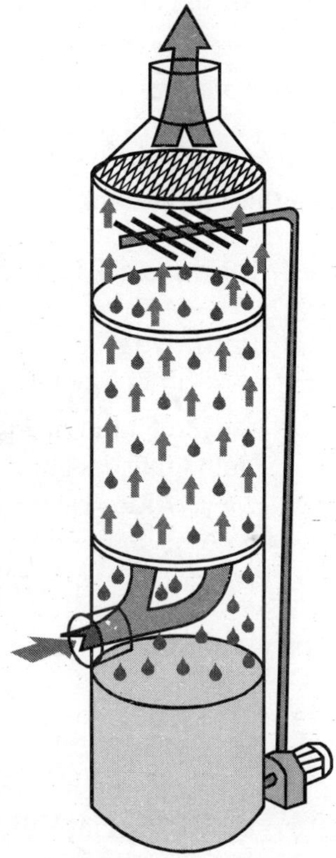

Fig. 4.5. Wet Scrubber

4.4.5. Fabric Filter

Filtration is the outlet and generally one of the most versatile and efficient method for removing particulate matter from industrial gases. Fabric filters are used to remove particles from a gas stream. Fabric filters are made of a woven or felted material in the shape of a cylindrical bag or a flat supported envelope.

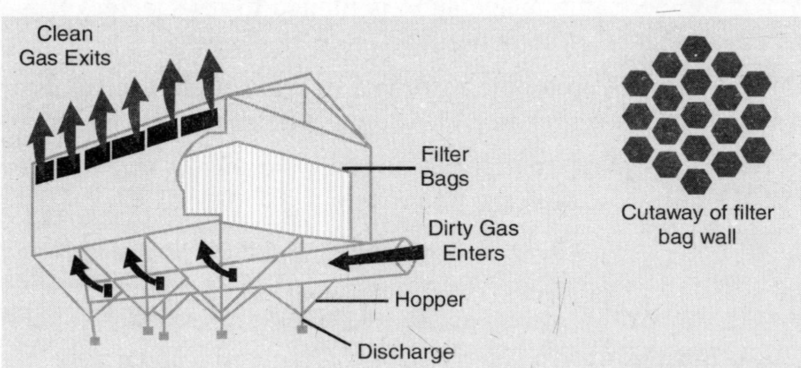

Fig. 4.6. Fabric Filter

These elements are contained in a housing which has gas inlet and outlet connections, a dust collection hopper; and a cleaning mechanism for periodic removal of the collected dust from the fabric. In operation, dust laden gas flows through the filters, which remove the particles from the gas stream.

In principle, the dust-laden gas passes though a fabric in which the particles are trapped due to different physical mechanism. The important mechanisms by which the particles are separated from the gas stream are usually inertial impaction, direct interception and diffusion.

Advantages and Disadvantages

(a) Advantages

(1) Very high collection efficiencies possible (99.9 + percent) with a wide range of inlet grain loadings and particle size variations. Within certain limits fabric collectors have a constancy of static pressure and efficiency, for a wider range of particle sizes and concentrations than any other type of single dust collector.

(2) Collection efficiency not affected by sulfur content of the combustion fuel as in ESPs.

(3) Reduced sensitivity to particle size distribution.

(4) No high voltage requirements.

(5) Flammable dust may be collected.

(6) Use of special fibers or filter aids enables submicron removal of smoke and fumes.

(7) Collectors available in a wide range of configurations, sizes, and inlet and outlet locations.

(b) Disadvantages

(1) Fabric life may be substantially shortened in the presence of high acid or alkaline atmospheres, especially at elevated temperatures.

(2) Maximum operating temperature is limited to 550 degrees Fahrenheit, unless special fabrics are used.

(3) Collection of hygroscopic materials or condensation of moisture can lead to fabric plugging, loss of cleaning efficiency, large pressure losses.

(4) Certain dusts may require special fabric treatments to aid in reducing leakage or to assist in cake removal.

(5) Fabric bags tend to burn or melt readily at temperature extremes.

4.5. MONITORING

Air pollution monitoring activities are typically separated into two classifications: source monitoring and ambient air monitoring. Monitoring can be made directly using continuous measurement instrumentation or manual methods, or remotely using optical sensing systems. Source monitoring involves the measurement of emissions directly from a fixed or mobile emission source, typically in a contained duct, vent, stack or chimney. Stationary source data is used to determine control technology performance, confirm established permit limits are being met, and as input to ozone and/or health risk prediction models. Major stationary sources may have continuous emissions monitors (CEMs) installed to report real-time emissions based on pre-established reporting cycles. Ambient air monitoring involves the measurement of specific pollutants present in an immediate surrounding atmosphere. Most Major urban areas often operate several ambient air monitoring instruments, each dedicated to measuring specific target pollutants.

Ambient Air Monitoring

For ambient air monitoring, specially designed High Volume samplers (HVS) having mass flow controller and cyclone separator are used. These HVS can accommodate both normal size filter paper and a PUF (Poly Urethane Foam) filter fitted at the trunk. High Volume Sampler instruments available in Indian market will require intricate modifications to suit sampling of dioxin and furan from ambient air. Anderson' make Hi-Vol sampler or Dicotomous sampler available in international market are suitable for ambient dioxin monitoring.

Stack/Source Monitoring

For source emission monitoring of dioxin and furan, an intricate flue gas sampling train and procedures are required to be adopted. The integrated stack gas sample is iso-kinetically withdrawn from selected traverse points along the stack cross-section. Semi-volatile organic compounds associated with particulate matter are collected in the front-half components of the sampling train. Semi-volatile organic compounds not collected by high efficiency glass or quartz fiber are adsorbed on porous, polymeric resin, Amberlite XAD-2. The borosilicate or quartz liner, encased in a stainless steel tube is required to perform stack monitoring. This stainless steel tube is capable of maintaining the exit gas temperature at 120 ± 14°C or at required temperature necessary to prevent condensation during sampling.

Air Quality Monitoring

Ambient air quality monitoring carried out at various cities/towns in the country, under National Air Monitoring Programme (NAMP) provide air quality information that form the basis for identifying areas with high air pollution levels and subsequently, for planning the strategies for control and abatement of air pollution. Data generated over the years reveal that Suspended Particulate Matter (SPM) and Respirable Suspended Particulate Matter (RSPM/PM10) exceed permissible levels at many locations, particularly in urban areas. Air pollution problem becomes complex due to multiplicity and complexity of air polluting source mix (e.g. industries, automobiles, generator sets, domestic fuel burning, road side dusts, construction activities, etc.). Ambient air quality monitoring is required to determine the existing quality of air, evaluation of the effectiveness of control programme and to identify areas in need of restoration and their prioritization.

A cost-effective approach for improving air quality in polluted areas involves : (*i*) identification of emission sources; (*ii*) assessment of extent of contribution of these sources to ambient air; (*iii*) prioritization of sources that need to be addressed; (*iv*) evaluation of various options for controlling the sources with regard to feasibility and economic viability; and (*v*) formulation and implementation of appropriate action plans.

Most frequently occurring pollutants in an urban environment are particulate matters (suspended particulate matter *i.e.* SPM and respirable suspended particulate matter *i.e.* RSPM), carbon monoxide (CO), hydrocarbons (HC), sulfur dioxide (SO_2), nitrogen dioxide (NO_2), ozone (O_3) and photochemical oxidants.

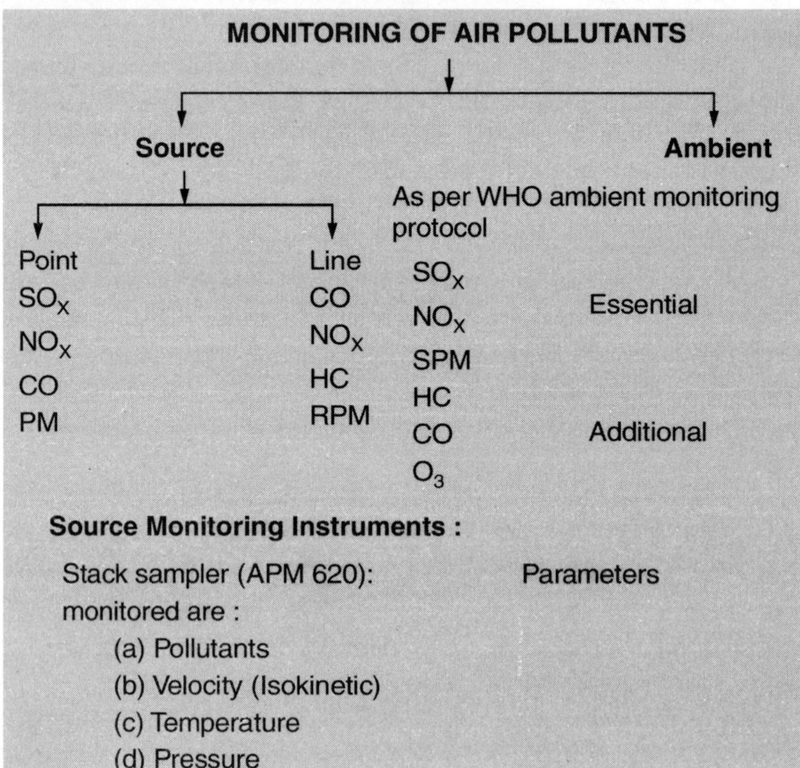

Fig. 4.7. Monitoring of Air Pollutants.

Objectives of Air Quality Monitoring

The major objectives for air quality monitoring are as below:

(i) Background Data

In order to generate background data, air quality monitoring is conducted to assess existing level of contamination and to asses possible effects of air contamination occurring in future.

(ii) Status and Trend Evaluation

The objective is to determine air pollution status and trend information from any continuous air quality monitoring programme. The information is used to determine, whether pollution control strategies as advised by implementing authority are giving acceptable values that is lowering of pollution levels or new or additional control are required to achieve acceptable levels.

(iii) Environment Exposure Level Determination

The air quality monitoring and survey concern itself with systematic study of considerable segment of environment to define inter-relationship of source of pollution, atmospheric parameter and measurable manifestations in order to evaluate the character and magnitude of existing problem.

(iv) Scavenging Behaviour of Environment

To understand natural scavenging or cleansing process undergoing in the environment through pollution dilution, dispersion, wind movement, dry deposition, precipitation and chemical transformation of pollutants generated.

(v) Air Quality Management

To assess the present status to judge effectiveness of air pollution control strategies and long term management of air quality.

The recommended criteria for siting the monitoring stations :

- The site is dependent upon the use/purpose of the results of the monitoring programs.
- The monitoring should be carried out with a purpose of compliance of air quality standards.
- Monitoring must be able to evaluate impacts of new/existing air pollution sources.
- Monitoring must be able to evaluate impacts of hazards due to accidental release of chemicals.
- Monitoring data may be used for research purpose.

Guidelines For Monitoring

For setting up of any ambient air quality monitoring station, the most important thing to be considered prior to commencement of actual monitoring is to collect its background information.

Background Information

The background information that needs to be collected includes details of sources and emissions, health status, demography, population growth, landuse pattern, epidemiological studies. Such prior information will provide immense help to identify the likely effects and in particular health impacts resulting from population exposure to air pollutants.

(i) Sources and Emissions

Sources in a city include vehicles, industries, domestic etc. In an industrial area, information should be obtained on the type of industries including their number, fuel used, composition of fuel, pollutants emitted etc. Information on number and distribution of sources should be collected. This information will help in identifying which pollutants can be expected in an area and thus should be measured. In case of industrial stacks, locations of maximum ground level concentrations should be determined by modeling.

The stations should be located at locations where maximum ground level concentrations are expected. Information on type and number of vehicles should be obtained. Information on domestic fuel that is used in household should be obtained.

Pollution load emanating from these sources should be estimated so as to identify sources that are generating significant amount of pollution.

(ii) Health and Demographic Information

Investigations shall be carried out based on the public complaints received from an area related to air pollution. If the results of such investigations reveal that the level are high that area can be considered for ambient air quality monitoring.

Areas where population density is high (more than one million) can be considered for locating monitoring stations. Information on age and socio-economic status of population is also important for making a decision on initiation of ambient air quality monitoring. Location of monitoring station in such areas will help in finding exposure levels to population which can be used further in epidemiological studies to evaluate health effects of air pollutants.

(iii) Meteorological Information

Meteorological data with respect to temperature, relative humidity, wind speed and direction should be collected. Predominant wind direction plays an important role in determining location of monitoring stations. Due to effects such as land and sea breezes, valley effects etc. it is important to collect local meteorological data specific to the site. The monitoring stations should be located in areas that are downwind from the sources. Mixing height data should also be collected. Mixing height data can be collected from Indian Meteorological Department. Information on duration of various seasons in a year is also important. Measurement frequency should be such that monitoring is done in all the seasons so that all seasonal variations are included in computing annual average.

(iv) Topographical Information

Local winds and stability conditions are affected by topography. In river valleys there is increased tendency of developing inversions. More number of monitoring stations should be located in areas where spatial variations in concentrations is large.

Mountains, hills, water bodies also affect dispersion of pollutants.

(v) Previous Air Quality Information

Any previous information collected on ambient air quality can serve as a basis for selecting areas where monitoring should be conducted and previous studies may include data collected for any health studies etc. Previous studies can be used to estimate the magnitude of the problem.

Once the background information is collected, the ambient air quality monitoring is to be initiated and selection of type of pollutant to be measured, number and distribution of monitoring stations etc. should be made.

Selection of Monitoring Location

Principal factors governing the locations of the sampling stations are the objectives, the particular method of instrument used for sampling, resources available, physical access and security against loss and tampering. Air quality monitoring should be done in areas where pollution problem exists or is expected *i.e.* mainly in industrial areas, urban areas, traffic intersections etc. One of the objective of monitoring is to determine status and trends and the air quality monitoring should be done in metropolitan cities and other urban areas so as to compare their levels and determine trends. Selection of site is very important as a incorrect location may result in data that may not meet the objectives of monitoring and will be of limited value. In general the following requirements should be satisfied for site selection.

(a) Representative Site

A site is representative if the data generated from the site reflects the concentrations of various pollutants and their variations in the area. The station should be located at a place where interferences are not present or anticipated. In general the following conditions should be met:

1. The site should be away from major pollution sources. The distance depends upon the source, its height and its emissions. The station should be at least 25 m away from domestic chimneys, especially if the chimneys are lower than the sampling point ; with larger sources the distance should be greater (WHO,1977).
2. The site should be away from absorbing surfaces such as absorbing building material. The clearance to be allowed will depend on the absorbing properties of the material for the pollutant in question, but it will normally be at least I m. (WHO, 1977).
3. The objective of monitoring is often to measure trends in air quality and measurements are to be conducted over a long time; thus the site should be selected such that it is expected to remain a representative site over a long time and no landuse changes, rebuildings etc. are foreseen in near future.

The instrument must be located in such a place where free flow of air is available. The instrument should not be located in a confined place, corner or a balcony.

(b) Comparability

For data of different stations to be comparable, the details of each location should be standardised. The following is recommended in IS 5182 (Part 14) 2000

(*i*) On all the sides it should be open, that is the intake should not be within a confined space, in a corner, under or above a balcony.

(*ii*) For traffic pollution monitoring the sampling intake should be 3 m above the street level. The height of 3m is recommended to prevent re-entrainment of particulates from the street, to prevent free passage of pedestrians and to protect the sampling intake from vandalism.

(*iii*) Sampling in the vicinity of unpaved roads and streets results in entrainment of dust into the samplers from the movement of vehicles. Samplers are therefore to be kept at a distance of 200 m from unpaved roads and streets.

(c) Physical Requirement of the Monitoring Site

Following physical aspects of the site must be met :

- The site should be available for a long period of time
- Easy access to the site should be there anytime throughout the year.
- Site sheltering and facilities such as electricity of sufficient rating, water, telephone connection etc. should be available.
- It should be vandal proof and protected from extreme weather highest concentrations and concentration gradients of carbon monoxide are likely to be in the vicinity of roads, highways. The gradients vary in both time and space on the micro and on the neighbourhood scale.

(d) Topographical and Meteorological Factors

Topographical and meteorological factors must also be considered for selecting a monitoring site. The topographical factors that must be considered are mountains, valleys, lakes, oceans and rivers. These factors cause a meteorological phenomena that may affects air pollutants distribution.

Winds caused by daytime heating and nightime cooling may affect pollutant transport causing either buildup of pollutants or dilution. Canyons or valleys may channel the local winds into a particular direction resulting in increase in wind speed. The presence of large water bodies may cause a land-sea breeze wind pattern which may determine pollutant transport. The mountain or hilly terrain may cause precipitation that may affect pollutant concentration.

In general the following requirement may be met for siting the monitoring station.

(*i*) Height of the inlet must be 3 – 10 m above the ground level.

(*ii*) The sampler must be more than 20 m from trees.

(*iii*) Distance of the sampler to any air flow obstacle i.e. buildings, must be more than two times the height of the obstacle above the sampler.

(*iv*) There should be unrestricted airflow in three of four quadrants

(*v*) There should be no nearby furnace or incinerator fumes.

Once a area has been selected for locating a monitoring stations, the site can be selected by finding maximum concentration using air quality modeling.

Modeling refers to the mathematical expression for the fate of pollutants when they are released into the atmosphere taking into consideration the various aspects of atmospheric effects such as dispersion, advection etc. Air quality models are capable of predicting the temporal and spatial distribution of pollutants for a given domain of interest. Air quality modelling can be applied to ground level sources, elevated points sources, line sources, areas sources, flying sources under unlimited mixing, limited mixing, inversion, fumigation, trapping and also on complex terrain, flat terrain and coastal areas. The methodology is different in each case. Maximum ground level concentrations can be calculated where the air quality monitoring station can be located.

Type of Ambient Monitoring Stations

Station type	Description
Type A	Downtown pedestrian exposure station- In central business districts, in congested areas, surrounding by buildings, many pedestrians, average traffic flow > 10000 vehicles per day. Location of station- 0.5 m from curve; height 2.5 to 3.5 m from the ground.
Type B	Downtown neighbor hood exposure stations- In central business districts but not congested areas, less high rise buildings, average vehicles < 500 vehicles per day. Typical locations like parks, malls, landscapes areas etc. Location of station- 0.5 m from curve; height 2.5 to 3.5 m from the ground.
Type C	Residential population exposure station – In the midst of the residential areas or sub-urban areas but not in central business districts. The station should be more than 100 m away from any street.
Type D	Mesoscale stations – At appropriate height to collect meteorological and air quality data at upper elevation; main purpose to collect the trend of data variations not human exposure.
Type E	Non-urban stations – In remote non-urban areas, no traffic, no industrial activity. Main purpose to monitor trend analysis. Location of station- 0.5 m from curve; height 2.5 to 3.5 m from the ground.
Type F	Specialized source survey stations – to determine the impact on air quality at specified location by an air pollution source under scrutiny. Location of station- 0.5 m from curve; height 2.5 to 3.5 m from the ground.

Number of Stations

- Minimum number is three.
- The location is dependent upon the wind rose diagram that gives predominant wind directions and speed.
- One station must be at upstream of predominant wind direction and other two must at downstream pre dominant wind direction.
- More than three stations can also be established depending upon the area of coverage.

Sampling Duration and Frequency

The period and frequency of sampling should be such that statistically reliable averages can be obtained with the data. National Ambient Air Quality Standards states that annual average should be computed of 104 measurements taken twice a week of 24 hours duration. One of the objective of monitoring under NAMP is to determine compliance to the NAAQS so monitoring should be done for 24 hours and minimum 104 days in a year.

The pollutants vary diurnally and seasonally and these variations should be taken into account for determining frequency of sampling. The precision required in the data is also important in determining frequency of sampling. Sampling should be more frequent than the frequency of variation of pollutants.

Particulate matter levels are lower during the monsoon months due to removal by wet deposition. Air pollutants such as CO levels are higher during winter months due to lower mixing heights resulting in less volume of troposphere available for mixing and hence higher concentrations. Thus measurements should be conducted in all the seasons so that in annual average all the seasons are represented equally. In general minimum 20% of the reading should be taken in each season.

Basic Considerations for Sampling

- Sample must be representative in terms of time, location, and conditions to be studied.
- Sample must be large enough for accurate analysis.
- The sampling rate must be such as to provide maximum efficiency of collection.
- Duration of sampling must accurately reflect the fluctuations in pollution levels *i.e.* whether 1-hourly, 4-hourly, 6-hourly, 8-hourly, 24-hourly sampling.
- Continuous sampling is preferred.
- Pollutants must not be altered or modified during collection.

High-Volume Sampler

Monitoring of Particulates in Ambient Air

Air-borne particulate matter is an ensmble of solid particles suspended and dispersed in air. The properties of these particles vary in terms of chemical composition, morphology (size/shape) optical parameters (colour/light scattering), and electrical characteristics (charge, resistance). Particles, which in general are non spherical are classified in terms of their aerodynamic diameter. This is defined as the diameter of a sphere of density (1g/cm^3 having a characteristic same as that of the particle in question.).

In ambient air dust particles have size range from a few nanometer to hundreds of micrometer (µm). In ambient air SPM has bimodal mass distribution with respect to size whose shape depends on the predominant physical and chemical formation processes of particles.

Characteristics for Ambient Air Sampling Systems

Five main characteristicss are :

- Collection efficiency
- Sample stability
- Recovery
- Minimal interference
- Understanding the mechanism of collection

The first three must be 100% efficient. For example, for SO_2, the sorbent should be such that at ambient temperature it may remove the SO_2 from ambient atmosphere 100%. Sample must be stabled during the time between sampling and analysis. Recovery *i.e.* the analysis of particular pollutant must be 100% correct.

Measurement of TSPM and Respirable Particulate Matter (PM_{10}) using High Volume Respirable Dust Sampler Equipped with Cyclone

PM_{10} and TSPM are measured by passing air at flow rate of about 1.1 m^3/min through high efficiency cyclone which retains the dust particles greater then 10 micron size and allow only fines (less than 10 micron particles) to reach the glass microfibre filter where these particles are retained. The instrument provides instantaneous flow rate and the period of operation (on-time) for calculation of air volume passed through the filter.

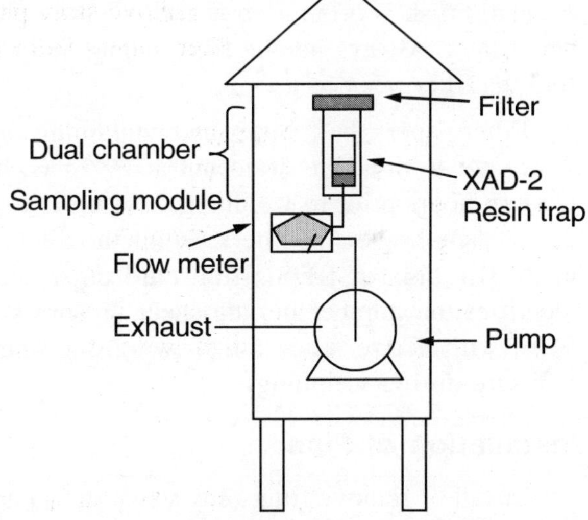

Fig. 4.8. High Volume Sampler

Amount of particulates collected is determined by measuring the change in weight of the cyclone cup and filter paper.

The passage of air entering in the cyclone is designed to prevent heavier settle able particles from reaching in the cyclone.

Selection of Sampling Site

Monitoring site need to be 3-10m high from ground level. This shall result elimination of local activities contribution if any. To get a representative sample the sampler must not be positioned near a wall or other obstruction that would prevent free airflow. In excessively turbulent conditions or in the presence of strong surface winds or otherwise inclement weather, the sampling rate is likely to decrease rapidly and perhaps in a nonlinear fashion due to filter choking. If the sampler is operated continuously, day-to-day variations in the measurement are expected due to varying meteorological conditions and changing atmospheric phenomena, like wind speed and direction, dispersion diffusion, etc. beside impact of emission sources.

Selection of Filter Medium

For most cases interest is limited to a gravimetric determination of the PM10 and total suspended particulate concentration. Glass Micro Fibre filters having low resistance to air flow, low affinity for moisture and 99% collection efficiency for particles of 0.3 microns or larger size are suitable for this purpose. However, where further analysis, of the particulates is to be attempted, to detect specific elements/radicals, care should be taken to choose special filter medium having a low background concentration of the substances of interest.

Preparing the Filter

Prior to use, expose each filter to a light source and inspect for pinholes, particles and other imperfections. Filters with visible defects should not be used. A small brush is often used to remove stray particles adhering to the surface of new filters. Always handle filter papers from their edges and do not crease or fold the filter prior to use.

Filter papers, both blank and containing samples should be conditioned in a desiccator with active desiccant at 20-25°C, humidity below 50% R.H., for at least 16 hours prior to weighing. It is usual to put an identification number and date of sampling on the filters. Weigh the filters to the nearest milligram (accurate up to 0.1 mg for RDS filters and cup) and record the weight and filters identification number in data sheet. Ensure keeping a perforated plastic bottle filled with active silica gel in weighing chamber for minimizing impact of moisture during weighing.

Installation of Filter

Install or remove filter only when the sampler is off. The numbered and the weighed filter is always kept with its rough side up. This increases the collection efficiency of the filter. The filter is placed in its proper position on the top of wire mesh. The sampler is started before putting back the top cover. This prevents the filter to shift from its place.

As a result, any undesirable leakage from top is prevented. The top cover is not to be under tightened as it will give way to leakage and over tightening will damage the rubber gasket and filter. It is not very difficult to ascertain the extent to which the top cover is to be tightened but, one learns through practice only. A very light application of talcum powder on the rubber gasket prevents the filter from sticking to it.

Errors in Sampling by HVS

- Particulates may be lost in sampling manifold – so not too long or too twisted manifold must be used.
- If 'isokinetic' conditioned are not maintained, biased results may be obtained for particulate matters.

Advantages of HVS

- High flow rate at low pressure drop
- High particulate storage capacity
- No moisture regain
- High collection efficiency
- Low coast
- Not appreciable increase in air flow resistance
- Filter is 99% efficient and can collect the particles as fine as 0.3 μm
- Absorption principle is 99% efficient in collecting the gases.

4.6. CLIMATIC CHANGES

The **climate** is the commonly considered to be the wether averaged over a long period of time, typically 30 years. Somewhat more precisely, the concept of "climate" also includes the statistics of the weather - such as the degree of day-to-day or year-to-year variation expected. The Intergovernmental Panel on Climate Change (IPCC) glossary definition is : *Climate in a narrow sense is usually defined as the "average weather", or more rigorously, as the statistical description in terms of the mean and variability of relevant quantities over a period of time ranging from months to thousands or millions of years. The classical period is 30 years, as defined by the World Meteorological Organization (WMO). These quantities are most often surface variables such as temperature, precipitation, and wind.*

The main causes for climatic changes are :

4.6.1. Global Warming

Global warming is a term used to describe the trend of increases in the average temperature of the Earth's atmosphere and oceans that has been observed in recent decades. The scientific opinion on climate change, as expressed in the UN Intergovernmental Panel on Climate Change (IPCC) Third Assessment Report in 2001 and explicitly endorsed by the national science academies of the G8 nations in 2005, is that the average global temperature has risen 0.6 ± 0.2 °C since the late 19th century, and that it is likely that "most of the warming observed over the last 50 years is attributable to human activities". The increased volumes of carbon dioxide and other greenhouse gases released by the burning of fossil fuels, land clearing and agriculture, and other human activities, are the primary sources of the human-induced component of warming. The natural greenhouse effect keeps the Earth about 33 °C warmer than it otherwise would be; adding carbon dioxide to a planet's atmosphere, with no other changes, will make that planet's surface warmer.

The **greenhouse effect** is a warming of the Earth's surface and lower atmosphere that tends to intensify with an increase in atmospheric carbon dioxide. The atmosphere allows a large percentage of the rays of visible light from the Sun to reach the Earth's surface and heat it. A part of this energy is reradiated by the Earth's surface in the form of long-wave infrared (IR) radiation, much of which is absorbed by molecules of carbon dioxide and water vapour in the atmosphere and which is reflected back to the surface as heat. This is roughly analogous to the effect produced by the glass panes of a greenhouse, which transmit sunlight in the visible range but hold in heat. The trapping of this infrared radiation causes the Earth's surface and lower atmospheric layers to warm to a higher temperature than would otherwise be the case. An increase in atmospheric concentrations of other trace gases such as chlorofluorocarbons (Freons), nitrous oxide, and methane, due again largely to human activity, may also aggravate greenhouse conditions and the IR active gases responsible for the effect are likewise referred to as greenhouse gases. They are mainly

$$CO_2 - 52\%,\ CH_4 - 24\%,\ N_2O - 18\%,\ CFC\text{'s} - 6\%.$$

Greenhouse Gases and Society

Greenhouse gases naturally blanket the Earth and keep it about 33 degrees Celsius warmer than it would be without these gases in the atmosphere. This is called the Greenhouse Effect. Over the past century, the Earth has increased in temperature by about 0.5 degrees Celsius and many scientists believe this is because of an increase in concentration of the main greenhouse gases: carbon dioxide, methane, nitrous oxide, and fluorocarbons.

Carbon Dioxide

Carbon Dioxide (CO_2) is a colorless, odorless non-flammable gas and is the most prominent Greenhouse gas in Earth's atmosphere. It is recycled through the atmosphere by the process photosynthesis, which makes human life possible. Photosynthesis is the process of green plants and other organisms transforming light energy into chemical energy. Light Energy is trapped and used to convert carbon dioxide, water, and other minerals into oxygen and energy rich organic compounds. Carbon Dioxide is emitted into the air as humans exhale, burn fossil fuels for energy, and deforests the planet. Every year humans add over 30 billion tons of carbon dioxide in the atmosphere by these processes, and it is up thirty percent since 1750. In 1996, carbon dioxide world emissions increased by 2.8 per cent. The U.S. reported a 3.3 per cent increase in CO_2 concentrations. The U.S. continues to emit more than any other country in the world, accounting for 25 per cent of all emissions. The European Union had an increase of 2.2 per cent, much larger than a small increase of 1.1 per cent in 1995. Eastern Europe had a decreasing rate of –2.4 per cent. China's increase in 1996 was 4.7 per cent.

Fossil Fuels were created chiefly by the decay of plants from millions of years ago. We use coal, oil and natural gas to generate electricity, heat our homes, power our factories and run our cars. These fossil fuels contain carbon, and when they are burned, they combine with oxygen, forming carbon dioxide. The two atoms of oxygen add to the total weight. The World Energy Council reported that global carbon dioxide emissions from buring fossil fuels rose 12 per cent between 1990 and 1995. The increase from developing countries was three times that from developed countries. Middle East carbon dioxide emissions from burning of fossil fuels increased 35 per cent, Africa increased 12 per cent, and Eastern Europe increased rates by 75 per cent from 1990-1995.

Deforestation is another main producer of carbon dioxide. The causes of deforestation are logging for lumber, pulpwood, and fuel wood. Also contributing to deforestation is clearing new land for farming and pastures used for animals such as cows. Forests and wooded areas are natural carbon sinks. This means that as trees absorb carbon dioxide, and release oxygen, carbon is being put into trees. This process occurs naturally by photosynthesis, which occurs less and less as we cut and burn down trees. As the abundance of trees declines, less carbon dioxide can be recycled. As we burn them down, carbon is released into the air and the carbon bonds with oxygen to form carbon dioxide, adding to the greenhouse effect.

Methane

Methane is a colorless, odorless, flammable gas. It is formed when plants decay and where there is very little air. It is often called *swamp gas* because it is abundant around water and swamps. Bacteria that breakdown organic matter in wetlands and bacteria that are found in cows, sheep, goats, buffalo, termites, and camels produce methane naturally. Since 1750, methane has doubled, and could double again by 2050. Each year we add 350-500 million tons of methane to the air by raising livestock, coal mining, drilling for oil and natural gas, rice cultivation, and garbage sitting in landfills. It stays in the atmosphere for only 10 years, but traps 20 times more heat than carbon dioxide.

Rice cultivation has developed into a large business; farmland has doubled in the past 45 years. It feeds 1/3 of the World's population. It grows mostly in flooded fields, where bacteria in waterlogged soil release methane.

Livestock such as cows, sheep, goats, camels, buffaloes, and termites release methane as well. Bacteria in the gut of the animal break down food and convert some of it to methane. When these animals belch, methane is released. In one day, a cow can emit ½ pound of methane into the air.

Nitrous Oxide

Nitrous oxide is another colorless greenhouse gas, however, it has a sweet odor. It is primarily used as an anesthetic because it deadens pain and for this characteristic is called laughing gas. This gas is released naturally from oceans and by bacteria in soils. Nitrous oxide gas raised by more than 15 per cent since 1750. Each year we add 7-13 million tons into the atmosphere by using nitrogen based fertilizers, disposing of human and animal waste in sewage treatment plants, automobile exhaust, and other sources not yet identified. It is important to reduce emissions because the nitrous oxide we release today will still be trapped in the atmosphere 100 years from now.

Nitrogen based fertilizer use has doubled in the past 15 years. These fertilizers provide nutrients for crops; however, when they breakdown in the soil, nitrous oxide is released into the atmosphere. In automobiles, nitrous oxide is released at a much lower rate than carbon dioxide, because there is more carbon in gasoline than nitrogen.

Fluorocarbons

Fluorocarbons are a general term for any group of synthetic organic compounds that contain fluorine and carbon. Many of these compounds, such as chlorofluorocarbons (CFCs), can be easily converted from gas to liquid or liquid to gas. Because of these properties, CFCs can be used in aerosol cans, refrigerators, and air conditioners. Studies in the 1970s showed that when CFCs are emitted into the atmosphere, they break down molecules in the Earth's ozone layer. Since then, the use of CFCs has significantly decreased and they are banned from production in the United States.

The substitutes for CFCs are hydrofluorocarbons (HFC's). HFCs do not harm or breakdown the ozone molecule, but they do trap heat in the atmosphere, making it a greenhouse gas, aiding in global warming. HFCs are used in air conditioners and refrigerators. The way to reduce emissions of this gas is to be sure that in both devices the coolant is recycled and all leaks are properly fixed. Also, before throwing the appliances away, be sure to recover the coolant in each.

Effects of Global Warming on Environment

There are many environmental problems coming from the increase concentration of greenhouse gases in Earth's atmosphere. As some news reported, Several signs indicate that we've begun changing Earth's climate: increased water vapor in the atmosphere, glaciers and polar ice caps appear to be melting, floods and droughts are becoming more severe, and sea levels have risen, on average, between 4 and 10 inches since 1990. These rises in sea level can increase the salinity of freshwater throughout the world, and cause coastal lands to be washed under the ocean. Warmer water and increased humidity may encourage tropical cyclones, and changing wave patterns could produce more tidal waves and strong beach erosion on the coasts.

Effects of Global Warming on Society

Agriculturally global warming is good for the human race, because it helps increase food production. Increasing amounts of greenhouse gases in the atmosphere and global warming could also lead to more health concerns. As temperatures increase towards the poles, similar to farmland, insects and other pests migrate towards Earth's poles. These insects and pests could be allowed to migrate up to 550 Km or 550 miles. Some insects carry diseases such as malaria and dengue fever. Thus, an increase in these particular insects and pests closer to the poles results in an increase in these diseases. This could lead to 50 to 80 million additional cases of Malaria annually, a 10-15 per cent increase.

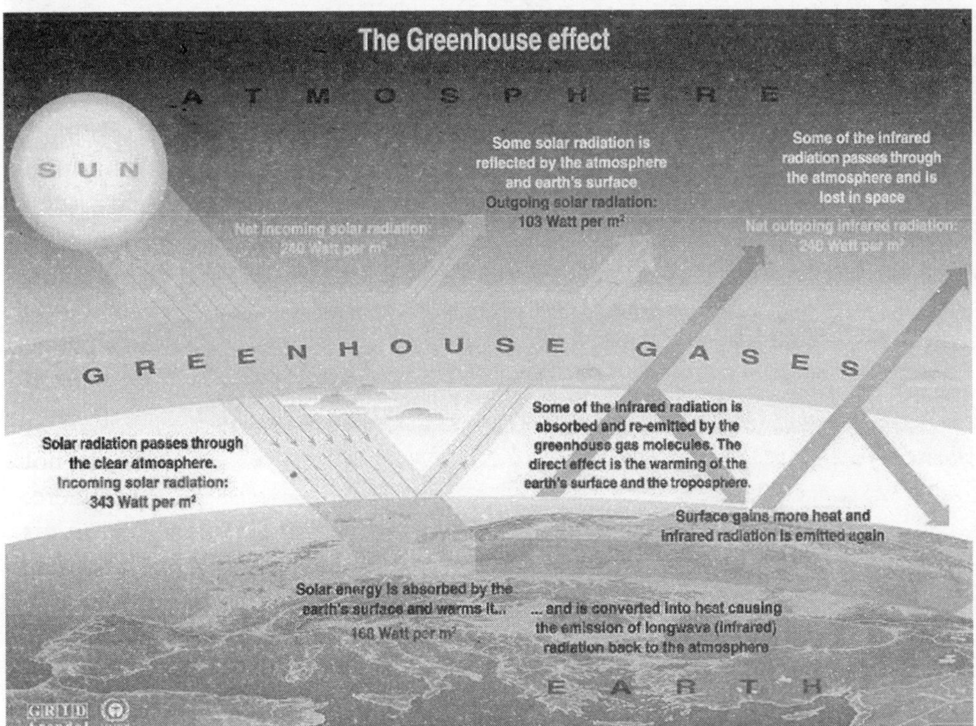

Fig. 4.9. Greenhouse Effect

The most obvious health effect is directly from the heat itself. With an increase in heat waves, there will be more people who will suffer from heatstroke, heart attacks and other ailments aggravated by the heat. Hot conditions could also cause smoke particles and noxious gases to linger in the air and accelerate chemical reactions that generate other pollutants. This leads to an increase in risk of respiratory diseases like bronchitis and asthma.

Global warming causes the oceans to warm and expand, inducing a rise in sea level. Eventually, the rising waters could take away land inhabited by people, forcing them to move. Warming of the oceans could also promote toxic algae which can lead to cholera.

4.6.2. Acid Rain

The term acid rain refers to what scientists call acid deposition. It is caused by airborne acidic pollutants and has highly destructive results. Acid rain, one of the most important environmental problems of all, cannot be seen. The invisible gases that cause acid rain usually come from automobiles or coal-burning power plants.

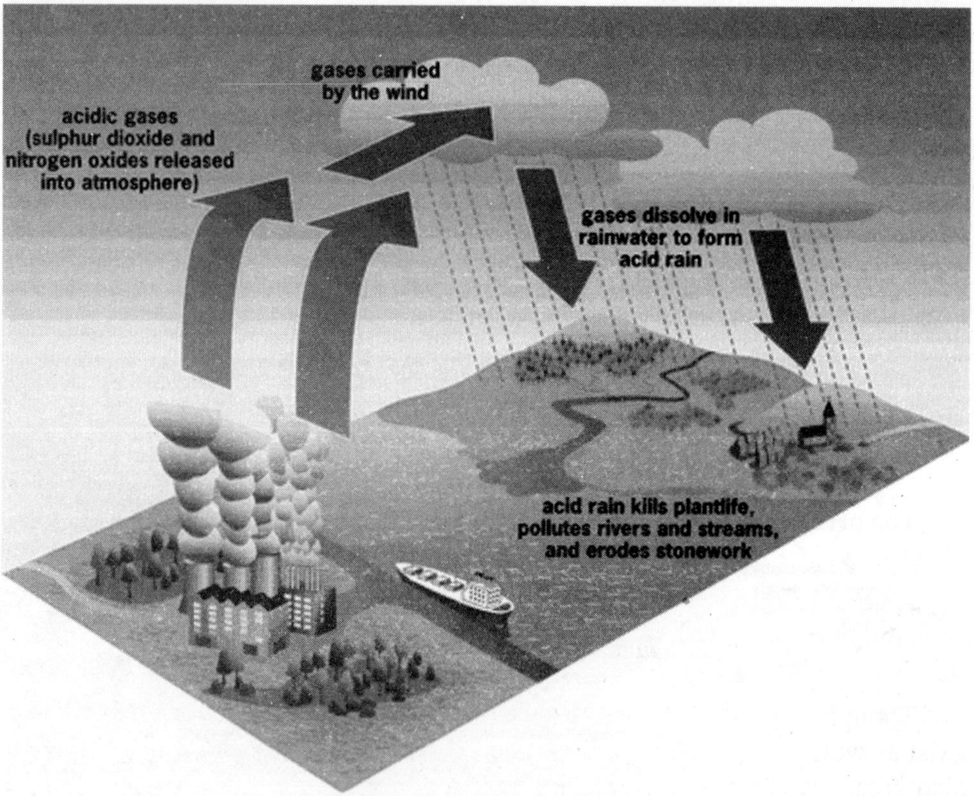

Fig. 4.10. Acid Rain

Scientists first discovered acid rain in 1852, when the English chemist Robert Agnus invented the term. From then until now, acid rain has been an issue of intense debate among scientists and policy makers.

Acid rain is one of the most dangerous and widespread forms of pollution. Sometimes called "the unseen plague," acid rain can go undetected in an area for years. Technically, acid rain is rain that has a larger amount of acid in it than what is normal.

The levels of acid rain vary from region to region. In Third World nations without pollution restrictions, acid rain tends to be very high. In Eastern Europe, China, and the Soviet Union, acid rain levels have also risen greatly. It is now common in many places for rain to be ten to seventy times more acid than unpolluted rain. Many living and non-living systems become harmed and damaged as a result of acid rain.

Causes of Acid Rain

The primary causes of acid rain are sulfur dioxide and nitrogen oxides. These chemicals are released by certain industrial processes, and as a result, the more industrialized nations of Europe as well as the US suffer severely from acid rain.

Most sulfur dioxide comes from power plants that use coal as their fuel. These plants emit 100 million tons of sulfur dioxide, 70 per cent of that in the world.

Acid rain is caused by smoke and gases that are given off by factories and cars that run on fossil fuels. When these fuels are burned to produce energy, the sulfur that is present in the fuel combines with oxygen and becomes sulfur dioxide; some of the nitrogen in the air becomes nitrogen oxide. These pollutants go into the atmosphere, react with rain water and form acid and produce acid rain. The main reactions are as follows :

$$2SO_2 + O_2 \longrightarrow 2SO_3$$
$$2NO + O_2 \longrightarrow 2NO_2$$
$$SO_3 + H_2O \longrightarrow H_2SO_4$$
$$NO_2 + H_2O \longrightarrow HNO_2 + HNO_3$$

The pH of natural rain water is near about 5.6 but the presence of H_2SO_4 and HNO_3 lowers down its pH below 5.6.

Automobiles produce about half of the world's nitrogen oxide. As the number of automobiles in use increases, so does the amount of acid rain. Power plants that burn fossil fuels also contribute significantly to nitrogen oxide emission.

Though human causes are primarily responsible for acid rain, natural causes exist as well. Fires, volcanic eruptions, bacterial decomposition, and lightening also greatly increase the amount of nitrogen oxide on the planet.

Once the tiny pollutant molecules have entered the atmosphere, they can travel for thousands of miles. Eventually, the particles will combine with other compounds to produce new, often harmful, chemicals.

Acid rain comes down to the earth in the form of rain, snow, hail, fog, frost, or dew. Once it reaches the ground, the acidity in the substance can harm and even destroy both natural ecosystems and man-made products, such as car finishes.

Effect on Environment

Acid rain is an extremely destructive form of pollution, and the environment suffers from its effects. Forests, trees, lakes, animals and plants suffer from acid rain. Trees are extremely important natural resource. They provide timber, regulate local climate, and forests are homes to wildlife. Acid rain can make trees lose their leaves or needles. The needles and leaves of the trees turn brown and fall off. Trees can also suffer from stunted growth; and have damaged bark and leaves, which makes them vulnerable to weather, disease, and insects. Lakes are also damaged by acid rain.

Effect on Architecture

Architecture and artwork can be destroyed by acid rain. Acid particles can land on buildings, causing corrosion. When sulfur pollutants fall of the surfaces of buildings (especially those made out of sandstone or limestone), they react with the minerals in the stone to form a powdery substance that can be washed away by rain. This powdery substance is called gypsum. Acid rain can damage buildings, stained glass, railroad lines, airplanes, cars, steel bridges, and underground pipes.

Effect on Human

Humans can become seriously ill, and can even die from the effects of acid rain. One of the major problems that acid rain can cause in a human being is respiratory problems. Many can find it difficult to breathe, especially people who have asthma. Asthma, along with dry cough, headaches and throat irritations can be caused by the sulfur dioxides and nitrogen oxides from acid rain. Acid rain can be absorbed by both plants (through soil and/or direct contact) and animals (from things they eat and/or direct contact). When humans eat these plants or animals, the toxins inside of their meals can affect them. Brain damage, kidney problems, and Alzheimer's disease has been linked to people eating "toxic" animals/plants.

Solution of Acid Rain

Sulphur Dioxide

The sulphur which is present in nearly all fossil fuels combines with oxygen when the fuel is burnt and is released into the atmosphere as SO_2 gas. These emissions can be reduced by measures taken before, during, or after the combustion process.

One approach is to use fuels which naturally have little sulphur in them. The sulphur content of coal can vary considerably. Some fuels may be treated to reduce their sulphur content, but effective treatment is expensive. Demand for low sulphur fuels is increasing as more countries develop programmes to reduce sulphur pollution, so they are becoming more expensive. During combustion it is possible to reduce the eventual emissions of SO_2 by the introduction of a sorbent such as limestone. The potential for sulphur reduction by this approach depends on the type of furnace or boiler.

After combustion, sulphur can be removed from flue gases or 'scrubbed'. This process is known as the flue gas desulphurization (FGD). In most FGD system a mixture of limestone and water is sprayed into the flue gas. The SO_2 is converted to gypsum (calcium sulphate), which can be used in the manufacture of plaster products. However, FGD systems of this type are expensive and use considerable amounts of limestone. If all power stations were fitted with FGD, gypsum production would exceed requirements, leading to a waste disposal problem. Although such a programme would increase limestone extraction by about 5 per cent, there would be a useful reduction in gypsum quarrying.

An alternative to limestone FGD systems is the regenerative FGD approach in which SO_2 is captured by a substance which can be recycled. Sulphur or sulphuric acid is obtained as a by-product and can be used in the chemical industry. Again, there are limits to the amount of by-product which industry can use.

Although FGD can reduce sulphur emissions by up to 90 per cent, such systems use extra energy and, therefore, increase emissions of the greenhouse gas CO_2.

Nitrogen Oxides

NOx is produced partly from the oxidation of nitrogen contained in the fuel and partly as a result of high temperature and pressure combustion, which oxidizes nitrogen in the air. Furnace burners can be changed to reduce outputs of NOx by up to 40 per cent (low-NOx burners). NOx in flue gas can be reduced by adding ammonia and passing it over a catalyst to produce nitrogen and water. This process is called selective catalytic reduction (SCR) and can reduce NOx from combustion plant by 85 per cent, NOx produced by cars can also be treated by using catalysts; fitting a catalytic converter to the exhaust system reduces NOx emissions by up to 90 per cent, although it may increase emissions of CO_2.

Other Options

Since most acid pollution comes from burning fossil fuels, one way of reducing emissions is to reduce the overall demand for energy by encouraging energy conservation and improving the efficiency of electricity generation. Another option is to develop non-fossil fuel energy sources such as nuclear power or renewable energy (solar, wind, tidal power, etc.)

However these have their own environmental problems which must be balanced against those of fossil fuels.

4.6.3. Ozone Layer Depletion

The term **Ozone depletion** is used to describe two distinct, but related, observations: a slow, steady decline, of about 3 per cent per decade, in the total amount of ozone in the earth's stratosphere during the past twenty years, and a much larger, but seasonal, decrease in stratospheric ozone over the earth's polar regions during the same period. (The latter phenomenon is commonly referred to as the "ozone hole".) The detailed mechanism by which the polar ozone holes form is different from that for the mid-latitude thinning, but the proximate cause of both trends is believed to be catalytic destruction of ozone by atomic chlorine and bromine. The primary source of these halogen atoms in the stratosphere is photo-dissociation of chlorofluorocarbon (CFC) compounds, commonly called Freons, and bromofluorocarbon compounds known as Halons, which are transported into the stratosphere after being emitted at the surface. Both ozone depletion mechanisms strengthened as emissions of CFCs and Halons increased.

Since the ozone layer prevents most harmful UVB wavelengths (270-315 nm) of ultraviolet light from passing through the Earth's atmosphere, observed and projected decreases in ozone have generated worldwide concern, leading to adoption of the Montreal Protocol banning the production of CFCs and halons as well as related ozone depleting chemicals such as carbon tetrachloride and trichloroethane (also known as methyl chloroform). It is suspected that a variety of biological consequences, including, for example, increases in skin cancer, damage to plants, and reduction of plankton populations in the ocean's photic zone, may result from the increased UV exposure due to ozone depletion.

Ozone Creation

Three forms (or allotropes) of oxygen are involved in the ozone-oxygen cycle Oxygen atoms, O, or atomic oxygen, oxygen molecules, O_2 and ozone, O_3. Ozone is formed in the stratosphere when oxygen molecules photo-dissociate after absorbing an ultraviolet photon whose wavelength is shorter than 240 nm. This produces two oxygen atoms. The atomic oxygen then combines with O_2 to create O_3. Ozone molecules strongly absorb UV light between 310 and 200 nm, following which ozone splits into a molecule of O_2 and an oxygen atom. The oxygen atom then joins up with an oxygen molecule to regenerate ozone. This is a continuing process which terminates when an oxygen atom "recombines" with an ozone molecule to make 2 O_2 molecules. Prior to the beginning of the depletion trend, the amount of ozone in the stratosphere was kept roughly constant by a balance between the rates of creation and destruction of ozone molecules by UV light.

$$O_2 + hv \longrightarrow O + O$$

$$O_2 + O + M \longrightarrow O_3 + M$$

$$O_3 + hv \longrightarrow O_2 + O$$

Ozone Destruction

Ozone can be destroyed by a number of free radical catalysts, the most important of which are hydroxyl (OH), nitric oxide (NO) and atomic chlorine (Cl) and bromine (Br). All of these radicals have both natural and anthropogenic (manmade) sources. At the present time, most of the OH and NO in the stratosphere is of natural origin, but human activity has dramatically increased the chlorine and bromine. These elements are found in certain stable organic compounds, especially chlorofluorocarbons (CFCs), which may find their way to the stratosphere without being destroyed in the troposphere. Once in the stratosphere, the Cl and Br atoms are liberated from the parent compounds by the action of ultraviolet light, and can destroy ozone molecules in a catalytic cycle. In this cycle, a chlorine atom reacts with an ozone molecule, taking an oxygen atom with it (forming ClO) and leaving a normal oxygen molecule. A free oxygen atom then takes away the oxygen from the ClO, and the final result is an oxygen molecule and a chlorine atom, which then reinitiates the cycle. The chemical shorthand for these reactions are:

$$CF_2Cl_2 + hv \longrightarrow Cl + CF_2Cl$$

$$Cl + O_3 \longrightarrow ClO + O_2$$

$$ClO + O \longrightarrow Cl + O_2$$

$$ClO + NO_2 \longrightarrow ClONO_2$$

$$ClONO_2 + H_2O \longrightarrow HOCl + HNO_3$$

$$HOCl + hv \longrightarrow Cl + OH^0$$

$$Cl + O_3 \longrightarrow ClO + O_2$$

$$OH^0 + O_3 \longrightarrow HO_2^0 + O_2$$

$$ClO + HO_2^0 \longrightarrow HOCl + O_2$$

For this mechanism to operate there must be a source of O atoms, which is primarily the photodissociation of O_3.

A single chlorine atom would keep on destroying ozone for up to two years (the time scale for transport back down to the troposphere) were it not for reactions that remove them from this cycle by forming reservoir species such as hydrochloric acid and chlorine nitrate. On a per atom basis, bromine is even more efficient than chlorine at destroying ozone, but there is much less bromine in the atmosphere at present. As a result, both chlorine and bromine contribute significantly to the overall ozone depletion.

Case Study

Antarctica Ozone Hole

In the area over Antarctica, there are stratospheric cloud ice particles that are not present at warmer latitudes. Reactions occur on the surface of the ice particles that accelerate the ozone destruction caused by stratospheric chlorine. Polar Regions get a much larger variation in sunlight than anywhere else, and during the 3 months of winter spend most of time in the dark without solar radiation. Temperatures stay close around or below –80°C for much of the winter and the extremely low antarctic temperatures cause cloud formation in the relatively "dry'' stratosphere. These Polar Stratospheric Clouds (PSC's) are composed of ice crystals that provide the surface for a multitude of reactions, many of which speed the degradation of ozone molecules. This phenomenon has caused documented decreases in ozone concentrations over Antarctica.

In fact, ozone levels drop so low in spring in the Southern Hemisphere that scientists have observed what they call a "hole" in the ozone layer. The ozone destruction process requires conditions cold enough for stratospheric clouds to form. Once these stratospheric clouds form the process can take place, even in warmer conditions.

It was first noticed by a research group from The British Antarctic Survey in the 1970's. Joseph Farman, Brian Gardiner and Jonathan Shanklin, are the BAS scientists who discovered the Antarctic ozone hole. In the 1980's the first measurements of this loss were actually documented. In October 1984, when the British first reported their findings, ozone levels were about 35 percent lower than the average for the 1960s.

4.7. NOISE POLLUTION

Sound is mechanical energy from vibrating source. Noise can be defined as an unpleasant and unwanted sound and has become a part of urban life and industrial centre in this century. Whenever a given sound is as pleasant as music or as unpleasant as noise depends on its duration, rhythm, loudness and mood of the person. Noise is a physical form of pollutant and is not directly harmful to the life supporting system namely air, water and soil. Its effects are more directly on the receiver *i.e.* man.

The noise measurements are expressed as sound pressure level (SPL) and increase in sound pressure does not invoke linear response of human ear. SPL is logarithmic ratio of the sound pressure to a reference pressure. It is expresses as a dimensional unit, decibel (dB). The decibel scale is logarithmic; each 10 decibel increase represents a tenfold increase in noise intensity. Human perception of loudness also conforms to a logarithmic scale; a 1'0 decibel increase is perceived as roughly a doubling of loudness. The central pollution control board (CPCB) committee has recommended permissible noise level for different locations.

Category of area	Noise level in dB	
	Day (6 AM – 9 PM)	Night (9 PM – 6 AM)
Industrial	75	70
Commercial	65	55
Residential	55	45
Sensitive	50	40

4.7.1. Sources of Noise Pollution

Noise pollution may come from loudspeakers, factories, aero-planes, moving trains, construction activity or even a radio. The sources are categorized as follows :

1. Industrial operation-textile mills, printing presses, metal works and engineering establishment
2. Transport vehicles-air, road and rail transportation
3. Household-electric home appliances, entertainment equipments, loud conversation and moving of furniture
4. Celebrations-social/religious functions, elections, demonstration and commercial advertising.
5. Agricultural machines- tractors, thrashers, harvesters, tube-wells.
6. Construction- Noise from construction equipments is a major source of noise pollution like mixers etc.
7. Defense equipment-artillery, tanks, launching of rockets, explosions, exercising of military airplanes, shooting practices, screams of jet engines and sonic booms.
8. Miscellaneous sources-automobile repair ship, construction works, bulldozing, stone crushing, blasting etc.

4.7.2. Effects of Noise Pollution

Noise level of 80 decibels or more for more than 8 hours a day increases tension and changes in breathing patterns. Continued exposure of noise results high levels fatigue, temporary or permanent shift of the hearing threshold of a person (hearing loss or even total loss of hearing), changes in blood circulation, changes in breathing, etc. Noise pollution above 120 decibels can cause many adverse biochemical changes. Cholesterol levels in the blood and white cell counts increase, besides causing hypertension and chances of heart diseases. Other effects are :

1. Exposure to sufficiently intense noise for a long duration results in permanent damage to the delicate tympanic membrane or the ear drum.
2. Noise can also cause some physiological effects like headache by deleting blood vessels of the brain, Cardiovascular and gastric effects.

Noise Level in decibels	Source	Psychological and physical Effect
135	pneumatic drill	painful
110	rock band	pneumatic drill
88	industry / city traffic	hearing impairment on prolonged exposure
80	alarm clock	annoying
65	average city traffic	intrusive

3. Muscular strain and nervous breakdown.
4. Noise can disturb sleep.
5. Noise pollution can have a considerable effect on communication.
6. lack of concentration.
7. Efficiency of the person is reduced.
8. Noise affects the quality of life.

4.7.3. Control of Noise Pollution

Noise pollution could be controlled by either reducing the noise at the source or by preventing its transmission or by protecting the receiver.

At the source :

- Reduction in sources of noise.
- Proper oiling will reduce the noise from machinery.
- Tightening the loose nuts.
- Decibel metres should be installed along highways and in factories to check and control the intensity of noise pollution.

In the path :

- A green-belt effectively reduces the noise.
- A 20 foot wide plantation inside the compound protects the house from the noise of vehicular traffic.
- Noise making machine should be kept in containers with sound absorbing media.
- Use of sound absorbing silencers.
- Through laws.

At the receivers end :

If by the above methods we are not able to bring down the noise level up to required level the only alternative is to provide ear plugs while working or moving in a noisy area.

Case Studies

Bhopal Gas Tragedy

The Bhopal disaster is an example of industrial pollution accidents. Bhopal provides an example of the failure in handling a toxic material in large scale chemical processing and the consequences that have to be faced by a city in a short time. The disaster struck Bhopal on December 3, 1984 at mid-night when pesticide manufacturing plant (Union Carbide factory) released a potent toxicant methyl isocynate gas (MIC) in to the atmosphere due to the alleged functional failure of vent scrubber outlet. About 30 tones of the gas escaped from underground storage tanks resulting in the deaths of about 1,00,000 people and severe disability of an additional 1,00,000 of the city's population. The chief causes of deaths and casualties were violent coughing, eye infection (chemical conjunctivitis), suffocation, cardiac failure and pulmonary disorders. The full consequences are not yet known.

London Smog (Chemical Smog)

This type of smog is thought to be caused by sulphur dioxide. It is formed by combination of smoke, dust, fog and SO_2. Chemical smog is usually formed around some industrial and power plants. It is generally bad in early morning hours and become worse after sunrise, due to sunlight induced oxidation of SO_2 to SO_3 followed by reaction with humidity. Sulphuric acid is formed in this type of smog, which is very harmful to people suffering from bronchial diseases such as asthma or emphysema.

On Thursday, December 4, 1952, a high temperature air mass moved over Southern England and thus creates a temperature inversion and caused white fog to settle over London. As the particulate and sulphur dioxide levels rose, due to the extensive use of coal-fired heating and power production system, the fog began to blacken. The high pressure area stalled and the pollutant build-up worsened further, since there were no air currents to disperse the smog. By Friday morning, visibility was zero. According to an eye-witness such was the state that white collar became almost black within 20 minutes.

The smog so formed was intensely irritating to the human respiratory system. As a result, most of the people soon developed red eyes, hurning throats and nagging coughs. Soon, reports of smog-related deaths began to come in. the elderly and the people with chronic respiratory problems became the first casualty; followed by young and otherwise healthy persons. More than 4,000 people died before the fog disappeared/lifted on December 9, 1952.

Los Angeles Smog (Photochemical Smog)

Photochemical smog is associated with adverse meteorological conditions (inversion conditions) when air movement is restricted in summers in highly motorized areas. The first noticed incidence of Photochemical smog was is Los Angeles, USA, in the mid 1940's and thus, it is sometimes referred as Los Angeles smog.

Photo-chemical smog is oxidizing smog having a high concentration of oxidants. Photochemical smog is a mixture of reactants and products that result from the interaction between volatile organic compounds (VOCs) and oxidants (such as NO_x, O_3, CO etc.) the primary pollutants in photochemical smog are nitric oxide and hydrocarbons. These primary pollutants convert rapidly to the secondary pollutants like ozone, organic nitrates (peroxyacetyl nitrate), oxidized hydrocarbons and photochemical aerosols in the presence of sunlight. It is characterized by brown and hazy fumes which reduce visibility, cause eye and lung irritation, damage crops, leads to the cracking of rubber, damage of plants, etc.

In India, Bombay and Calcutta are ideal conditions for the formation of photochemical smog, but it may be masked by smoke and SO_2.

In a vastly oversimplified from, the production of Photochemical smog can be represented by the following equation:

$$HCs + NO_x \xrightarrow{\text{Sunlight}} \text{Photochemical Smog}$$

But in actual practice, there are many complex reactions which take place in the formation of photochemical smog.

Normal NO_2 photolytic cycle:

$$NO_2 + h\nu \longrightarrow NO + O$$

$$O + O_2 \longrightarrow O_3$$ where M is a third molecule (usually N_2/O_2)

$$NO + O_3 \longrightarrow NO_2 + O_2$$

Production of hydroxyl radical:

$$O_3 + h\nu \longrightarrow O^+ + O_2$$

$$O^+ + H_2O \longrightarrow 2OH''$$

Chain and disturbance reactions:

$$\left[\begin{array}{l} CO + OH^\bullet \rightarrow CO_2 + H^\bullet \\ H^\bullet + O_2 \rightarrow HO_2^\bullet \\ HO_2^\bullet + NO \rightarrow NO_2 + OH^\bullet \end{array}\right.$$

$$\left[\begin{array}{l} RCHO + OH^\bullet \rightarrow RC(O)^\bullet + H_2O \\ RC(O)^\bullet + O_2 \rightarrow RC(O)O_2^\bullet \\ RC(O)O_2^\bullet + NO_2 \rightarrow RC(O)O_2NO_2 \end{array}\right.$$

$$\left[\begin{array}{l} RC(O)O_2'' + NO \rightarrow NO_2 + RC(O)O'' \\ RC(O)O'' + O_2 \rightarrow RO_2'' + CO_2 \\ RO_2'' + NO \rightarrow RNO_3 \end{array}\right.$$

$$\left[\begin{array}{l} RO_2'' + NO \rightarrow NO_2 + RO'' \\ RO'' + O_2 \rightarrow R'CHO + HO_2'' \\ HO_2'' + NO \rightarrow NO_2 + OH'' \end{array}\right.$$

$$\left[\begin{array}{l} RH + OH'' \rightarrow R'' + H_2O \\ R'' + O_2 \rightarrow RO_2'' \end{array}\right.$$

$$\left[\begin{array}{l} RCHO + h\nu \rightarrow R'' + HCO'' \\ HCO'' + O_2 \rightarrow CO + HO_2'' \end{array}\right.$$

Terminating Reaction:

$$\begin{cases} OH'' + NO_2 \rightarrow HNO_3 \\ RO_2'' + NO \rightarrow RNO_3 \\ RC(O)O_2'' + NO_2 \rightarrow RC(O)O_2NO_2 \end{cases}$$

Summary

- Pollution is the release of chemical, physical, biological or radioactive contaminants to the environment.
- Pollution is direct or indirect change in any environmental components that is causing harmful effect adversely the industrial progress, culture or natural assets or general environment.
- Air pollution may be defines as the presence of any contaminant in the atmosphere that tends to be injurious to human health, animal or plants or properties or would unreasonable interfere with enjoyment of life and property.
- Primary pollutants are those which are emitted directly into the atmosphere and Secondary pollutants are pollutants formed by the photochemical reaction of primary pollutants.
- The air pollution may be due to natural disasters like dust storms, earthquakes and volcanic eruptions or man-made causes like toxic emissions from factories, power plants, vehicular traffic, etc.
- Depending on the pollutant, the concentration and the duration of exposure, they may cause irritation and effects mucous membranes and respiratory organs.
- When the temperature increases with altitude, called an inversion. This may be of three type subsidence inversion, radiation inversion and advection inversion.
- Climate is the statistical description over a period of time ranging from months to thousands or millions of years.
- The main causes for climatic changes are global warming, acid rain, ozone depletion and nuclear accidents.
- Global warming is a term used to describe the trend of increases in the average temperature of the Earth's atmosphere and oceans.
- Acid rain is rain that has a larger amount of acid in it than what is normal.
- The Ozone depletion is a slow, steady decline in the total amount of ozone in the earth's stratosphere.
- Sound is mechanical energy from vibrating source.
- Noise can be defined as an unpleasant and unwanted sound and has become a part of urban life and industrial centre in this century.

- Whenever a given sound is as pleasant as music or as unpleasant as noise depends on its duration, rhythm, loudness and mood of the person.
- Noise is a physical form of pollutant and is not directly harmful to the life supporting system namely air, water and soil.
- The effects of noise pollution are more directly on the receiver i.e. man.
- The noise measurements are expressed as sound pressure level (SPL) and increase in sound pressure does not invoke linear response of human ear.
- SPL is logarithmic ratio of the sound pressure to a reference pressure.
- It is expresses as a dimensional unit, decibel (dB).
- The decibel scale is logarithmic; each 10 decibel increase represents a tenfold increase in noise intensity.
- Noise pollution could be controlled by either reducing the noise at the source or by preventing its transmission or by protecting the receiver.
- Photochemical smog is oxidizing smog and it should be clearly distinguished from the London smog which forms due to the combination of smoke and fog.

Descriptive Type Questions

1. What do you understand by the term pollution and pollutants? What are the different types of pollution and how you will define them?
2. Enumerate various air pollutants for which National Ambient Air Quality Standards have been given in India. Explain the various non-point sources contributing to air pollution.
3. What is air pollution? What are its effects on human health? Discuss the measures used for controlling air pollution.
4. Explain four major air pollutants and their consequences.
5. What is inversion? Define all its types and explain the causes and effects of inversion in atmosphere.
6. What is photochemical smog? How does if different from London smog? What are its effects?
7. The industrial developments have caused positive as well as negative effects on human health – comment.
8. Explain water conservation, global warming, acid rain and ozone layer depletion.
9. Discuss the phenomenon of 'Greenhouse Effect'. What are its effects? What remedial measures you suggest?
10. Mention the major sources of carbon emission.
11. What do you understand by Chlorofluorocarbons? What are their uses? What are the effects on global warming and ozone layer depletion?
12. What is acid rain? What are the causes and effects of acid rain? How can be problem be overcome?

13. Explain the problems associated with ozone layer depletion.
14. What is ozone hole? What are the causes of ozone hole formation. Discus the effects of ozone hole layer depletion and its remedial measures.

Short Answer Type Questions

1. Name the gases and reactions causing acid rain.
2. How ozone layer does protect us?
3. What do you understand by ozone layer?
4. 'Ozone is a pollutant in troposphere while life saver in stratosphere'. Justify this statement.
5. Mention the causes of Acid rain
6. Write Short notes on Ozone layer depletion
7. Name different types of pollutions. Define them.
8. What do you understand by secondary pollutants? Give the reaction for their formation.
9. Explain the following terms :
   (*i*) PAN (*ii*) CFC's
   (*iii*) MIC
10. Write short notes on Sources, effects and control measures of noise pollution
11. What are the effects of ultraviolet radiations on human being?
12. Name the major air pollutants and their sources.
13. Differentiate the following:
    (*i*) Natural and man-made sources of pollution (with example)
    (*ii*) London smog and Los Angeles smog
    (*iii*) Fog and smog
    (*iv*) Noise and sound
    (*v*) Primary and secondary air pollutants
14. Write short notes :
    (*a*) Bhopal gas tragedy
    (*b*) Point, area and line source of air pollutants
    (*c*) Threshold limit of hearing

Objective Type Questions

Multiple Choice Questions

1. Global atmospheric temperatures are likely to be increased due to :
   (*a*) Burning of fossil fuel (*b*) Water pollution
   (*c*) Soil erosion (*d*) None of the above
2. Which of the following is not a 'green house gas'?
   (*a*) Oxygen (*b*) Carbon dioxide
   (*c*) Chiorofluoro carbons (*d*) Methane

3. Greenhouse effect is related to :

(*a*) Green trees on house (*b*) Global warming

(*c*) Grasslands (*d*) Greenery in country

4. Global Warming could affect :

(*a*) Climate (*b*) Food production

(*c*) Melting of glaciers (*d*) All of the above

5. Which of the following gases contributes maximum to the 'Greenhouse effect' on earth?

(*a*) Carbon dioxide (*b*) Methane

(*c*) Chlorofluoro carbons (*d*) Freon

6. Which of the following is not a solution of global warming?

(*a*) Reducing fossil fuel consumption

(*b*) Planting more trees

(*c*) Deforestation

(*d*) None of the above

7. Acid rain is caused by increase in the atmospheric concentration of?

(*a*) Ozone and dust (*b*) SO and NO

(*c*) SO and CO (*d*) CO and CO

8. The primary cause of acid rain around the world is :

(*a*) Carbon dioxide (*b*) Sulphur dioxide

(*c*) Carbon monoxide (*d*) Ozone

9. Ozone layer of upper atmosphere is being destroyed by?

(*a*) Ozone (*b*) PAN

(*c*) Aldehydes (*d*) All of the above

10. Increasing skin cancer and high mutation rate are the result of?

(*a*) Ozone depletion (*b*) Acid rain

(*c*) CO pollution (*d*) CO pollution

11. Ozone layer is present in :

(*a*) Troposphere (*b*) Mesosphere

(*c*) Stratosphere (*d*) Thermosphere

12. Formation of hole in Ozone is maximum over :

(*a*) India (*b*) Antarctica

(*c*) Europe (*d*) Africa

13. Which of the following statements about ozone is true?

(*a*) Ozone is a major constituent of photochemical smog

(*b*) Ozone protects us from the harmful UV radiation of sun

(*c*) Ozone is highly reactive

(*d*) All of the above

14. Ozone layer thickness is measured in :

(*a*) Millimeter (*b*) Centimeter

(*c*) Decibels (*d*) Dobson units

15. Which of the following is an air pollutant?
(*a*) Nitrogen (*b*) Carbon dioxide
(*c*) Carbon mono-oxide (*d*) Oxygen

16. Which of the following is a secondary air pollutant?
(*a*) Ozone (*b*) Carbon dioxide
(*c*) Carbon mono-oxide (*d*) Sulphur dioxide

17. Which of the following on inhalation dissolved in the blood hemoglobin more rapidly than oxygen?
(*a*) Sulphur dioxide (*b*) Carbon monoxide
(*c*) Ozone (*d*) Nitrous oxide

18. Smog is :
(*a*) A natural phenomenon (*b*) A combination of smoke and fog
(*c*) Is colourless (d) All of the above

19. Which of the following are likely to be present in photochemical smog?
(*a*) Sulphur dioxide (*b*) Photochemical oxidants
(*c*) Chlorofluorocarbon (*d*) Smog

20. Taj Mahal at Agra may be damaged by?
(*a*) Sulphur dioxide (*b*) Chlorine
(*c*) Hydrogen (*d*) Oxygen

21. During Inversion :
(*a*) Temperature increases with altitude
(*b*) Temperature decreases with altitude
(*c*) Temperature remains constant
(*d*) None of the above

22. Gas leaked in Bhopal tragedy :
(*a*) Methyl isocyanate (*b*) Potassium isothiocynate
(*c*) Ethyl isocyanate (*d*) Sodium isothiocyanate

23. Which of the following devices is suitable for the removal of gaseous pollutants?
(*a*) Cyclone separator (*b*) Electrostatic precipitator
(*c*) Fabric filter (*d*) Wet scrubber

24. Which of the following air pollution control devices is suitable for the removing the finest dust from the air?
(*a*) Cyclone separator (*b*) Electrostatic precipitator
(*c*) Fabric filter (*d*) Wet scrubber

25. Air pollution from automobiles can be controlled by fitting?
(*a*) Cyclone separator (*b*) Electrostatic precipitator
(*c*) Catalytic converter (*d*) Wet scrubber

26. Noise is :
(*a*) Loud sound (*b*) Sound of high frequency
(*c*) Unwanted sound (*d*) Constant sound

27. Short term properties of the atmosphere at a given place and time is referred as :
(*a*) Climate (*b*) Microclimate
(*c*) Season (*d*) Weather

28. Which of the following is a management option for air pollution?

(*a*) Regulations and standards (*b*) Emission changes
(*c*) Transport planning (*d*) All of the above

Fill in the Blanks

1. The thinning of stratospheric ozone layer during the spring time is called ....
2. With increasing height, air temperature drops uniformly with altitude, this phenomenon is commonly called the ....
3. The global warming contributes to raise in sea level due to .... of Ocean and melting of ....
4. The active chlorine catalytically destroys .... by converting it into ....
5. Automobiles are largest sources of .... pollution in cities.
6. Increase in the concentration of soluble salts in soil is called ....
7. PAN is formed by the interaction of oxides of nitrogen and .... in presence of sunlight.

True/False Questions

1. All pollutants are degradable
2. Smog causes cracking of rubber
3. PAN is a primary pollutant
4. The noise is measured Richter scale.

ANSWERS

Multiple Choice Questions

1. (*a*)	**2.** (*a*)	**3.** (*b*)	**4.** (*d*)	**5.** (*a*)
6. (*c*)	**7.** (*b*)	**8.** (*b*)	**9.** (*c*)	**10.** (*a*)
11. (*c*)	**12.** (*b*)	**13.** (*d*)	**14.** (*d*)	**15.** (*c*)
16. (*a*)	**17.** (*b*)	**18.** (*b*)	**19.** (*c*)	**20.** (*a*)
21. (*a*)	**22.** (*a*)	**23.** (*d*)	**24.** (*b*)	**25.** (*c*)
26. (*c*)	**27.** (*d*)	**28.** (*d*)		

Fill in the Blanks

1. Ozone hole
2. Inversion
3. Thermal expansion, glaciers
4. Ozone, oxygen
5. Air
6. Salinization
7. Hydrocarbons

True/False Questions

1. False **2.** False **3.** False **4.** False

❑❑❑

CHAPTER

5

Water Pollution and Disaster Management

5.1. WATER POLLUTION

INSIDE THIS CHAPTER

Like clean air, fresh water is also becoming a scarcity. The limited availability of fresh water and its unequal distribution make water pollution a matter of great concern. Water pollution is generally localized and confined, making it more severe. The pollutants undergo many reactions and can become hazardous. 70 per cent of India's fresh water is polluted, including several high altitude lakes. While water pollution is easier to study and manage, its control is highly complex and very costly.

Water pollution is defined as presence of any foreign substance (organic, inorganic, biological or radiological) in water which tends to degrade the quality so as to constitute a hazard, or impair the usefulness of water.

It contains various types of impurities such as dissolved gases, dissolved minerals and suspended matters and even microbes. The presence of water pollutants alters the amount dissolved oxygen and biochemical oxygen demand. Dissolved oxygen (DO) is the amount of oxygen present in the aquatic system that is available for aquatic life and is essential for sustaining the plant and animal life in any aquatic system. The biochemical oxygen demand (BOD) is a measure of the oxygen required by aerobic micro-organisms to bio-chemically oxidize the organic matter present in the waste and is expressed in mg/l. Another quick measure of organic matter present in waste waste is chemical oxygen demand (COD) and in this case chemical oxidants (potassium dichromate in acidic conditions) are used instand of aerobic mocro-organisms for the oxidation of organic matter. Generally COD of a waste will be higher than its BOD because more organic compound can be chemically oxidized than bio-chemically.

5.1.1. Types of Water Pollution

Natural water pollution : Leaching of rocks, decaying of organic or dead matter, soil erosion and silting etc.

Man-made water pollution : Deforestation, industrial setup near water bodies, industrial waste, domestic sewage radio-active waste, synthetic chemicals, fertilizers and pesticides etc.

Water Pollutants

Categories	Nature	Examples
Physical	Temperature	Waste heat from industries
	Turbidity and color	Dyes and pigments
	Suspended and floating matter	Sand, silt, metal pieces, plastic, rubber, wood chips, paper, foam sewage etc
Chemical	Organic	Synthetic detergents, tars, plastic, fertilizers, organic effluent from industries etc.
	Inorganic	Nitrates, phosphates, chlorides, fluorides, trace elements etc.
Biological	Pathogenic	Bacteria, fungi, virus, worms etc.

5.1.2. Sources of Water Pollution

In underdeveloped countries, sewage is a major source of water pollution. Human excreta contain 400 different species of bacteria and viruses. Even well-treated sewage contains pathogenic bacteria and virus, unless properly chlorinated before being discharged into any water course. Sewage is a major contributor to water-borne diseases and affects the health of people and other organisms in the environment in many ways.

Industrial effluents from sugar factories, distilleries, tanneries and paper industries are accompanied by very high organic loads. By-products of paper and pulp industry cause depletion of fish up to as far as 40 km downstream. The wastes from oil refineries and steel industries contain phenol which imparts a strong odour, apart from poisoning the water body. Fertilizer industry wastes contain ammonia, urea, phosphate and sulphate which, in water, cause algal bloom and are toxic to aquatic fauna and flora. Alkaline industry wastes contain mercury which can kill human beings who consume mercerized fishes.

Lead generated from battery, printing, petrol and paste-processing industries, trace and toxic elements such as zinc, copper etc., and effluents from mining industries are injurious to aquatic organisms.

Effects of Water Pollution on Environment and Human Health

Major Water Pollutants and Their Effect on Human Health

Pollutant	Origin	Effects
Arsenic	Pesticide industries and agriculture	Toxic, eruptions on skin, carcinogenic
Chlorine	Industries using chlorine bleach	Physiological
Chromium	Industrial waste, especially from tanneries	Toxic and carcinogenic
Cyanide	Pesticides	Toxic and fatal
Lead	Industrial, mine discharges, water pipes	Causes a disease called plumbism
Nitrates and Nitrites	Fertilizer industries	Cause metaemo globinaemia in infants
Mercury	Pesticides	Toxic, causes brain damage
Bacteria	Sewage	Cholera, typhoid, hepatitis, dysentery
Sulphates	Leaching of gypsum or industrial waste discharges	Have laxative effect in combination with magnesium or sodium in drinking water, may form hydrogen sulphide gas which is fatal for workers in sewers

Other effects are :

- Objectionable odor.
- Increase in turbidity.
- Destruction in aesthetic value.
- Increase in temperature of water body.
- Due to increase in turbidity of water, the rate of photosynthesis decrease in the water bodies as aquatic plants didn't get sufficient sun light.
- As due to run off with water the nutrients reaches in the water body and that becomes rich nutrient (process is known as eutrophication) and the rate of growth of aquatic plants increases very rapidly and the penetration of sun light to the bottom of the water body in not possible and amount of dissolved oxygen decreases that effects aquatic life.

5.1.3. Control of Water Pollution

The water pollution can be controlled by treating polluted water by removing the contaminants up to acceptable quality standards, so that water can be reused or discharged into water bodies. Available waste water treatment process can be broadly classified as physical, chemical or biological.

Physical processes are the simplest forms of treatment and are depends on the removal of physical pollutants and also known as primary treatment. They remove identifiable suspended solid and floating matter. These principally comprise screening, sedimentation, filtration and flotation.

In biological treatment, also known as secondary treatment, organic matter that is soluble or in the colloidal form is removed. Secondary treatment is done with the help of aerobic (works in the presence of oxygen) and anaerobic (works in the absence of oxygen) micro-organism. Biological process utilizes biochemical reactions; typical examples are biological filtration (trickling filters), oxidation ponds and the activated sludge process.

Chemical processes utilize the chemical properties of the impurities or of the added reagents. Commonly used chemical processes are adsorption, precipitation, coagulation and disinfection (the removal of nitrate and phosphate bacteria is carried out by passing chlorine/ozone gas)

Advance waste treatment may involve physical, chemical or biological processes or their various combinations depending on the impurities to be removed. These processes are employed to remove residual soluble non-biodegradable organic compounds, including surfactants, inorganic nutrients and salts, trace elements and dissolved inorganic salts. The advance waste treatment process are expensive and are used only when water produced is required to be of higher quality.

After these treatment stages, the water is clean to supply in house-hold and discharge in water bodies. The sludge is the by-product of the treatment process and it can use in agriculture as manure.

Sewage treatment is done with the help of septic tank, which is an anaerobic suspended growth biological treatment system. They are made up of concrete and situated under the ground, having distributor box and porous rods. In septic tank anaerobic bacteria bio-chemically oxidize the organic matter. Septic tanks are usually adopted for small communities of population up to 300.

5.2. SAMPLING TECHNIQUE

Wastewater sampling is generally performed by one of two methods, grab sampling or composite sampling. Grab sampling is just what it sounds like; all of the test material is collected at one time. As such, a grab sample reflects performance only at the point in time that the sample was collected, and then only if the sample was properly collected.

Grab sampling is the simple process of bringing up surface sediments from the seafloor. It cannot be used to characterize different sedimentary layers since a mixture of sediments is produced when they are brought up, therefore, it is only used for the surface. This method has neither the ability nor the ergonomic qualities to penetrate the ground to depth.

Once it is launched, the jaws of the grab sampler open and it descends to the seafloor. A spring closes the jaws, and they trap sediments or loose substrate. The grab sampler is then brought up to the surface where its contents are studied in detail.

Composite sampling consists of a collection of numerous individual discrete samples taken at regular intervals over a period of time, usually 24 hours. The material being sampled is collected in a common container over the sampling period. The analysis of this material, collected over a period of time, will therefore represent the average performance of a wastewater treatment plant during the collection period.

Numerous industry references list various parameters for wastewater testing and whether samples should be collected using grab sampling or composite sampling methods. For example, grab sampling allows the analysis of specific types of unstable parameters such as pH, dissolved oxygen, chlorine residual, nitrites and temperature. However, the most widely used indicators of treatment plant performance, including CBOD5 (five day carbonaceous biochemical oxygen demand), TSS (total suspended solids) and TN (total nitrogen) require the use of composite sampling techniques.

Routine variations in the volume and strength characteristics of incoming wastewater create fluctuations in the quality of treatment plant effluent. Therefore, an effluent grab sample taken at one specific time throughout the daily flow pattern will not be representative of system performance over the entire day.

While the limitations of determining system performance by grab sampling are apparent, the use of grab samples for the evaluation of a residential treatment unit is further compromised if the sample is collected from a location where the effluent does not have sufficient velocity to keep the effluent solids in suspension.

Composite samples of effluent collected, stored, analyzed, tabulated and averaged over an extended period of time provide the only verifiable indication of treatment plant performance. Collecting and analyzing these composite samples is often an expensive and time-consuming process. For these reasons, most regulatory organizations recognize independent third-party certifiers, who use composite sampling methods to conduct performance evaluation and accurately measure system performance in a standardized, reproducible setting. Attempting to evaluate a residential treatment system in the field by analyzing a grab sample taken from a sump or any other containment vessel provides a compound degree of error and will yield erroneous conclusions about system performance.

5.3. SOIL POLLUTION

Soil is the thin layer of organic and inorganic material that coverts the earth's rocky surface. Several factors contribute to the formation of soil from parent material. This includes the mechanical weathering of rocks due to temperature changes and abrasion, wind, moving water, glaciers, chemical weathering activities and lichens. The surface litter layer consists of freshly-fallen and partially decomposed leaves, twigs, animal waste, fungi and other organic materials are known as 'topsoil'.

Soil pollution is the introduction of substances, biological organisms or energy into the soil, resulting in a change of the soil quality, which is likely to affect the normal use of the soil or endangering public health and the living organism. Soil pollutants have an adverse effect on the physical, chemical and biological properties of soil and reduce its productivity.

Land is a very valuable but limited resource, as the population increases rapidly. Many highly urbanized cities are faced with acute space problems, as in Calcutta or Bombay. Besides the limited availability of land, 175 million hectares of land are becoming less productive every year. India loses 20 tons of topsoil per hectare in a year due to floods, rainfall and deforestation. 20 per cent to 50 per cent of lands under irrigation can go out of cultivation at this rate because of water logging and salinity.

5.3.1. Causes of Soil Pollution

- Indiscriminate discharge and dumping of industrial effluents on land, thus contaminating them.
- An increase in the use of pesticides and fertilizers for agriculture increases soil's toxicity
- Open defecation by animals and human beings
- Accumulation of solid waste; this is a major problem in developed countries like India where the garbage and refuse products are not degraded
- Radioactive substances from nuclear plants which are released into the soil. These substances when reach to the soil and persist there for a long time and keep on emitting radiation.
- Nitrification process where nitrates are leached out from the soil by plants in presence of nitrifying bacteria.
- Acid rain increases the normal pH of soil and converts neutral soil to acidic one.
- Soil erosion of causes the loss of top soil and makes the soil less fertile and reduces its water holding capacity, hence productivity. It also contributes to water pollution by clogging lakes and increasing the turbidity of the water and ultimately leads the loss of aquatic life.
- Water logging and salinization increases the soluble salts in soil and makes the soil toxic.

5.3.2. Effects of Soil Pollution

- Effects human health
- Reduces soil productivity
- Decomposition of organic matter by micro-organism releases unpleasant odor
- Contaminates ground water
- Radio-active isotopes replace the essential elements of body and causes abnormalities.

5.3.3. Control of Soil Pollution

- Effluents should be properly treated before discharging them on soil.
- Solid waste should be properly collected and disposed off by appropriate method.
- From the waste, recovery of useful products should be done.
- Biodegradable organic waste should be used for generation of bio-gas.
- Cattle dung should be used for methane generation.
- There should be optimum use of fertilizers and pesticides.
- By bioremediation. Bioremediation is a treatment process that uses micro organisms (yeast, fungi or bacteria) to break down or degrade hazardous substances into less toxic or non toxic substances (carbon dioxide and water).

5.4. MARINE POLLUTION

Marine pollution can be defined as the direct or indirect introduction of humans of substances or energy into the marine environment (including estuarine), resulting in harm to living resources, hazards to human health, hindrance to marine activities including fishing, impairment of the quality of sea water. A big problem is that many poisons and toxic components adhere to tiny particles which are taken up within a few days by aquatic plants and aquatic small animals, most of which concentrating upward within ocean food chains. Because most animal feeds contain high fish meal and fish oil content, toxins can be found a few weeks later in commonly consumed food items derived from livestock and animal husbandry such as meat, eggs, milk and butter.

5.4.1. Sources of Marine Pollution

1. Sewage, runoff from forestry, farming and other land use.
2. Erosion from mining and forestry
3. Barges and ships also dump industrial waste, sludge from sewage plant and materials dredged or scraped from the bottoms of harbours and streams to maintain shipping channels into the ocean.
4. Rivers which bring pollutants from their drainage basin
5. Catchment's area i.e. coastline where human settlement in the form of hotels, industries and agricultural practices have been established
6. Crude petroleum and refined petroleum are accidentally or deliberately released into the environment from a number of sources.
7. Dumping of discarded radio-active waste and nuclear submarine.

5.4.2. Effects of Marine Pollution

1. Causes algal booms in coastal water.
2. Volatile organic hydrocarbons in oil immediately kill a number of aquatic organisms and oil coating destroy the animals' natural insulation and buoyancy.

3. Affects recreational facilities on the beaches.
4. Decrease in aesthetic value.
5. Causes radio active radiation.
6. Oil on beaches can have serious economic effects on coastal residents, who lose income from fishery and tourist activities.

5.4.3. Control of Marine Pollution

1. Toxic pollutants from industries and sewage treatment plants should not be discharged in coastal water.
2. Run-off from non-point sources (rivers) should be prevented to reach coastal area.
3. Dumping of toxic, hazard and radio-active waste should be banned.
4. Developmental activities on coastal area should be minimized.
5. Oil and greases from service station should be processed for reuse.
6. By legal regulations.

5.5. DISASTER

What is Disaster ?

Disaster is a sudden, calamitous event bringing great damage, loss, and destruction and devastation to life and property. The damage caused by disasters is immeasurable and varies with the geographical location, climate and the type of the earth surface/degree of vulnerability. This influences the mental, socio-economic, political and cultural state of the affected area. Generally, disaster has the following effects in the concerned areas :

1. It completely disrupts the normal day to day life;
2. It negatively influences the emergency systems; and
3. Normal needs and processes like food, shelter, health, etc. are affected and deteriorate depending on the intensity and severity of the disaster.

It may also be termed as "a serious disruption of the functioning of society, causing widespread human, material or environmental losses which exceed the ability of the affected society to cope using its own resources."

Thus, a disaster may have the following main features :

- Unpredictability
- Unfamiliarity
- Speed
- Urgency
- Uncertainty
- Threat

Thus, in simple terms we can define disaster as a hazard causing heavy loss to life, property and livelihood, *e.g.* a cyclone killing 10,000 lives and a crop loss of one crore can be termed as disaster.

5.5.1. Types of Disaster

Generally, disasters are of two types : **Natural** and **Manmade.** A natural disaster is the effect of a natural hazard that affects the environment, and leads to financial, environmental and/or human losses. The resulting loss depends on the capacity of the population to support or resist the disaster, and their resilience.

Various disasters like earthquake, landslides, volcanic eruptions, fires, flood and cyclones are natural hazards that kill thousands of people and destroy billions of dollars of habitat and property each year. The rapid growth of the world's population and its increased concentration often in hazardous environment has escalated both the frequency and severity of natural disasters. With the tropical climate and unstable land forms, coupled with deforestation, unplanned growth proliferation non-engineered constructions which make the disaster-prone areas mere vulnerable, tardy communication, poor or no budgetary allocation for disaster prevention, developing countries suffer more or less chronically by natural disasters. Asia tops the list of casualties due to natural disaster. Among various natural hazards, earthquakes, landslides, floods and cyclones are the major disasters adversely affecting very large areas and population in the Indian sub-continent. These natural disasters are of (*i*) geophysical origin such as earthquakes, volcanic eruptions, land slides and (*ii*) climatic origin such as drought, flood, cyclone, locust, forest fire.

Man-made disasters are events which, either intentionally or by accident cause severe threats to public health and well-being. Because their occurrence is unpredictable, man-made disasters pose an especially challenging threat that must be dealt with through vigilance, and proper preparedness and response. Information on the major sources of man-made disasters is provided here to help educate the public on their cause and effects as they relate to emergency planning.

Based on the devastation, these are further classified into major/minor natural disaster and major/minor manmade disasters. Some of the disasters are listed below

Major natural disasters :

- Flood
- Cyclone
- Drought
- Earthquake

Minor natural disasters :

- Cold wave
- Thunderstorms
- Heat waves
- Mud slides
- Storm

Major manmade disaster :

- Setting of fires
- Epidemic
- Deforestation
- Pollution due to prawn cultivation
- Chemical pollution
- Wars

Minor manmade disaster:

- Road/train accidents, riots
- Food poisoning
- Industrial disaster/crisis
- Environmental pollution

Risk is a measure of the expected losses due to a hazardous event of a particular magnitude occurring in a given area over a specific time period. Risk is a function of the probability of particular occurrences and the losses each would cause. The level of risk depends on :

- Nature of the Hazard
- Vulnerability of the elements which are affected
- Economic value of those elements

Vulnerability

It is defined as *"the extent to which a community, structure, service, and/or geographic area is likely to be damaged or disrupted by the impact of particular hazard, on account of their nature, construction and proximity to hazardous terrain or a disaster prone area"*

Hazards

Hazards are defined as ***"Phenomena that pose a threat to people, structures, or economic assets and which may cause a disaster. They could be either manmade or naturally occurring in our environment."***

The extent of damage in a disaster depends on :

(*i*) The impact, intensity and characteristics of the phenomenon and

(*ii*) How people, environment and infrastructures are affected by that phenomenon.

This relationship can be written as an equation:

Disaster Risk = Hazard + Vulnerability

5.6. EARTHQUAKES

Perhaps the Earthquake is the most **fearful natural phenomenon** in the human life. It is more so, because it is unpredictable and arrives without notice or without announcing it's vigor and strength. Scientists are constantly in search of this unknown.

An earthquake is a sudden movement of the earth's surface. An **earthquake** (also known as a **tremor** or **temblor**) is the result of a sudden release of energy in the Earth's crust that creates seismic waves.

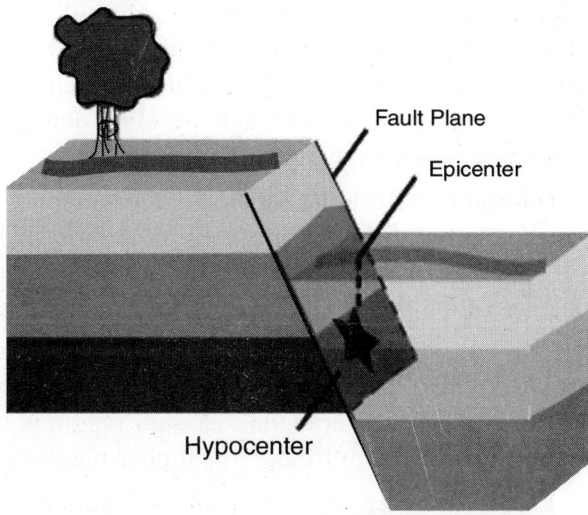

Fig. 5.1. Epicenter and Hypocenter

An ***earthquake*** is what happens when two blocks of the earth suddenly slip past one another. The surface where they slip is called the ***fault*** or ***fault plane***. The location below the earth's surface where the earthquake starts is called the ***hypocenter***, and the location directly above it on the surface of the earth is called the ***epicenter***.

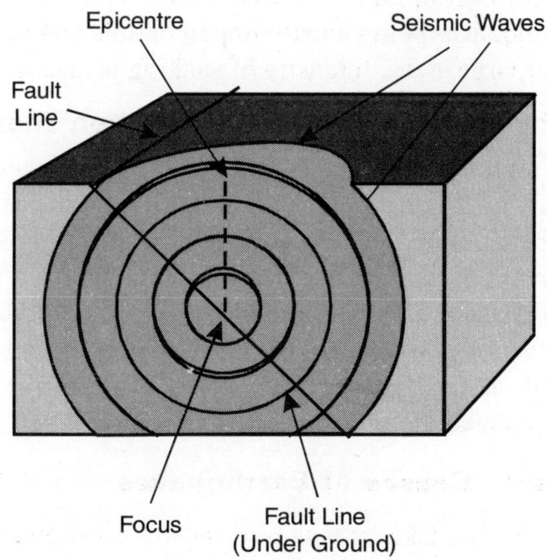

Fig. 5.2. Seismic Waves and Fault Line

When the earth shakes due to the movements of plates below the earth's crust it is known as Earthquake. Earthquakes are natural disasters, which kill thousands of people in an instant and can destroy cities and countries-human habitation across miles. The vibration during an earthquake has the potential to wreak havoc and destruction, which is beyond imagination. The destruction is the maximum near the epicenter, the place from where the vibrations arise and spread. Of late there have been many such natural disasters, which can be associated with earthquakes like the Tsunami that hit the shores of India, Thailand and razed down buildings and annihilated many nm/.hjlives. The shaking in earthquakes can also trigger landslides and occasionally volcanic activity.

In a simpler way it can be narrated that the earth as such consists of different rock layers of decreasing densities right from the center towards it's surface. Deep inside at the center, the earth is hot and molten. Because of Earth's rotation and other energy factors different shells or the rock layers constantly move or slid past each other. As a result, the different continental mass fragments of lesser densities float and move overriding the denser rock layers or the plates, either towards or away from each other. This is happening since long geological past. This movement acts like a conveyor belt and during the journey the plates meet each other or gets obstacle by other denser plate and the rock layers start descending when we call a subduction Zone.

Naturally in the adjacent sides of the subduction zone the rock layers get up heaved resulting folding, thrusting and faulting. The second belt in parts of India *i.e.* Andaman- Great Nicobar-Java-Sumatra zone is actually is a sub ducting zone for which the long island chains have emerged.

This zone, in fact, is very dynamic and active giving rise deep to intermediate earthquake foci. The sub ducting force in this part, is directed from west to east while the same in Uttaranchal - Assam region is from south to north. Entire this zone is susceptible to tremors of higher magnitude.

Earthquakes are recorded with a seismometer, also known as a seismograph. The moment magnitude (or the related and mostly obsolete Richter magnitude) of an earthquake is conventionally reported, with magnitude 3 or lower earthquakes being mostly imperceptible and magnitude 7 causing serious damage over large areas. Intensity of shaking is measured on the modified Mercalli scale.

Why Does the Earth Shake When There is an Earthquake?

While the edges of faults are stuck together, and the rest of the block is moving, the energy that would normally cause the blocks to slide past one another is being stored up! When the force of the moving blocks finally overcomes the *friction* of the jagged edges of the fault and it unsticks, all that stored up energy is released! The energy radiates (moves) outward from the fault in all directions in the form of *seismic waves* like ripples on a pond. The seismic waves shake the earth as they move through it, and when the waves reach the earth's surface, they shake the ground and anything on it.

5.6.1. Causes of Earthquakes

Earthquakes are caused mostly by rupture of geological faults, but also by volcanic activity, landslides, mine blasts, and nuclear experiments. An earthquake's point of initial rupture is called its focus or hypocenter. The term epicenter refers to the point at ground level directly above the hypocenter.

They may be caused by various activities at the earth's surface such as ebb and the flow of the tides, the rush of traffic in the city streets, the tumbling of streams over high falls, magma tic outbursts from within the interior of the earth or explosion of high power nuclear or atomic bomb and so and so forth.

If seen broadly we can say that earthquakes are caused due to two major reasons. The first reason is the eruption of volcanoes, which are sudden, and as is known volcanoes are seat of inner disturbance and can effect the plates which is the second cause of earthquakes. Earthquakes are caused due to disturbance in the movement of plates, which again can be caused due to various reasons like under crust waves or cracks in the plates.

5.6.2. Earthquake Fault Types

There are three main types of fault that may cause an earthquake:

1. Strike-slip Fault

The fault is vertical, and the blocks slide past one another horizontally. This occurs in areas where the crustal blocks are sliding past one another. Strike-slip faults are steep structures where the two sides of the fault slip horizontally past each other; transform boundaries are a particular type of strike-slip fault.

2. Normal Fault

The fault is at an angle, and the block above the fault (the *hanging wall*) moves down relative to the block below the fault (the *foot wall*). This occurs in areas where there is extension or pulling of the crustal blocks. Normal faults occur mainly in areas where the crust is being extended such as a divergent boundary.

Fig. 5.3. Fault Types

3. Thrust (Reverse) Fault

The fault is at an angle, and the *hanging wall* moves up relative to the *foot wall*. This occurs in areas where the crustal blocks are being pushed together. Reverse faults occur in areas where the crust is being shortened such as at a convergent boundary.

Normal and reverse faulting are examples of dip-slip, where the displacement along the fault is in the direction of dip and movement on them involves a vertical component. Many earthquakes are caused by movement on faults that have components of both dip-slip and strike-slip; this is known as oblique slip.

Safety Tips

Before the Earthquake

If we wait until the earth starts to shake, it may be too late. Consider the following safety measures:

- Always keep the following in a designated place: bottled drinking water, non-perishable food (chura, gur, etc), first-aid kit, torchlight and battery-operated radio with extra batteries.
- Identify places in the house that can provide cover during an earthquake.
- It may be easier to make long distance calls during an earthquake. Identify an out-of-town relative or friend as family's emergency contact.

If the family members get separated after the earthquake and are not able to contact each other, they should contact the designated relative/ friend. The address and phone number of the contact person/relative should be with all the family members.

Safeguard House :

- Consider retrofitting our house with earthquake-safety measures. Reinforcing the foundation and frame could make our house quake resistant.
- Kutchha buildings can also be retrofitted and strengthened.

During Quake

Earthquakes give no warning at all. Sometimes, a loud rumbling sound might signal its arrival a few seconds ahead of time. Those few seconds could give us a chance to move to a safer location. Here are some tips for keeping safe during a quake.

- Take cover. Go under a table or other sturdy furniture; kneel, sit, or stay close to the floor. Hold on to furniture legs for balance. Be prepared to move if your cover moves.
- If no sturdy cover is nearby, kneel or sit close to the floor next to a structurally sound interior wall. Place hands on the floor for balance.
- Do not stand in doorways. Violent motion could cause doors to slam and cause serious injuries.
- Move away from windows, mirrors, bookcases and other unsecured heavy objects.
- If you are in bed, stay there and cover yourself with pillows and blankets
- Do not run outside if you are inside. Never use the lift.
- If you are living in a kutcha house, the best thing to do is to move to an open area where there are no trees, electric or telephone wires.

If outdoors :

- Move into the open, away from buildings, streetlights, and utility wires. Once in the open, stay there until the shaking stops.
- Avoid places where there are loose electrical wires and do not touch metal objects that are in touch with the loose wires.
- Do not re-enter damaged buildings and stay away from badly damaged structures.

If in a moving vehicle :

- Move to a clear area away from buildings, trees, overpasses, or utility wires, stop, and stay in the vehicle. Once the shaking has stopped, proceed with caution. Avoid bridges or ramps that might have been damaged by the quake.

After the Quake

Here are a few things to keep in mind after an earthquake. The caution you display in the aftermath can be essential for your personal safety.

- Wear shoes/chappals to protect feet from debris
- After the first tremor, be prepared for aftershocks. Though less intense, aftershocks cause additional damages and may bring down weakened structures. Aftershocks can occur in the first hours, days, weeks, or even months after the quake.
- Check for fire hazards and use torchlights instead of candles or lanterns.
- If the building you live in is in a good shape after the earthquake, stay inside and listen for radio advises. If you are not certain about the damage to your building, evacuate carefully. Do not touch downed power line.
- Help injured or trapped persons. Give first aid where appropriate. Do not move seriously injured persons unless they are in immediate danger of further injury. In such cases, call for help.
- Remember to help your neighbours who may require special assistance-infants, the elderly, and people with disabilities.
- Listen to a battery-operated radio for the latest emergency information.
- Stay out of damaged buildings.
- Return home only when authorities say it is safe. Clean up spilled medicines, bleaches or gasoline or other flammable liquids immediately. Leave the area if you smell gas or fumes from other chemicals. Open closet and cupboard doors cautiously.
- If you smell gas or hear hissing noise, open windows and quickly leave the building. Turn off the switch on the top of the gas cylinder.
- Look for electrical system damages - if you see sparks, broken wires, or if you smell burning of amber, turn off electricity at the main fuse box. If you have to step in water to get to the fuse box, call an electrician first for advice.
- Check for sewage and water lines damage. If you suspect sewage lines are damaged, avoid using the toilets. If water pipes are damaged, avoid using water from the tap.
- Use the telephone only for emergency calls.
- In case family members are separated from one another during an earthquake (a real possibility during the day when adults are at work and children are at school), develop a plan for reuniting after the disaster. Ask an out of state / district relative or friend to serve as the "family contact". Make sure everyone in the family knows the name, address, and phone number(s) of the contact person (s).

Case Study

Significantly, most of the earthquakes in this belt are associated with volcanic activities. The violent outbursts of Kraktao in 1883 were accompanied by severe shocks and about 35,000 people died instantly and the whole village was displaced beneath the ocean bottom.

With intermittent volcanic activity and earth tremors, this region recently, on 26th December 2004 was subjected to catastrophic earthquake of 9 magnitudes in Richter scale, which again recurred on 28th March 2005. The previous one led tsunami causing severe damage to both life and property. The other parts of India, was used to be considered as earthquake resistant areas or the shield areas. Oldest rocks of more than 3000 m years form the foundations of the continental mass compare to much younger rock sequence in the seismically active belt just described.

Therefore earthquakes causing loss of life and property around Latur or Koyna raised a serious concern among the geoscientists. One important aspect in this regard should always be kept in mind that the tremors had hit mostly the coastal points excepting a few areas in the hinterland.

Recently, a light earthquake (m.l=2.8) struck the coastal Kunnakulam region in the northern Kērala on 20.12.2006 at 19:19 hours local time. The adjoining districts of Mallapuram, Pallakad and Trissur also experienced the same tremors. People became panicky particularly when the famous festival of Trissur Pooram was at the peak.

5.7. CYCLONE

A cyclone is a storm accompanied by high speed whistling and howling winds. It brings torrential rains. A cyclonic storm develops over tropical oceans like the Indian Ocean and Bay of Bengal and the Arabian Sea. Its strong winds blow at great speed, which can be more than 118 kilometers per hour. When a cyclonic storm approaches, the skies begin to darken accompanied by lightening and thunder and a continuous downpour of rain.

In meteorology, a **cyclone** is an area of closed, circular fluid motion rotating in the same direction as the Earth. This is usually characterized by inward spiraling winds that rotate counter clockwise in the Northern Hemisphere and clockwise in the Southern Hemisphere of the Earth.

It is often difficult to predict where a cyclone will strike. When it starts moving from oceans towards the land area, a cyclone can change track and hit areas other than those anticipated earlier.

Large-scale cyclonic circulations are almost always centred on areas of low atmospheric pressure. The largest low-pressure systems are cold-core polar cyclones and extratropical cyclones which lie on the synoptic scale. Warm-core cyclones such as tropical cyclones, mesocyclones, and polar lows lie within the smaller mesoscale. Subtropical cyclones are of intermediate size. Cyclones have also been seen on other planets outside of the Earth, such as Mars and Neptune.

Cyclogenesis describes the process of cyclone formation and intensification. Extratropical cyclones form as waves in large regions of enhanced midlatitude temperature contrasts called baroclinic zones. These zones contract to form weather fronts as the cyclonic circulation closes and intensifies. Later in their life cycle, cyclones occlude as cold core systems. A cyclone's track is guided over the course of its 2 to 6 day life cycle by the steering flow of the cancer or subtropical jetstream.

Weather fronts separate two masses of air of different densities and are associated with the most prominent meteorological phenomena. Air masses separated by a front may differ in temperature or humidity. Strong cold fronts typically feature narrow bands of thunderstorms and severe weather, and may on occasion be preceded by squall lines or dry lines. They form west of the circulation center and generally move from west to east. Warm fronts form east of the cyclone center and are usually preceded by stratiform precipitation and fog. They move poleward ahead of the cyclone path. Occluded fronts form late in the cyclone life cycle near the enter of the cyclone and often wrap around the storm center.

Tropical cyclogenesis describes the process of development of tropical cyclones. Tropical cyclones form due to latent heat driven by significant thunderstorm activity, and are warm core. Cyclones can transition between extratropical, subtropical, and tropical phases under the right conditions. Mesocyclones form as warm core cyclones over land, and can lead to tornado formation. Waterspouts can also form from mesocyclones, but more often develop from environments of high instability and low vertical wind shear.

How does a Cyclone Affect Us ?

- A cyclone causes heavy floods.
- It uproots electricity supply and telecommunication lines. Power supply shuts down and telephones stop functioning.
- Road and rail movements come to halt because floods damage rail tracks and breach roads. Rail movements are also disrupted because of communication failure.
- The inclement weather conditions also disrupt Air services. Seaports stop work due to high winds, heavy rains and poor visibility. Sometimes ships overturn or are washed ashore. The high speed winds bends and plucks out trees and plants.
- A cyclone tears away wall sidings and blows off roofs of houses.
- Houses collapse and people are rendered homeless. In villages kachha houses get blown away. The speeding winds cause loose metal and wooden sheets to fly turning them to potential killers. Broken glass pieces can cause serious injuries.
- The floodwaters can take time to recede.
- The floodwaters can turn the fields salty.

- Bridges, dams and embankments suffer serious damages.
- Floods wash away human beings and animals and make water unfit for drinking. There can be outbreak of diseases like Cholera, Jaundice or Viral fever due to intake of impure water. Water gets contaminated because of floating corpses of animals and human beings and mixing of sewage stored food supplies, get damaged.

Safety Tips

Before the Cyclone Season

- Keep watch on weather and listen to radio or TV. Keep alert about the community warning systems – loudspeakers, bells, conches, drums or any traditional warning system.
- Get to know the nearest cyclone shelter / safe houses and the safest route to reach these shelters.
- Do not listen to rumours.
- Prepare an emergency kit containing:
  (*i*) A portable radio, torch and spare batteries;
  (*ii*) Stocks dry food – Chura, Chhatua, Mudhi, gur, etc.
  (*iii*) Matches, fuel lamp, portable stove, cooking utensils, waterproof bags.
  (*iv*) A first aid kit, manual, etc.
  (*v*) Katuri, pliers, small saw, axe and plastic rope
- Check the roof and cover it with net or bamboo. Check the walls, pillars, doors and windows to see if they are secure. If not, repair those at the earliest. In case of tin roofs, check the condition of the tin and repair the loose points. Cover the mud walls with polythene or coconut leaves mats or straw mats on a bamboo frame. Bind each corner of the roof with a plastic rope in case of thatched roof.
- Trim dry tree branches, cut off the dead trees and clear the place/courtyard of all debris, including coconuts and tree branches.
- Clear your property of loose materials that could blow about and cause injury or damage during extreme winds.
- If your area is prone to storm surge, locate safe high ground or shelter.
- Keep important documents, passbook, etc. in a tight plastic bag and take it along with your emergency kits if you are evacuating.
- Identify the spot where you can dig holes to store food grains, seeds, etc. in polythene bags.
- Keep a list of emergency addresses and phone numbers on display. Know the contact telephone number of the government offices /agencies, which are responsible for search, rescue and relief operations in your area.

Upon a Cyclone Warning

- Store loose items inside. Put extra agricultural products/ stock like paddy in plastic bags and store it by digging up a hole in the ground, preferably at a higher elevation and then cover it properly. Fill bins and plastic jars with drinking water.

- Keep clothing for protection, handy
- Prepare a list of assets and belongings of your house and give information to volunteers and other authorities about your near and dear ones.
- Fill fuel in your car/motorcycle and park it under a solid cover. Tie bullock carts, boats securely to strong posts in an area, which has a strong cover and away from trees. Fallen trees can smash boats and other assets.
- Close shutters or nail all windows. Secure doors. Stay indoors, with pets.
- Pack warm clothing, essential medications, valuables, papers, water, dry food and other valuables in waterproof bags, to be taken along with your emergency kit.
- Listen to your local radio / TV, local community warning system for further information.
- In case of warning of serious storm, move with your family to a strong pucca building. In case of warning of cyclones of severe intensity, evacuate the area with your family, precious items and documents and emergency kit. Take special care for children, elders, sick, pregnant women and lactating mothers in your family. Do not forget your emergency food stock, water and other emergency items. GO TO THE NEAREST CYCLONE SHELTER.
- Do not venture into the sea for fishing.

On Warning of Local Evacuation

Based on predicted wind speeds and storm surge heights, evacuation may be necessary. Official advice may be given on local radio / TV or other means of communication regarding safe routes and when to move.

- Wear strong shoes or chappals and clothing for protection.
- Lock your home, switch off power, gas, water, and take your emergency kit.
- If evacuating to a distant place take valuable belonging, domestic animals, and leave early to avoid heavy traffic, flooding and wind hazards.
- If evacuating to a local shelter or higher grounds carry the emergency kit and minimum essential materials.

What One should not do During a Cyclone Attack?

- During the storm do not venture out unless advised to evacuate.
- If you have a vehicle and wish to move out of your house, leave early before the onset of a cyclone. It is often best to stay at home
- Avoid remaining on the top floor of dwellings. Stay close to the ground.
- Fishermen are advised not to venture out into the sea. They should keep boats and rafts tied up in a safe place.

- Avoid taking shelters near old and damaged buildings or near trees.
- Do not touch power lines. One may get electrocuted.

When the cyclone strikes :

- Disconnect all electrical appliances and turn off gas.
- If the building starts crumbling, protect yourself with mattresses, rugs or blankets under a strong table or bench or hold on to a solid fixture (e.g. a water pipe)
- Listen to your transistor radio for updates and advice.
- Beware of the calm 'eye'. If the wind suddenly drops, don't assume the cyclone is over; violent winds will soon resume from the opposite direction. Wait for the official "all clear".
- If driving, stop – but well away from the sea and clear of trees, power lines and watercourses. Stay in the vehicle.

After the Cyclone

- Do not go outside until officially advised it is safe.
- Check for gas leaks. Do not use electric appliances, if wet.
- Listen to local radio for official warnings and advice.
- If you have to evacuate, or did so earlier, do not return until advised. Use a recommended route for returning and do not rush.
- Be careful of snake bites and carry a stick or bamboo
- Beware of fallen power lines, damaged bridges, buildings and trees, and do not enter the floodwaters.
- Heed all warnings and do not go sightseeing.

CASE STUDY : Orissa Over the Bay of Bengal (NOAA)

October 29th, 1999 : The Orissa cyclone was the deadliest Indian Ocean tropical cyclone since the 1991 Bangladesh cyclone, and deadliest Indian storm since 1971. The storm made landfall just weeks after a Category 4 storm hit the same general area. On October 28th it became a severe cyclone with a peak of 160 mph (260 km/h) winds. It hit India the next day as a 155 mph (250 km/h) cyclone. It caused the deaths of 20,000 people and 700,000 cattle. The storm also left 20 million people homeless.

5.8. FLOODS

A **flood** is an overflow or accumulation of an expanse of water that submerges land. In the sense of "flowing water", the word may also be applied to the inflow of the tide. Flooding may result from the volume of water within a body of water, such as a river or lake, which overflows or breaks levees, with the result that some of the water escapes its normal boundaries. While the size of a lake or other body of water will vary with seasonal changes in precipitation and snow melt, it is not a significant flood unless such escapes of water endanger land areas used by man like a village, city or other inhabited area.

Fig. 5.4. Flooded Area

Floods can also occur in rivers, when the strength of the river is so high it flows out of the river channel, particularly at bends or meanders and causes damage to homes and businesses along such rivers.

5.8.1. Principal Types of Flood

Based on Geography

Riverine Floods

- **Slow kinds :** Runoff from sustained rainfall or rapid snow melt exceeding the capacity of a river's channel. Causes include heavy rains from monsoons, hurricanes and tropical depressions, foreign winds and warm rain affecting snow pack. Unexpected drainage obstructions such as landslides, ice, or debris can cause slow flooding upstream of the obstruction.
- **Fast kinds :** include flash floods resulting from convective precipitation (intense thunder storms) or sudden release from an upstream impoundment created behind a dam, landslide, or glacier.

Estuarine Floods

- Commonly caused by a combination of sea tidal surges caused by storm-force winds. A storm surge, from either a tropical cyclone or an extratropical cyclone, falls within this category.

Coastal Floods

- Caused by severe sea storms, or as a result of another hazard (e.g. tsunami or hurricane). A storm surge, from either a tropical cyclone or an extratropical cyclone, falls within this category.

Catastrophic Floods

- Caused by a significant and unexpected event e.g. dam breakage, or as a result of another hazard (e.g. earthquake or volcanic eruption).

Muddy Floods

- A muddy flood is generated by run off on crop land.

A muddy flood is produced by an accumulation of runoff generated on cropland. Sediments are then detached by runoff and carried as suspended matter or bedload. Muddy runoff is more likely detected when it reaches inhabited areas.

Muddy floods are therefore a hillslope process, and confusion with mudflows produced by mass movements should be avoided.

Other

- Floods can occur if water accumulates across an impermeable surface (e.g. from rainfall) and cannot rapidly dissipate (i.e. gentle orientation or low evaporation).
- A series of storms moving over the same area.
- Dam-building beavers can flood low-lying urban and rural areas, often causing significant damage.

Based on Speed

Snow Melt Floods

Precipitation in the form of snow does not produce runoff or infiltrate into the soil, but waits until the snow melts, which means that several months of precipitation can accumulate above the soil surface. Rain on the snow pack or water from the melting snow can be held within the pack until the high liquid water content finally causes the pack to collapse, releasing water catastrophically and causing a very large runoff very rapidly. Warm rainwater a large cold spell may cause the snow pack to melt while the underlying ground is still frozen, which prevents any infiltration. Because of these reasons, snowmelt floods can be very large.

Storm Surge

Floods in coastal areas and in river estuaries are usually due to storm surges, which result from the sea being driven on to the land by meteorological forces. Here to physical forces act together. A storm with intense low pressure causes the level of sea to rise because of barometric effects and strong winds associated with this storm, if directed on shore, drive the sea on the land. Storm surges are thus, commonly associated with tropical cyclones. The east coast of India is particularly prone to storm surges. The storm that produces the surge can also give rise to heavy rainfall inland so that the estuary region can be subject simultaneously to river flooding and storm surge.

Flash Floods

Flash floods are defined floods of short duration with a relatively high peak discharge. They arise from local precipitation of extremely high intensity, typical of thunderstorms. The high concentration of rainfall on a small area can have devastating effects as the river flow can rise to several hundred times the normal flow in the space of a few hours. Flash floods are common in arid and semi-arid areas. In these areas, what little rainfall there is usually occurs in short, intense storms. The intensity of the storms and the poor absorptive capacity of arid zone soils lead to much to the annual runoff occurring as flash floods. Which can also occur following thunderstorms in more humid regions. Mountainous areas are prone to thunderstorms and the steep terrain and thin soils in the mountains assure high runoff with a short delay time.

5.8.2. Causes of Floods

Flooding conditions may occur due to:

(*i*) River in spate
(*ii*) Snowmelt
(*iii*) Storm surges
(*iv*) Short intense storm causing flash floods

Flooding in rivers in mainly caused by:

(*i*) Inadequate capacity within the banks of the rivers to contain high flows.
(*ii*) River bank erosion and silting of riverbeds.
(*iii*) Landslides leading to obstruction of flow and change in the river course.
(*iv*) Synchronization of flood in the main tributary rivers.
(*v*) Flow retardation due to tidal and backwater effects.
(*vi*) Poor natural drainage.
(*vii*) Cyclone and heavy rainfall.

5.8.3. Effects of Flood

Primary Effects

(*i*) **Physical Damage :** Can range anywhere from bridges, cars, buildings, sewer systems, roadways, canals and any other type of structure.

(*ii*) **Casualties :** People and livestock die due to drowning. It can also lead to epidemics and waterborne diseases.

Secondary Effects

(*i*) **Water Supplies :** Contamination of water. Clean drinking water becomes scarce.

(*ii*) **Diseases :** Unhygienic conditions. Spread of water-borne diseases.

(*iii*) **Crops and Food Supplies :** Shortage of food crops can be caused due to loss of entire harvest. However, lowlands near rivers depend upon river silt deposited by floods in order to add nutrients to the local soil.

(*iv*) **Trees :** Non-tolerant species can die from suffocation.

Tertiary/Long-Term Effects

(*i*) **Economic :** Economic hardship, due to: temporary decline in tourism, rebuilding costs, food shortage leading to price increase etc.

5.8.4. Floods in India

The rapid and durable recovery which does not reproduce the original vulnerable conditions The country receives an annual precipitation of 400 million - hectare meters. Of the annual rainfall, 75 per cent is received during four months of monsoon (June- September) and, as a result, almost all the rivers carry heavy discharge during this period. The flood hazard is compounded by the problems of sediment deposition, drainage congestion and synchronization of river floods with sea tides in the coastal plains. The area vulnerable to floods is 40 million hectares and the average area affected by floods annually is about 8 million hectares. The Table below depicts the average annual loss in the 1953 onwards.

Average Annual Loss Due to Floods

S. No.	Items	Loss
1.	Area affected	7.351 million hectare
2.	Population affected	40.967 million
3.	Human lives lost	1793 number
4.	Cattle lost	85599 number
5.	Houses damaged	1452904 number
6.	Houses damaged	370.607crore
7.	Crop area damaged	3.725 million hectare
8.	Crop damaged	1095.132crore
9.	Public Utilities damaged	1186.456crore
10.	Total losses	2706.243crore

(**Source:** Central Water Commission, Ministry of Water Resources, Government of India.)

5.8.5. Safety Tips

This guide lists simple things you and your family can do to stay safe and protect your property from floods.

Before Flooding Occurs

(*i*) All your family members should know the safe route to nearest shelter/ raised pucca house.

(*ii*) If your area is flood-prone, consider alternative building materials. Mud walls are more likely to be damaged during floods. You may consider making houses where the walls are made of local bricks up to the highest known flood level with cement pointing.

(*iii*) Have an emergency kit on hand which includes a:

(*a*) A portable radio, torch and spare batteries;

(*b*) Stocks of fresh water, dry food (chura, mudi, gur, biscuits), kerosene, candle and matchboxes;

(c) Waterproof or polythene bags for clothing and valuables, an umbrella and bamboo stick (to protect from snake), salt and sugar.

(d) A first aid kit, manual and strong ropes for tying things.

When you hear a flood warning or if flooding appears likely :

(i) Tune to your local radio/TV for warnings and advice.

(a) Keep vigil on flood warning given by local authorities.

(b) Don't give any importance to rumours and don't panic.

(c) Keep dry food, drinking water and clothes ready.

(ii) Prepare to take bullock carts, other agricultural equipments, and domestic animals to safer places or to higher locations.

(iii) Plan which indoor items you will raise or empty if water threatens to enter your house.

(iv) Check your emergency kit.

During Floods

(i) Drink boiled water.

(ii) Keep your food covered, don't take heavy meals.

(iii) Use raw tea, rice-water, tender coconut-water, etc. during diarrhoea; contact your ANM/AWW for ORS and treatment.

(iv) Do not let children remain on empty stomach.

(v) Use bleaching powder and lime to disinfect the surrounding.

(vi) Help the officials/volunteers distributing relief materials.

If you Need to Evacuate

(i) Firstly pack warm clothing, essential medication, valuables, personal papers, etc. in waterproof bags, to be taken with your emergency kit.

(ii) Take the emergency kit.

(iii) Inform the local volunteers (if available), the address of the place you are evacuating to.

(iv) Raise furniture, clothing and valuables onto beds, tables and to the top of the roof (electrical items highest).

(v) Turn off power.

(vi) Whether you leave or stay, put sandbags in the toilet bowl and over all laundry/bathroom drain-holes to prevent sewage back-flow.

(vii) Lock your home and take recommended/known evacuation routes for your area.

(viii) Do not get into water of unknown depth and current.

If you stay or on your return :

(i) Stay tuned to local radio for updated advice.

(ii) Do not allow children to play in, or near, flood waters.

(*iii*) Avoid entering floodwaters. If you must, wear proper protection for your feet and check depth and current with a stick. Stay away from drains, culverts and water over knee-deep.

(*iv*) Do not use electrical appliances, which have been in floodwater until checked for safety.

(*v*) Do not eat food, which has been in floodwaters.

(*vi*) Boil tap water (in cities) until supplies have been declared safe. In case of rural areas, store tube well water in plastic jars or use halogen tablets before drinking.

(*vii*) Be careful of snakes, snakebites are common during floods.

5.8.6. Flood Disaster Management

The various measures adopted for flood mitigation may be categorized into two groups :

(*i*) Structural

(*ii*) Non- structural

The general approach was aimed at preventing floodwaters from reaching the potential damage centers, as a result of which a large number of embankments came up along the various flood prone rivers. The main thrust of the flood protection programme undertaken in the country so far in the form of structural measures may be grouped into the following:

(*a*) Dams and Reservoirs

(*b*) Embankments, flood walls, sea wall

(*c*) Natural detention basin

(*d*) Channel improvement

(*e*) Drainage improvement

(*f*) Diversion of floodwaters.

The effective functioning of all the physical measures taken, it is necessary that pre- and post-monsoon checks must be made special repairs must be carried out prior to flood period.

The non-structural measures, on the other hand, aim at modifying the susceptibility to flood damage as well as modifying the loss burden. The various non-structural measures being implemented in the country are:

Modifying the susceptibility to flood damages through :

(*i*) Flood plain management

(*ii*) Flood proofing including disaster preparedness, and response planning and

(*iii*) Flood forecasting and Warning

Modifying the flood loss burden through :

(*i*) Disaster Relief

(*ii*) Flood fighting including Public Health Measures

Setting up of flood forecasting and warning services is one of the most cost-effective non- structural measures available.

5.9. LANDSLIDE

Landslide is a general term for a wide variety of downslope movements of earth materials that result in the perceptible downward and outward movement of soil, rock, and vegetation under the influence of gravity. The materials may move by falling, toppling, sliding, spreading, or flowing. Some landslides are rapid, occurring in seconds, whereas others may take hours, weeks, or even longer to develop.

Mudflows (or debris flows) are fluid mass of rock, earth, and other debris saturated with water. Mudflows are characteristics of steep, scanty vegetated slopes on which heavy rainfall initiates movement in a thick layer of weathered material. They develop when water rapidly accumulates in the ground, such as during heavy rainfall or rapid snowmelt, changing the earth into a flowing river of mud or slurry. Slurry can flow rapidly down slopes or through channels, and can strike with little or no warning at avalanche speeds.

5.9.1. Causes of Landslides

Many factors contribute to slides, including geology, gravity, weather, groundwater, wave action, and human actions. Although landslides usually occur on steep slopes, they also can occur in areas of low relief. Landslides can occur as ground failure of river bluffs, cut and-fill failures that may accompany highway and building excavations, collapse of mine-waste piles, and slope failures associated with quarries and open-pit mines. Underwater landslides usually involve areas of low relief and small slope gradients in lakes and reservoirs or in offshore marine settings. Typically, a landslide occurs when several of these factors converge.

I. Natural Factors

(*i*) **Gravity :** Gravity works more effectively on steeper slopes, but more gradual slopes may also be vulnerable.

(*ii*) **Geological factors :** Many slides occur in a geologic setting that places permeable sands and gravels above impermeable layers of silt and clay, or bedrock. Water seeps downward through the upper materials and accumulates on the top of the underlying units, forming a zone of weakness.

(*iii*) **Heavy and prolonged rainfall :** Water is commonly the primary factor triggering a landslide. Slides often occur following intense rainfall, when storm water runoff saturates soils on steep slopes or when infiltration causes a rapid rise in groundwater levels. Groundwater may rise as a result of heavy rains or a prolonged wet spell. As water tables rise, some slopes become unstable.

(*iv*) **Earthquakes :** Seismic activities have always been a main cause of landslides throughout the world. Any time plate tectonics move the soil that covers moves with it. When earthquakes occur on areas with steep slopes, many times the soil slips causing landslides. Furthermore, ashen debris flows caused by earthquakes can also trigger mass movement of soil.

(*v*) **Forest fire :** Fires cause soil erosion and induce floods and landslides due to the destruction of the natural vegetation. (eg. ridges of Manipur-Nagaland border).

(*vi*) **Volcanoes :** Strato volcanoes are prone to sudden collapse, especially during wet conditions. The conditions commonly prevail after volcanic eruptions that kill vegetation over extensive areas and spread loose volcanic rocks over the landscape. During subsequent rainy seasons, swollen rivers will erode the new deposits and sometimes generate lahars that are dangerous to people downstream. (E.g. lahars at Mount Pinatubo, Philippines in 1990)

(*vii*) **Waves :** Wave action can erode the beach or the toe of a bluff, cutting into the slope, and setting the stage for future slides.

II. Anthropogenic Factors

Human actions most notably those that affect drainage or groundwater, can trigger landslides *e.g.* are Inappropriate drainage system, change in slope/land use pattern, deforestation, agricultural practices on steep slopes, cutting and deep excavations on slopes for buildings, roads, canals and mining ,inappropriate disposal of debris after excavations are examples.

(*i*) **Inappropriate drainage system :** Natural drainage lines on slopes are blocked by terracing/ contour bounding adopted to prevent soil erosion and to enhance percolation during dry season for cultivation, without adequate provision for surface drainage of excess storm water during high intensity rains increase the landslide vulnerability.

(*ii*) **Cutting and deep excavations on slopes for buildings, roads, canals and mining :** Developmental activities like construction of buildings, road cutting, embankments, cut and fill structures causes modification of natural slopes, blocking of surface drainage, loading of critical slopes and withdrawal to toe support promoting vulnerability of critical slopes.

(*iii*) **Change in slope/land use pattern, deforestation, agricultural practices on steep slopes :** Deforestation and cultivation of seasonal crops and increase in settlements. Improper land use practices such as heavy tilling, agricultural practices and settlement patterns have contributed to creep and withdrawal of toe support in many cases.

5.9.2. Types of Landslides

The common types of landslides are described below. These definitions are based mainly on the work of Varnes (Varnes, D.J., 1978) (*i*) **Falls :** Abrupt movements of materials that become detached from steep slopes or cliffs, moving by free-fall, bouncing, and rolling.

(*ii*) **Flows :** General term including many types of mass movement, such as creep, debris flow, debris avalanche, lahar, and mudflow.

(*iii*) **Creep :** Slow, steady downslope movement of soil or rock, often indicated by curved tree trunks, bent fences or retaining walls, tilted poles or fences.

(*iv*) **Debris Flow :** Rapid mass movement in which loose soils, rocks, and organic matter combine with entrained air and water to form slurry that then flows downslope, usually associated with steep gullies.

(*v*) **Debris avalanche :** A variety of very rapid to extremely rapid debris flow.

(*vi*) **Lahar :** Mudflow or debris flow that originates on the slope of a volcano, usually triggered by heavy rainfall eroding volcanic deposits, sudden melting of snow and ice due to heat from volcanic vents, or the breakout of water from glaciers, crater lakes,or lakes dammed by volcanic eruptions

(*vii*) **Mudflow :** Rapidly flowing mass of wet material that contains at least 50 per cent sand-, silt-, and clay-sized particles.

(*viii*) **Lateral spreads :** Often occur on very gentle slopes and result in nearly horizontal movement of earth materials. Lateral spreads usually are caused by liquefaction, where saturated sediments (usually sands and silts) are transformed from a solid into a liquefied state, usually triggered by an earthquake.

(*ix*) **Slides :** Many types of mass movement are included in the general term "landslide." The two major types of landslides are rotational slides and translational landslides.

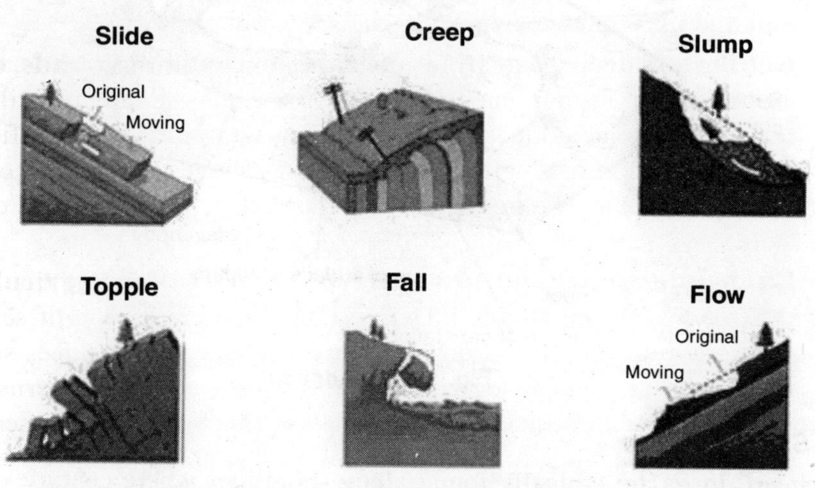

Fig. 5.5. Common Landslide Types

(*a*) ***Rotational Landslide :*** The surface of rupture is curved concavely upward (spoon shaped), and the slide movement is more or less rotational. A slump is an example of a small rotational landslide.

(*b*) ***Translational Landslide :*** The mass of soil and rock moves out or down and outward with little rotational movement or backward tilting.

Translational landslide material may range from loose, unconsolidated soils to extensive slabs of rock and may progress over great distances under certain conditions.

(*c*) ***Topple :*** A block of rock that tilts or rotates forward and falls, bounces, or rolls down the slope.

5.9.3. Where Does Landslides Mostly Occur?

Where They have Occurred Before

Large, deep-seated slides tend to be a reactivation of existing landslide complexes. Slope stability maps can provide an excellent indication of unstable areas. A competent geological analysis can usually provide an estimate of stability of problem areas on a site.

It cannot reliably provide a probability of failure or an exact map of the area to be affected.

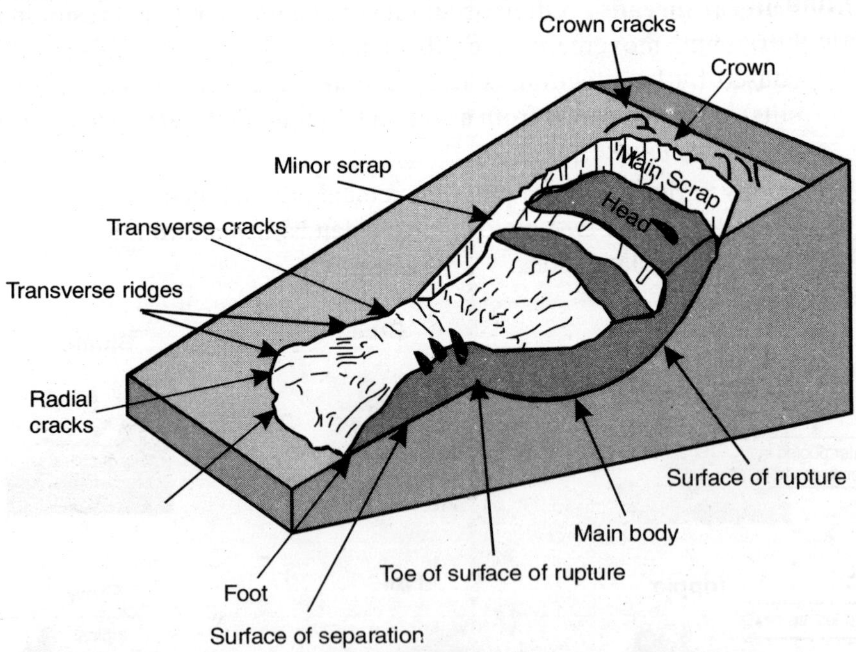

Fig. 5.6. Landslide

On Steep Slopes

Steep slopes are typically found along shorelines where centuries of wave or river currents have eroded the toe of the slope. Most steep slopes experience sliding.

On Benches

Relatively level benches on an otherwise steep slope often indicate areas of past slope movement.

Where Drainage is Causing a Problem

Landslides are often triggered by the failure of drainage systems. Large amounts of water flowing from driveways, roof areas, roads and other impermeable surfaces can cause slides.

Where certain geologic conditions exist

Landslides occur where certain combinations of soils are present. When layers of sand and gravel lie above less permeable silt and clay layers, groundwater can accumulate and zones of weakness can develop.

5.9.4. How do Landslides Affect us?

In hilly terrains of India, Himalayan mountains, western ghats and northeastern part, landslides have been a major natural disasters that strike life and property almost perennially. These landslides, year after year, bring about untold misery to human settlements apart from causing devastating damages to transportation and communication network.

Landslides cause property damage, injury and death and adversely affect a variety of resources. For example, water supplies, fisheries, sewage disposal systems, forests, dams and roadways can be affected for years after a slide event.

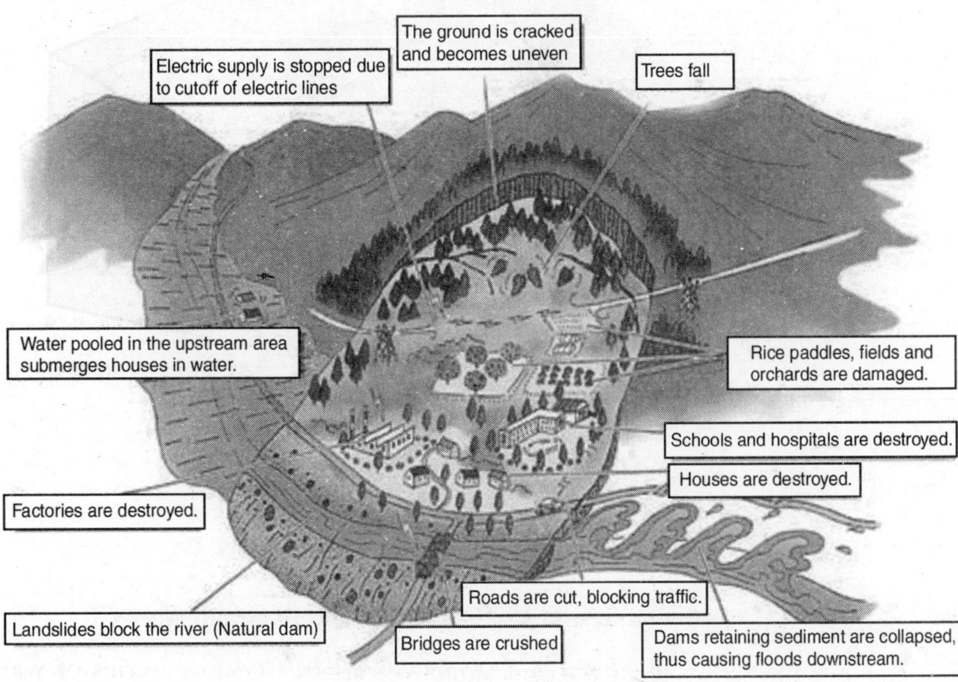

Fig. 5.7. Causes of Landslide

The negative economic effects of landslides include the cost to repair structures, loss of property value, disruption of transportation routes, medical costs in the event of injury, and indirect costs such as lost timber and lost fish stocks.

Water availability, **quantity and quality** can be affected by landslides. Geotechnical studies and engineering projects to assess and stabilize potentially dangerous sites can be costly. Large, infrequent landslides contribute less to personal and property losses than do the smaller, more frequent slides and debris torrents in populated areas.

Every year in India especially in mountain regions like Himalaya landslides damage many houses and cause millions of rupees damage to buildings, roads, railways, pipelines, agricultural land and crops.

5.9.5. Preventive Measures Against Landslides

A landslide is caused by a combination of various factors (topography, geology, geological structure, ground water, etc.). Accordingly, measures to be taken for landslide prevention come in a variety of types. Broadly the landslide preventive measures are classified into two types of works: control works and restraint works. The control works are intended to remove or mitigate factors which may lead to the occurrence of a landslide. On the other hand, the restraint works aim at stabilizing a slope by the construction of structures. Measure to control land slide :

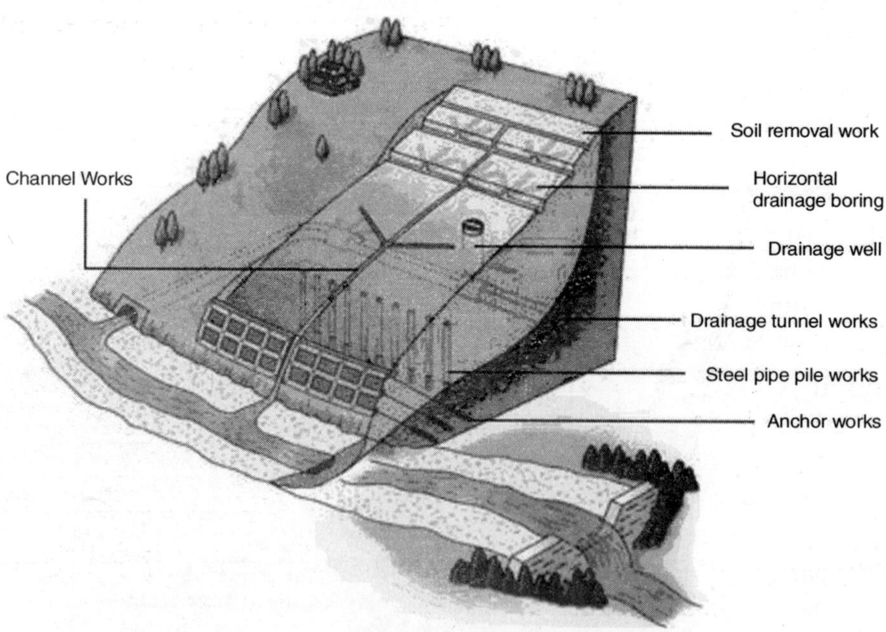

Fig. 5.8. Schematic View of Control Works and Restraint Works

- Plant trees over slope so the roots can stabilize the soil.
- Cover slope s with concrete, while putting pipes in it to let the water from the top out.

- Don't live in landslide hazard area.
- Put steel net that are designed to stop falling rocks. This net is use to distort the impact from the free falling rocks so it wouldn't cause so much damage. The nets are made up of circular rings to provide more strength.
- Don't construct anything in an area where landslide has occurred before.
- Make sure that all water pipes do not leak.
- Build a strong wall in front of the bottom of the slope to prevent any lahars or flow to run rapidly, which also change the direction of flow.
- Make sure to stay indoors when the landslide warning is given.
- Stay away from mountainous area where there is heavy rain.
- Water pipes could also be put into slope so the water can be taken out, preventing it from flowing too high which provides a slimmer chance of a landslide.

During a Landslide

(*i*) Stay alert and awake. Many debris-flow fatalities occur when people are sleeping. Listen to a Weather Radio or portable, battery-powered radio or television for warnings of intense rainfall. Be aware that intense, short bursts of rain may be particularly dangerous, especially after longer periods of heavy rainfall and damp weather.

(*ii*) If you are in areas susceptible to landslides and debris flows, consider leaving if it is safe to do so. Remember that driving during an intense storm can be hazardous. If you remain at home, move to a second story if possible. Staying out of the path of a landslide or debris flow saves lives.

(*iii*) Listen for any unusual sounds that might indicate moving debris, such as trees cracking or boulders knocking together. A trickle of flowing or falling mud or debris may precede larger landslides. Moving debris can flow quickly and sometimes without warning.

(*iv*) If you are near a stream or channel, be alert for any sudden increase or decrease in water flow and for a change from clear to muddy water. Such changes may indicate landslide activity upstream, so be prepared to move quickly. Don't delay! Save yourself, not your belongings.

(*v*) Be especially alert when driving. Embankments along roadsides are particularly susceptible to landslides. Watch the road for collapsed pavement, mud, fallen rocks, and other indications of possible debris flows.

What to Do if You Suspect Imminent Landslide Danger

(*i*) Contact your local fire, police, or public works department. Local officials are the best persons able to assess potential danger.

(*ii*) Inform affected neighbors. Your neighbors may not be aware of potential hazards. Advising them of a potential threat may help save lives. Help neighbors who may need assistance to evacuate.

(*iii*) Evacuate. Getting out of the path of a landslide or debris flow is your best protection.

After the Landslide

(*i*) Stay away from the slide area. There may be danger of additional slides.

(*ii*) Check for injured and trapped persons near the slide, without entering the direct slide area. Direct rescuers to their locations.

(*iii*) Help a neighbor who may require special assistance - infants, elderly people, and people with disabilities. Elderly people and people with disabilities may require additional assistance. People who care for them or who have large families may need additional assistance in emergency situations.

(*iv*) Listen to local radio or television stations for the latest emergency information.

(*v*) Watch for flooding, which may occur after a landslide or debris flow. Floods sometimes follow landslides and debris flows because they may both be started by the same event.

(*vi*) Look for and report broken utility lines to appropriate authorities. Reporting potential hazards will get the utilities turned off as quickly as possible, preventing further hazard and injury.

(*vii*) Check the building foundation, chimney, and surrounding land for damage. Damage to foundations, chimneys, or surrounding land may help you assess the safety of the area.

(*viii*) Replant damaged ground as soon as possible since erosion caused by loss of ground cover can lead to flash flooding.

(*ix*) Seek the advice of a geotechnical expert for evaluating landslide hazards or designing corrective techniques to reduce landslide risk. A professional will be able to advise you of the best ways to prevent or reduce landslide risk, without creating further hazard.

Case Study Landslide Disasters-India

1. Thangi Slide

The Thangi slide is located at km 389.2 on H.T. road (NH-22) on the right bank of Sutluj river opposite the confluence with Tirung Khad. The rocks here are highly jointed, low dipping mica schist of lower Haimanta Group. The schists are interbanded with quartzite layers. Numerous apophyses and veins of granite and pegmatite occur along joints and foliation.

It is a recurring landslide disrupting the traffic and transport on national highway after heavy snowfall and its melting. It is a composite slide-cum-wedge

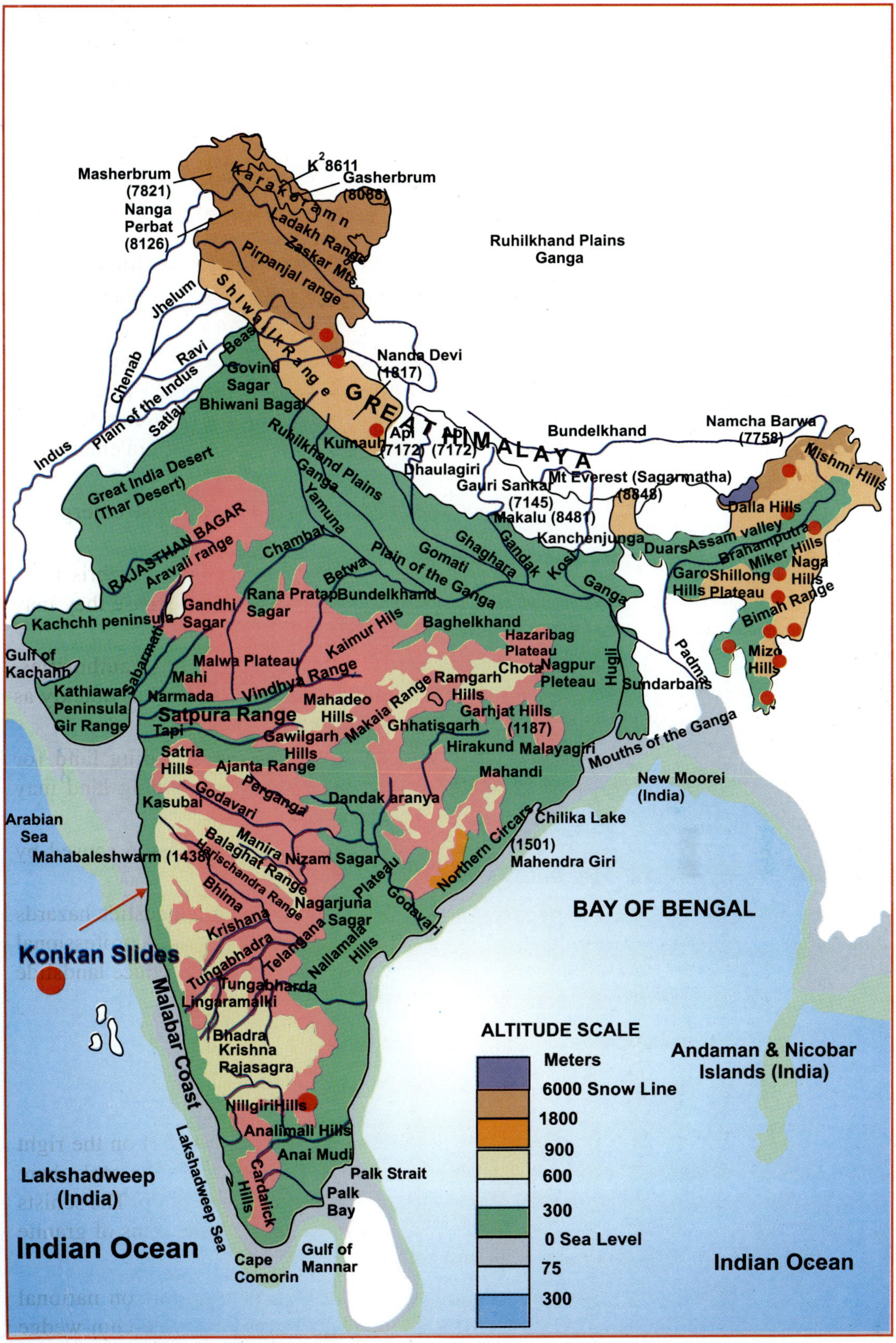
Masherbrum (7821)
Nanga Perbat (8126)
K2 8611
Gasherbrum (8068)
Karakoram
Ladakh Range
Zaskar Mts
Pirpanjal range
Shiwalik Range
Ruhilkhand Plains Ganga
Jhelum
Chenab
Ravi
Beas
Satlaj
Indus
Plain of the Indus
Govind Sagar
Bhiwani Bagal
Nanda Devi (7817)
GREAT HIMALAYA
Kumauh
Api (7172)
Api (7172)
Dhaulagiri
Gauri Sankar (7145)
Makalu (8481)
Mt Everest (Sagarmatha) (8848)
Kanchenjunga
Bundelkhand
Namcha Barwa (7758)
Mishmi Hills
Dalla Hills
Assam valley
Brahamputra
Duars
Miker Hills
Naga Hills
Garo Hills
Shillong Plateau
Bimah Range
Mizo Hills
Great India Desert (Thar Desert)
RAJASTHAN BAGAR
Aravali range
Ruhilkhand Plains
Ganga
Yamuna
Chambal
Betwa
Gomati
Ghaghara
Gandak
Kosi
Plain of the Ganga
Ganga
Gandhi Sagar
Rana Pratap Sagar
Bundelkhand
Kachchh peninsula
Gulf of Kachahh
Kathiawar Peninsula
Gir Range
Sabarmati
Mahi
Malwa Plateau
Kaimur Hills
Baghelkhand
Hazaribag Plateau
Chota Nagpur Pleteau
Hugli
Padma
Sundarbans
Narmada
Vindhya Range
Satpura Range
Tapi
Mahadeo Hills
Makaia Range
Ramgarh Hills
Garhjat Hills (1187)
Ghhatisgarh
Gawilgarh Hills
Hirakund
Malayagiri
Mouths of the Ganga
Satria Hills
Ajanta Range
Perganga
Mahandi
New Moorei (India)
Kasubai
Godavari
Dandak aranya
Chilika Lake
Arabian Sea
Manira
Nizam Sagar
Northern Circars
(1501) Mahendra Giri
Mahabaleshwarm (1438)
Balaghat Range
Harischandra Range
Plateau
Godavari
Bhima
Nagarjuna Sagar
BAY OF BENGAL
Krishana
Telangana
Nallamala Hills
Konkan Slides
Tungabhadra
Tungabharda
Lingaramalki
Malabar Coast
Bhadra
Krishna
Rajasagra
ALTITUDE SCALE
Meters
6000 Snow Line
1800
900
600
300
0 Sea Level
75
300
Andaman & Nicobar Islands (India)
NillgiriHills
Analimali Hills
Anai Mudi
Cardalick Hills
Lakshadweep Sea
Palk Strait
Palk Bay
Lakshadweep (India)
Indian Ocean
Cape Comorin
Gulf of Mannar
Indian Ocean

failured rockfall in the part and debris slide in the lower part.The thickness of the soil and weathered mantle along the road cut is 0.5 m and upto 1 to 3.5 respectively. The chemical weathering evidenced by white and yellow encrustation has made the rocks very weak and friable. The lower and middle slopes are covered with old glacial debris. The upslope area is covered with open forest and middle slopes are occupied by fields and orchards. The lower slopes are barren rock cliffs.

The barren and steep lower slope had become unstable due to toe erosion by Satluj river and also due to cutting in the road widening. High density of joints and fractures and the contact between unconsolidated and morainic debris and parent rocks have created channels for water seepage which has moistened the slopes and in turn reduced shear strength of the slope material. In the initial stages, a rock fall had occurred in the upper part of this slide along two wedge fractures dipping 80° towards SE 72° towards SW direction. Falling of boulders from rockfall on the debris cone coupled with undercutting of already moistened slope by the Satluj river had reactivated the debris cone lying below the rockfall on the steep slope to slide along slip plane dipping 56° towards downslope.

The high discharge of Tirung Khad has forced the Satluj river towards west to undercut the slope.

Khadra Dhang

Khadra Dhang slide zone (75 × 1300 × 10 m), was activated in the early sixties and is still active. It is located on the old Hindustan - Tibet road along the right bank of river Satluj, opposite to Ribba. The valley slopes here are mainly covered with semi-consolidated old glacial debris resting on weathered granitic gneisses. The slide is a translational debris slide which occurred in lower and middle slopes. The upper slopes are covered with open forest and middle slopes by agricultural land. Though slope material is generally dry but seepage zones and springs present in the lower stretches glacial debris, this build up the pore water pressure and thus reducing the shear resistance of the material. The sparsely vegetated semi-consolidated old glacial debris had become dangerously unstable following the initiation of road cuttings. The seepage had converted the glacial debris into slurry and excavation of the semi-consolidated unstable slope for road construction have disturbed the critical shear stress equilibrium.

This coupled with over steepening of the already steep slope (48° dipping SSW) due to toe cutting by river Satluj have contributed to translational debris slide along a slip plane dipping 70° towards SSW. Though the H.T. road has now been diverted to the left bank of the river, but toe cutting by river Satluj during peak discharge continuously causes active sliding.

Pangi Slide

Pangi slide (measurring 65 × 300 × 6 m) is located on Pangi Kalpa link road its junction with NH-22 (371.25 km stone). It occurred in biotite gneiss and garnetiferous mica schist of Vaikrita Group covered with old scree and boulders of gneisses. The slide is located in embankment part of the lower valley slopes, dipping 48° NW, along Pangi Nala.

The slide is a rotational debris slide in the upper part and translational planner slide in the lower part along the alip plane dipping 70° towards SW which is parallel to the dipping 60°, 70° and 55° towards NNE, SE, SSW respectively, have converted the rock mass into blocks. Freezing and thawing action in the joints has further reduced the cohesive strength of the rocks. A number of cracks have also developed along the flank of the slide during road construction. The presence of seasonal seepages within the sear indicate building up of pore-water pressure and toe cutting by the Pangi Nala have rendered the slopes unstable. This together with uncontrolled blasting for the construction of NH- 22 – Kalpa link road has also contributed to the generation of this slide.

Powari Landslide

The Powari landslide zone (measuring 1000 × 500 × 25 m), is one of the largest landslides located on the lower slope along the right bank of Satluj river between Powari and Peo in a terrain composed by highly jointed weathered and fractured, mica gneisses of Vaikrita formation.

The 1962 survey of India map shows the existence of a small landslide. The major slide occured about 15 years ago after the construction of approach road from Powari to Kalpa.

This road climbs through six hair pin bends and three stretches have been affected badly by the slides.

The landslide is acomposite slide which represts a combination of debris fall along upper part, rotational slump in middle slope and debris slide at the toe. At least four landslides scars are visible. The first scar formed due to debris fall, is a small one present below fifth bend of road and the second one is a wedge rockfall along joints dipping 70° towards SE. The third scar is formed due to planner debris slide along foliation plane and the fourth scar is formed due to debris slide and slumps. A number of slumps scars and seepages are present on each bend. The dip of the failure plane and slip surface varies between 61° and 70° and dipping towards SSE and SE direction. Foliation planes (dipping 34° and NE) of gneiss and 3 set of prominent joints and fractures dipping in the slope direction.The saturated overburden consisting of old glacial debris,rich in mica ,are responsible for the slope failure. An irrigation canal passing along the upper part of the slope is contributing water to anumber of seepage zones located within the slumped mass.

This together with snow melt has resulted in the build up of pore water pressure over a long period of time and gradual decrease in shear resistance and have also pulverized overlaying micaceous rich old glacial debris and slope material. This together with road, excavation of slopes for the hair pin band (five loops in slope length of 500m) road construction from Powari to Peo and continuous toe erosion by Satluj river has triggered the slide. The presence of

NE-SW oriented transverse cracks (4-5 m scar length) on the flank of fourth scar, convex slope in toe portion, tilted trees, continuous debris slide sinking of road due to active slumping are indicative of continued downward movements or creep within the slumped mass. The U shape road bends in a stretch of 500m slope length. Continuous seepage of water at the H.T. Road and along road to Peo at different levels indicates that the water from irrigation canals needs to be diverted from controlling the slide. The road has subsided by 3-4 meters along lower three bends.

Barua Slide

The Barua landslide (measuring 60 × 100 × 15 m) is located on the left middle slope in the Bapsa valley about 5km southeast of Karcham. Though, the slide was triggered in 1987-88 it has been repeatedly activated. The thickness of the soil - mantle along the road cut is of the order of 1.5-2 m. The location of the landslide is close to the Karcham and Vaikrita thrusts. The sparsely vegetated steep lower valley slope dipping 73° towards west 10° south, zig-zag road and moderate upslope (25° towards WSW) covered with old glacial material has become unstable following the widening the road. Seasonal snow melt from upper slope in old glacial material has build up pore water pressure and thus reduced the shear strength. The removal of toe to widen the road and already moistened glacial material and seasonal seepage from upslope has contributed to the occurance of this landslide along the sliding plane dipping 58° due WSW. The road to village Shong from Karcham has subsided by 5-7m and is not in use any more. A number of houses as well as orchards and fields have also been damaged.

Urni Rockfall

The Urni rockfall (500 × 250 × 5 m) is located near village Tapri along the lower slope, on the right bank of river Satluj on Hindustan-Tibet road (NH-22) (347.00 km) in highly jointed Wangtu gneisses. While the upperb slopes are covered with old glacial debris and open forest, the lower slope are barren. The thickness of soil and weathered mantle along the road cut is 0.5 to 0.8m. Though no seepage or springs are visible in the slide zone, seasonal seepage from snow melt does take through the joints.

Over steepening of slopes due to toe cutting by the Satluj river has been further intensified by rockfall on the opposite bank, forcing the river towards present slide. This over together with opening of joints due to uncontrolled blasting for the widening of NH-22 have reduced the shear strength of the rock mass and led to rockfall along wedge joints.

The fallen rock block range from 7-8 m across. Five person died due to reactivation of this slide in 1992.

Nachar Slide

The Nachar landslide (350 × 200 × 35 m) in the lower middle slope on left bank of Satluj river is located on NH-22 road in highly jointed and sheared granite-gneiss of central Crystallines(Jutogh Groups). The gneisses have interbanded amphibolities and are cut across by a network of the pegmatic veins. Numerous shear zones(1-2m thick), parallel to the foliation planes are traversing the host rock. The thickness of the soil and weathered mantle along the road cut is of the order 50-70 cm. The upper slope (40° due North) is covered with old rockfall and glacial deposits resting on dip slopes and joints dipping 54° towards NNE. Village Nachar and Paunda are situated on these slopes. The upper is occupied by fields and orchard. The lower slopes are moderately dipping at 43° towards north. There is one seepage zone present in the slide zone and one seasonal stream on the other flank of the slide. Five sets of closely spaced fractures and joints developed due to tectonic activity and uncontrolled blasting have broken the rock into blocks and act as easy passages for watre seepage. The seepage through these joints and fracture and upper slope debris over a long period of time has reduced the shear strength of the rock.

5.10. VOLCANIC ERUPTION

Volcano is an opening in the earth's crust through which molten lava, ash, and gases are ejected or a mountain formed by the materials ejected from a volcano. Volcano is vents or fissures in the earth's crust through which gases,

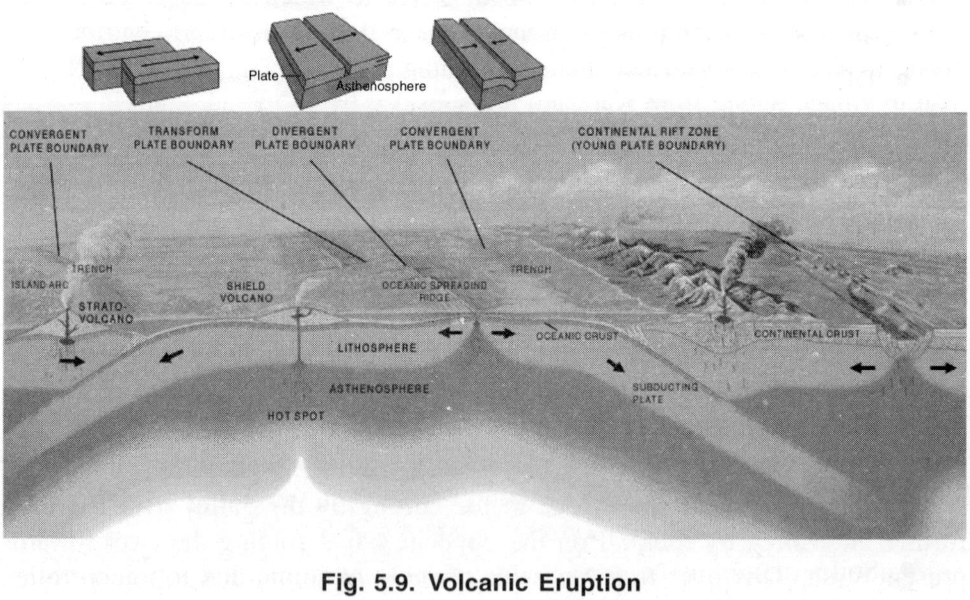

Fig. 5.9. Volcanic Eruption

molten rock, or lava and solid fragments are discharged. Volcanoes are described as active, dormant, or extinct. The soil resulting from decomposition of volcanic materials is extremely fertile, and the ash itself is a good polishing and cleansing agent. It is known that an eruption usually follows movement of magma (a mixture of finely divided solids with enough liquid to produce a pasty mass) upwards into the solid layer (the earth's crust) beneath a volcano and occupying a magma chamber (a magma chamber is the chamber beneath a volcano where the magma is held before an eruption.). Eventually, magma in the chamber is forced upwards and flows out across the planet surface as lava, or the rising magma can heat water in the surrounding landform and cause explosive discharges of steam; either this or escaping gases from the magma can produce forceful ejections of rocks, cinders, volcanic glass, and/or volcanic ash also known as tephra. While always displaying powerful forces, eruptions can vary from effusive to extremely explosive.

Increases in temperature and pressure with depth cause formation of magma (melt). This melt is less dense than the surrounding mantle and will consequently raise though the mantle to the overlying crust. As the magma (melt) rises through the crust it may melt and assimilate some of the surrounding crust, it may cool and begin to grow crystals, and it may exsolve gas. If the magma reaches the surface, it will generate a volcanic eruption.

Summary

- Water pollution is defined as presence of any foreign substance (organic, inorganic, biological or radiological) in water which tends to degrade the quality and impair the usefulness of water.
- The presence of water pollutants alters the amount of dissolved oxygen and biochemical oxygen demand.
- Dissolved oxygen (DO) is the amount of oxygen present in the aquatic system that is available for aquatic life.
- The biochemical oxygen demand (BOD) is a measure of the oxygen required by aerobic micro-organisms to bio-chemically oxidize the organic matter present in the waste.
- Another quick measure of organic matter present in waste is chemical oxygen demand (COD) and in this case chemical oxidants are used instant of aerobic micro-organisms for the oxidation of organic matter.
- Sewage and Industrial waste are the major sources of water pollution.
- The water pollution can be controlled by treating polluted water by physical, chemical or biological means and removing the contaminants up to acceptable quality standards, so that water can be reused or discharged into water bodies.
- Soil pollution is the introduction of substances, biological organisms or energy into the soil, resulting in a change of the soil quality, which is likely to affect the normal use of the soil or endangering public health and the living organism.
- Marine pollution can be defined as the direct or indirect introduction of substances or energy into the marine environment (including estuarine), resulting in harm to living resources, hazards to human health, hindrance to marine activities including fishing, impairment of the quality of sea water.

- Thermal pollution can be defined as presence of waste heat in the water which causes undesirable changes in the natural environment.

Descriptive Type Questions

1. What do you mean by self-purification of rivers? Explain the various natural mechanisms involved in this process.
2. What do you understand by water pollution? Suggest various remedial and control measures to minimize water pollution.
3. What do you mean by Water Pollution? Discuss its major sources of water treatment with layout of a water treatment plant.
4. Define water pollution? Discuss its major sources and effects.
5. What is BOD? How is domestic discharge responsible for increasing the BOD level into the stream?
6. What is thermal pollution? Discuss its causes and effects. State the approaches to control it?
7. Define 'Marina Pollution' and 'Thermal Pollution' and discuss the various sources contributing it. How can we prevent pollution of our oceans?

Short Answer Type Questions

1. Differentiate the following:
   (*a*) Wet cooling tower and dry cooling tower
   (*b*) BOD and COD
2. Explain the following terms:
   (*i*) Eutrophication (*ii*) Sanilization (*iii*) Biomagnification

Objective Type Questions

Multiple Choice Questions

1. BOD is :
   (*a*) A measure of the organic matter present in water
   (*b*) Usually less than COD
   (*c*) Biochemical oxygen demand
   (*d*) All of the above
2. Biochemical Oxygen Demand measures :
   (*a*) Industrial pollution
   (*b*) Air pollution
   (*c*) Pollution capacity of effects
   (*d*) Dissolved O_2 needed by microbes to decompose organic waste.
3. Which of the following is a non-point source of water pollution?
   (*a*) Factories (*b*) Sewage treatment plants
   (*c*) Urban and suburban lands (*d*) All of the above

4. Septic tank is :
   (*a*) An aerobic attached growth treatment system
   (*b*) An aerobic suspended growth biological treatment system
   (*c*) An anaerobic attached growth biological treatment system
   (*d*) An anaerobic suspended growth treatment system
5. Which of the following is not a marine pollutant?
   (*a*) Oil (*b*) Plastics
   (*c*) Dissolved oxygen (*d*) All of the above
6. Which of the following is a major source of thermal pollution in water bodies?
   (*a*) Sewage treatment plant (*b*) Solid waste disposal sites
   (*c*) Thermal power plant (*d*) All of the above
7. Disease caused by eating fish inhabiting mercury contaminated water is :
   (*a*) Bright's disease (*b*) Hiroshima episode
   (*c*) Mina-mata disease (*d*) Osteosclerosis
8. The surface litter layer consists of freshly-fallen and partially decomposed leaves, twigs, animal waste, fungi and other organic materials are known as :
   (*a*) topsoil (*b*) Bottom soil
   (*c*) humas (*d*) Mist
9. It causes algal booms in coastal water :
   (*a*) Air pollution (*b*) Noise pollution
   (*c*) Soil pollution (*d*) Marine pollution
10. A treatment process that uses micro organisms to break down or degrade hazardous substances into less toxic or non toxic substances :
    (*a*) Bio-magnification (*b*) Bioremediation
    (*c*) Eutrophication (*d*) Photosynthesis
11. Biodegradable organic waste should be used for generation of bio-gas :
    (*a*) Bio-gas (*b*) Natural gas
    (*c*) Coal (*d*) Oil

ANSWERS

Multiple Choice Questions

1. (*d*)	**2.** (*d*)	**3.** (*c*)	**4.** (*d*)	**5.** (*c*)
6. (*c*)	**7.** (*c*)	**8.** (*a*)	**9.** (*d*)	**10.** (*b*)
11. (*a*)				

❑❑❑

CHAPTER

6

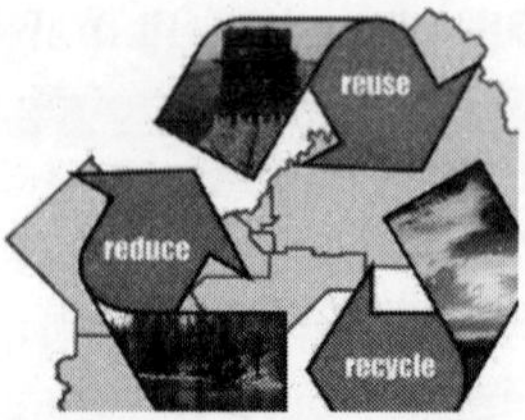

Society and Ethics

INSIDE THIS CHAPTER

6.1. ENVIRONMENTALLY OPTIMUM SUSTAINABLE POPULATION

An 'optimum' population, in dictionary terms, is the 'best or most favourable' population. an optimum population means, at its simplest, a population size which is environmentally sustainable in the long term, affords people a good quality of life, has adequate renewable and non-renewable resources necessary for its long-term survival and consumes or recycles them to ensure it will not compromise the long-term survival of its progeny.

The **Optimum population** for a region or country is a level of (human) population which is ecologically sustainable. It is less than the carrying capacity (the amount before a country would theoretically sink under the weight of its population), so the amount of resources available can support the country's population needs indefinitely, which maximizes the overall welfare (happiness, standard of living) of the population. There are different views on the appropriate measure of welfare, and this is an area of political disagreement.

- If the population is **below** its optimum, it can increase with a net social benefit without damaging the natural resources needed for the population.
- If it is **above** the optimum, the negative effects of overcrowding (pollution, crime, social dysfunction, etc) and environmental damage

reduces welfare or there are too few resources to maintain the population at its current level.

To reach optimum population, a country must be able to measure human welfare and environmental impacts, as well as having policies in place to influence the fertility rate and net migration. This implies agreement on welfare, environmental issues, accurate projections of immigration, age distribution and changes in lifespan among other factors. Reaching optimum population is an important objective in the wider project of creating a sustainable society.

The main difference is that here we take *sustainable* to mean the maximum sustainable values of populations in simple calculational terms and which are often used in such discussions. This is dealt with in the first part of this briefing.

Optimum, on the other hand, goes on develop a more realistic set of numbers to explore and include margins for unknowable but assessable future events.

Given OPT's definition of a sustainable population, it follows that a government or other body elected to protect, administer and ensure the long term good of such a population has the duty to take all steps necessary to bring about its sustainability.

Sustainability has become an increasingly important issue over the last two or three decades. It has progressed from an esoteric subject within a few scientific and sociological communities to mainstream national and international political agendas and debates. Such interest has been triggered by:

(*i*) The continuing rapid growth of populations during the 20th century and beyond;

(*ii*) The increasing rate of pollution of the land and waters of the earth through excessive and ever-rapid exploitation of the world's biological and geological assets;

(*iii*) The now-generally-accepted view that global warming—and thus climate change—is a direct result of human activity, and threatens the future of the human race and other species;

(*iv*) The growing realisation that collective human consumption has:

- (*a*) exceeded the renewable resources available to it and that the human race is, as a result, in [F3] danger of a catastrophic collapse;
- (*b*) by its sheer magnitude caused irreversible damage to many ecosystems and other species.

Although not generally accepted, OPT considers that it is indisputable that (*ii*), (*iii*), (*iv*) are all a direct consequence of (*i*).

The *sustainable population* is, in mathematical terms, a dependent variable—which is why it is represented on the vertical axis of the graph. The independent variables on which it depends are *biocapacity* and *footprint* (rate of consumption). Increase the footprint—and/or decrease biocapacity—and the maximum sustainable population decreases—and vice-versa.

It is clearly neither moral nor desirable to instantly reduce today's world population of 6.7 billion to a sustainable level of 5.1 billion by, for example, dispatching 24% of each country's population. It is, therefore, not helpful to discuss sustainable populations in *today's terms*. Assumptions need be made about conditions expected to exist when the sustainable level is deemed to be achieved. Thus, Earth's sustainable population in (say) 2050 will depend on how much world population will be consuming each year *at that time*. This is not an *a priori* definable quantity, since the global biocapacity, collective consumption, expectations and behaviour of individual countries and social groups in response to changes in the environment will all vary over time. Thus managing population decline is like trying to hit a moving target.

The future maximum *sustainable population* of the planet is only definable subject to qualifying statements regarding the projected global biocapacity and footprint of its human occupants.

6.1.1. Achieving Optimum Population

Although no country has yet formally adopted an optimum population target, many recognize the problems caused by population growth and overpopulation. Most of the effort has been focussed on reducing population growth. There are several policy areas that influence this:

Social and Cultural

Changing people's views and attitudes on religion to adjust it into a modern fashion, changing social attitudes, such as giving women more rights and thought in starting a large family than following tradition.

Economic

Increasing career opportunities will have peoples' minds set on education, career prospects, and maintaining their job, such that the immediate impulse to start a family might be delayed.

Medical and Scientific

Increasing the amount of contraception in LEDC (Less Economically Developed Country) educating adults and children about sexual education, on how to use contraception and the risks involved.

Political

Improving education to direct people into a career, this will have people concentrate on getting a stable job rather than planning ahead on starting a family.

Increasing emigration, and/or reducing immigration, within certain areas (especially those in which the residents have an atypically-high impact on the environment), is viewed by some as a key factor in stabilizing population and preserving natural resources.

6.1.2. Current Population of India

India, with 1,210,000,000 (1.21 billion) people is the second most populous country in the world, while China is on the top with over 1,350,044,605 (1.35

billion) people. The figures show that India represents almost 17.31% of the world's population, which means one out of six people on this planet live in India. Although, the crown of the world's most populous country is on China's head for decades, India is all set to take the numero uno position by 2030. With the population growth rate at 1.58%, India is predicted to have more than 1.53 billion people by the end of 2030.

Current Population of India in 2011	**1,210,000,000 (1.21 billion)**
Total Male Population in India	**623,700,000 (623.7 million)**
Total Female Population in India	**586,500,000 (586.5 million)**
Sex Ratio	**914 females per 1,000 males**
Age structure	
0 to 25 years	50% of India's current population
Currently, there are about 51 births in India in a minute.	
India's Population in 2001	1.02 billion
Population of India in 1947	350 million

More than 50% of **India's current population** is below the age of 25 and over 65% below the age of 35. About 72.2% of the population lives in some 638,000 villages and the rest 27.8% in about 5,480 towns and urban agglomerations. The birth rate (child births per 1,000 people per year) is 22.22 births/1,000 population (2009 est.) while death rate (deaths per 1000 individuals per year) is 6.4 deaths/1,000 population. Fertility rate is 2.72 children born/woman (NFHS-3, 2008) and Infant mortality rate is 30.15 deaths/1,000 live births (2009 estimated). India has the largest illiterate population in the world. The literacy rate of India as per 2001 Population Census is 65.38%, with male literacy rate at 75.96% and female at 54.28%. Kerala has the highest literacy rate at 90.86%, Mizoram (88.80%) is on the second position and Lakshadweep (86.66%) is on third.

Every year, India adds more people than any other nation in the world, and in fact the individual population of some of its states is equal to the total population of many countries. For example, Population of Uttar Pradesh (state in India) almost equals to the population of Brazil. It, as per 2001 Population Census of India, has 190 million people and the growth rate is 16.16%. The population of the second most populous state Maharashtra, which has a growth rate of 9.42%, is equal to that of Mexico's population. Bihar, with 8.07%, is the third most populous state in India and its population is more than Germany's. West Bengal with 7.79% growth rate, Andhra Pradesh (7.41%) and Tamil Nadu (6.07%) are at fourth, fifth and sixth positions respectively. The sex ratio of India stands at 933. Kerala with 1058 females per 1000 males is the state with the highest female sex ratio. Pondicherry (1001) is second, while Chhatisgarh (990) and Tamil Nadu (986) are at third and fourth places respectively. Haryana with 861 has the lowest female sex ratio.

Some of the reasons for India's rapidly growing population are poverty, illiteracy, high fertility rate, rapid decline in death rates or mortality rates and immigration from Bangladesh and Nepal. Alarmed by its swelling population, India started taking measures to stem the growth rate quite early. In fact India by launching the National Family Planning programme in 1952 became the first country in the world to have a population policy. The family planning programme yielded some noticeable results, bringing down significantly the country's fertility rate. In 1965-2009, the contraceptive usage more than tripled and the fertility rate more than halved. The efforts did produce positive results, however, failed to achieve the ultimate goal and the population of India since getting independence from Britain in 1947 increased almost three times. Whereas India has missed almost all its targets to bring the rate of population growth under control, China's 'One Child Policy' in 1978, has brought tremendous results for the latter. The policy claims to have prevented between 250 and 300 million births from 1978 to 2000 and 400 million births from 1979 to 2010.

6.2. POPULATION GROWTH

Population growth is the change in a population over time, and can be quantified as the change in the number of individuals of any species in a population using "per unit time" for measurement.

In demography, *population growth* is used informally for the more specific term population **growth rate**, and is often used to refer specifically to the growth of the human population of the world.

Simple models of population growth include the Malthusian Growth Model and the logistic model.

6.2.1. Population Growth Rate

In demographics and ecology, **Population growth rate** (PGR) is the fractional rate at which the number of individuals in a population increases. Specifically, PGR ordinarily refers to the change in population over a unit time period, often expressed as a percentage of the number of individuals in the population at the beginning of that period. This can be written as the formula:

$$\text{Growth rate} = \frac{(\text{Population at end of period} - \text{Population at beginning of period})}{\text{Population at beginning of period}}$$

(In the limit of a sufficiently small time period.)

The above formula can be expanded to: growth rate = crude birth rate - crude death rate + net immigration rate, or $\Delta P/P = (B/P) - (D/P) + (I/P) - (E/P)$, where P is the total population, B is the number of births, D is the number of deaths, I is the number of immigrants, and E is the number of emigrants.

This formula allows for the identification of the source of population growth, whether due to natural increase or an increase in the net immigration rate. Natural increase is an increase in the native-born population, stemming from either a higher birth rate, a lower death rate, or a combination of the two. Net immigration rate is the difference between the number of immigrants and the number of emigrants.

The most common way to express population growth is as a ratio, not as a rate. The change in population over a unit time period is expressed as a percentage of the population at the beginning of the time period. That is:

$$\text{Growth ratio} = \text{Growth rate} \times 100\%$$

A positive growth ratio (or rate) indicates that the population is increasing, while a negative growth ratio indicates the population is decreasing. A growth ratio of zero indicates that there were the same number of people at the two times—net difference between births, deaths and migration is zero. However, a growth rate may be zero even when there are significant changes in the birth rates, death rates, immigration rates, and age distribution between the two times. Equivalently, percent death rate = the average number of deaths in a year for every 100 people in the total population.

6.2.2. Excessive Growth and Decline

Population exceeding the carrying capacity of an area or environment is called overpopulation. It may be caused by growth in population or by reduction in capacity. Spikes in human population can cause problems such as pollution and traffic congestion, these might be resolved or worsened by technological and economic changes. Conversely, such areas may be considered "underpopulated" if the population is not large enough to maintain an economic system. Between these two extremes sits the notion of the optimum population.

6.2.3. The Rate of Natural Increase (r)

Birth rate (b) – death rate (d) = rate of natural increase (r).

- Birth rate expressed as number of births per 1000 per year (currently 14 in the U.S.);
- Death rate expressed as the number of deaths per 1000 per year (currently 8 in the U.S.);
- So the rate of natural increase is 6 per thousand (0.006 or 0.6%).

Although the value of r is affected by both birth rate and death rate, the recent history of the human population has been affected more by declines in death rates than by increases in birth rates. The introduction of public health measures, such as:

- Better nutrition
- Greater access to medical care
- Improved sanitation
- More widespread immunization

has produced a rapid decline in death rates, but until recently there was no corresponding decline in birth rates. In 2007, r is 1.7%. (Data from the Population Reference Bureau.)

Although death rates declined in all age groups, the reduction among infants and children had—and will continue to have—the greatest impact on population growth. This is because they will soon be having children of their own.

This situation, resulting in a rapid rate of population growth, is characteristic of many of the poorer regions of the world.

6.2.4. The Demographic Transition

Slowly declining birth rates following an earlier sharp decline in death rates are today characteristic of most of the less-developed regions of the world. The shift from high birth and death rates to low birth as well as death rates is called the demographic transition.

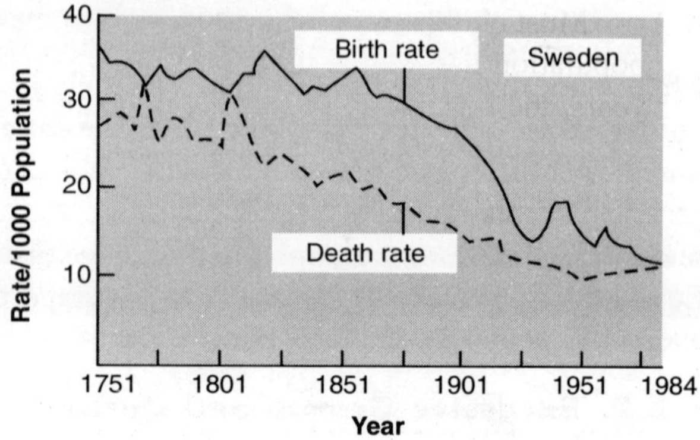

Fig. 6.1. Death Rate and Birth Rate

The birth rate in Sweden is (2007) 12/1000; the death rate 10/1000, giving a rate of natural increase (***r***) of 0.2%.

6.2.5. Doubling Times

When a population has doubled, $N = N_0 \times 2$.

Putting this in our exponential growth equation, $2N_0 = N_0 e^{rt}$ $e^{rt} = 2$

$$rt = \ln \text{ (natural logarithm) of } 2 = 0.69$$

doubling time, $t = 0.69 / r$

So Sri Lanka with an r of 1.3% (0.013) has a doubling time $t = 0.69/0.013 = 53$.

6.2.6. Total Fertility Rate (TFR)

The total fertility rate is the average number of children that each woman will have during her lifetime. The TFR is an average because, of course, some women will have more, some fewer, and some no children at all.

Theoretically, when the TFR = 2, each pair of parents just replaces itself.

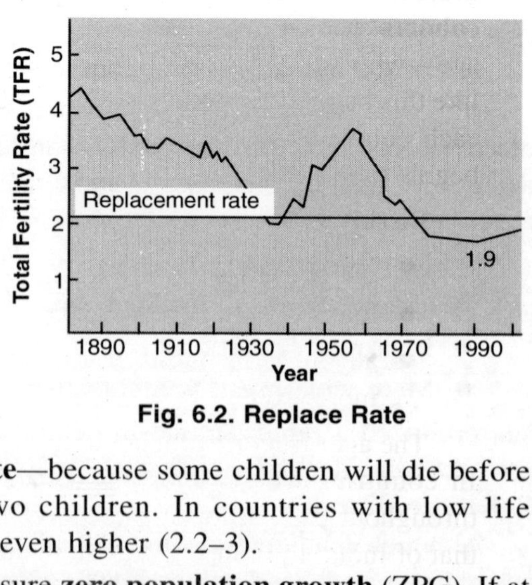

Fig. 6.2. Replace Rate

Actually it takes a TFR of 2.1 or 2.2 to replace each generation—this number is called the **replacement rate**—because some children will die before they grow up to have their own two children. In countries with low life expectancies, the replacement rate is even higher (2.2–3).

But even a TFR of 2.1 may not ensure **zero population growth** (ZPG). If at one period a population has an unusually large number of children, they will—

as they pass through their childbearing years—increase the *r* of the population even if their TFR goes no higher than 2.

6.2.7. Age Structure of Populations

Most childbearing is done by women between the ages of 15 and 49. So if a population has a large number of young people just entering their reproductive years, the rate of growth of that population is sure to rise.

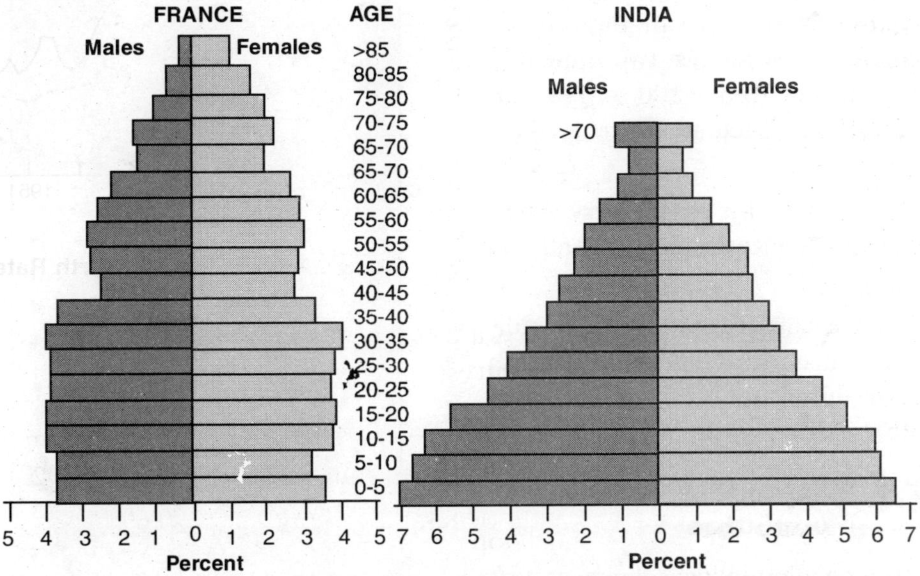

Fig. 6.3. Age Structure of France and India

These pyramids compare the age structure of the populations of France and India in 1984. The relative number (%) of males and females is shown in 5-year **cohorts**. Almost 20% of India's population were children—15 years or less in age—who had yet to begin reproduction. When the members of a large cohort like this begin reproducing, they add greatly to birth rates. In France, in contrast, each cohort is about the size of the next until close to the top when old age begins to take its toll.

Broad-based pyramids like India's are characteristic of populations

- with high birth rates;
- low life expectancies (where many people die before reaching old age);
- advances in public health have recently reduced infant and childhood mortality.

The age structure of a population also reflects the recent pattern of mortality. In countries where injuries, starvation, and disease, etc. take a heavy toll throughout life, a plot of the age cohorts produces a broad based pyramid like that of India. In France (and other countries in western Europe) almost everyone survives until old age, and a plot of the age cohorts is scarcely a pyramid at all.

So even if the TFRs were the same in both countries (they are not — in India it is 2.9, in France, 2.0), India is in for more years of rapid population growth, France is not.

6.2.8. The Population of the World

The solid line in the graph shows estimates of the size of the world's population over the last two millennia. The estimates from 1800 to 1991 are based on more accurate data than those before.

The dotted line shows what would happen if exponential growth continued to the year 2100.

As we can see, the world's population has been growing exponentially (except during the years of the black death).

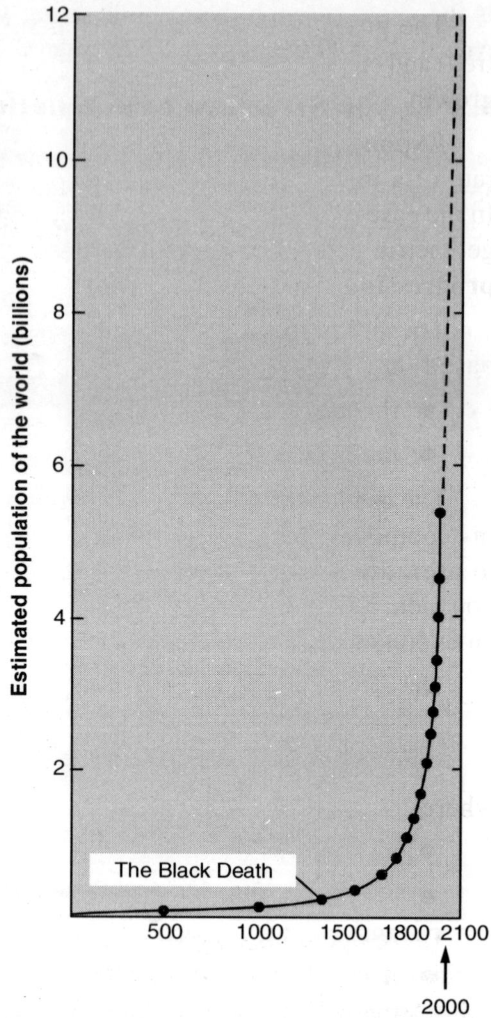

Fig. 6.4. Population of the World

6.2.9. Exponential Growth in Population

If a population has a constant birth rate through time and is never limited by food or disease, it has what is known as exponential growth. With exponential growth the birth rate alone controls how fast (or slow) the population grows.

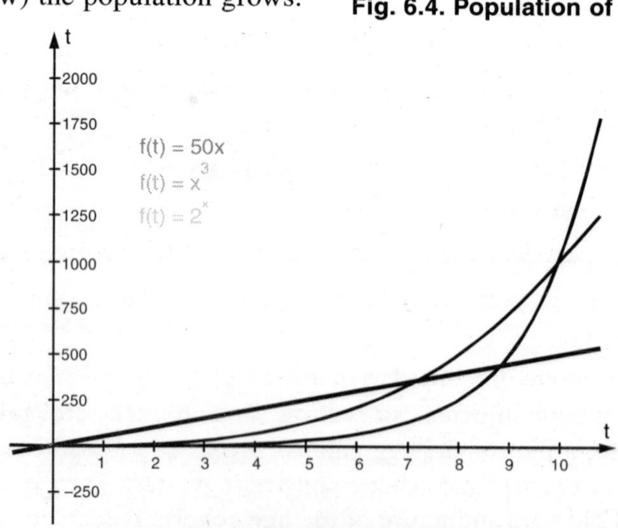

Fig. 6.5. Exponential Growth

The graph illustrates how exponential growth (green) surpasses both linear (red) and cubic (blue) growth. ■ Exponential growth ■ Linear growth ■ Cubic growth.

Exponential growth (including exponential decay) occurs when the growth rate of a mathematical function is proportional to the function's current value. In the case of a discrete domain of definition with equal intervals it is also called geometric growth or **geometric decay** (the function values form a **geometric progression**).

The prediction that Sri Lanka will double its population in 53 years is based on :

- the assumption that r will remain unchanged (which is surely false)
- the mathematics of exponential growth.

The product of growth grows itself. So the growth of populations is a problem in "compound interest". At the end of each year (or whatever period you choose to use), the base against which the rate is applied has grown. Whatever figures you pick, as long as r is positive, a plot of population as time elapses will produce an exponential growth curve like this one.

The rate of population growth at any instant is given by the equation

$$\frac{dN}{dt} = rt$$

where

- r is the rate of natural increase in
- t — some stated interval of time, and
- N is the number of individuals in the population at a given instant.

The algebraic solution of this differential equation is $N = N_0 e^{rt}$ where

- N_0 is the starting population
- N is the population after
- a certain time, t, has elapsed, and
- e is the constant 2.71828... (the base of natural logarithms).

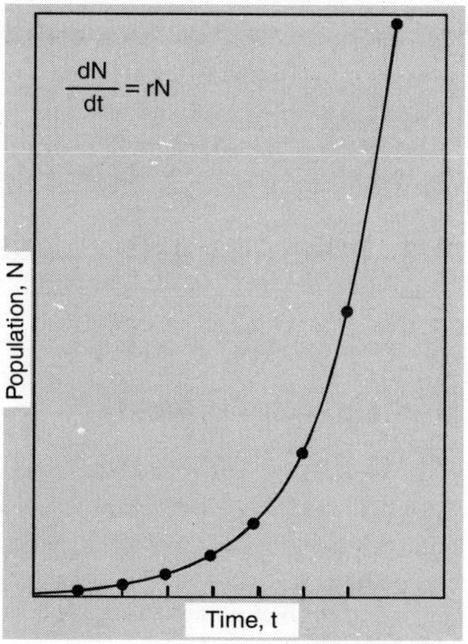

Fig. 6.6. Exponential Growth

Plotting the results gives this exponential growth curve, so-called because it reflects the growth of a number raised to an exponent (rt).

Equilibrium is difficult to achieve because exponential growth produces demographics such that they are substantially more people entering their reproductive years than those which are dying of old age.

6.2.10. Population Density

Population density is a measurement of population per unit area or unit volume. It is frequently applied to living organisms, and particularly to humans. It is a key geographic term.

Population density is population divided by total land area.

Low densities may cause an extinction vortex and lead to further reduced fertility. This is called the Allee effect after the scientist who identified it. Examples of the causes in low population densities include:

- Increased problems with locating mates
- Increased inbreeding

Different species have different expected densities. R-selected species commonly have high population densities, while K-selected species may have lower densities. Low densities may be associated with specialized mate location adaptations such as specialized pollinators; as found in the orchid family (*Orchidaceae*).

6.3. SOLID WASTE

All solid and semi-solid waste arising from human and animal activities, except human-excrete and sullage (liquid wastes from bathroom and kitchen etc.) are discarded as useless or unwanted are included in the term solid wastes. The term solid wastes encompasses the highly heterogeneous mass of discarded materials or throwaways from urban community as well as the more homogeneous accumulation of waste generated by agricultural, industrial and mining activities. Solid waste (waste other than liquid or gaseous) can be classified as municipal, industrial, agricultural, medical, mining waste and sewage sludge.

6.4. SOURCES AND CLASSIFICATION OF SOLID WASTES

Solid waste may be classified based party on content and partly on moisture and heating value. A typical classification is as follows:

Municipal Solid Waste

Municipal solid waste consists of household waste, construction and demolition debris, sanitation residue, and waste from streets. This garbage is generated mainly from residential and commercial complexes. They may be categories as:

Garbage

Refers to the putrescible solid waste constituents produced during the preparation or storage of meat, fruit, vegetable etc. these waste have a moisture content of about 70% and a heating value of around 6×10^6 J/kg.

Rubbish

Refers to non-putrescible solid waste constitute, either combustible or non-combustible. Combustible wastes would include paper, wood, scrap, rubber, leather etc. non-combustible waste are metals, glass, ceramics etc. these waste contain a moisture content of about 25% and the heating value of the waste is around 15×10^6 J/kg.

Hospital Waste

Hospital waste is generated during the diagnosis, treatment, or immunization of human beings or animals or in research activities in these fields or in the production or testing of biologicals. It may include wastes like sharps, soiled waste, disposables, anatomical waste, cultures, discarded medicines, chemical wastes, etc. These are in the form of disposable syringes, swabs, bandages, body fluids, human excreta, etc. The moisture content is 85% and there are 5% non-combustible solids. The heating value is around 2.5×10^6 J/kg. This waste is highly infectious and can be a serious threat to human health if not managed in a scientific and discriminate manner.

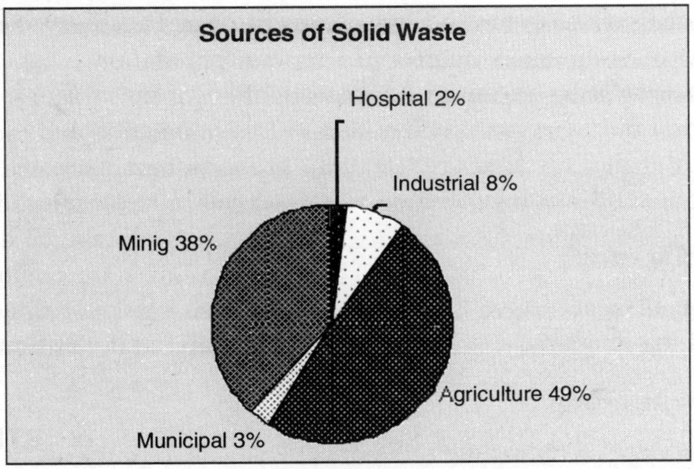

Fig. 6.7. Sources of Solid Waste

Industrial Waste

They include chemicals, paints, sand, metal ore processing, fly ash, sewage treatment sludge etc. Manufacturing industries produce wastes which are solid or semi-solid. This waste can be self-igniting, explosive, toxic or radio-active. Chemical process industries generate a variety of waste, both organic and inorganic, which are mixtures with wide range of component concentration.

Agricultural Waste

These include farm animal manure and crop residue etc. animal and vegetable waste contain valuable minerals and nutrients. Humus from agricultural wastes contains nitrogen, phosphorus, potash and trace elements which are vital to the fertility of the soil and optimum plant growth. Burning of wastes as fuel in the conventional manner makes poor use of the leaf content of the fuel burnt and further, leads to loss of valuable nutrients. In India, agricultural wastes contain around 2kg/person/day.

6.5. CAUSES OF SOLID WASTE

The main causes for the rapid growth in the quantity of solid waste are:

6.5.1. Overpopulation

Due to increasing industrialization and population large quantities of wastes are being generated in different forms such as- solid, liquid, sludge and gases. Each city produces tones of solid wastes daily from house holds, hospitals, industry offices, market centers etc., Some of these are biodegradable some are non biodegradable and Hazardous waste. Wastes are directly thrown away on the street roads, city garbage collecting places etc. Out of which most of it still remain there, which later pile up and chocks cities drainage lines.

6.5.2. Urbanization

Due to pressure of urbanization most of the cities are growing fast and sometimes they develop beyond the planned limits. Generally the unplanned areas of the city contain a quarter of the total population, where the spatial information is missing because of non-availability of up to date maps. Earlier waste disposal did not pose problem due to less population and lack of public awareness. But due to rapid urbanization, proportionate increase in domestic and industrial solid waste generation was observed in large cities.

6.5.3. Affluence

With the affluence, there is a tendency to declare items as being in or out of fashion and the items those are out of fashion results solid waste pollution.

6.5.4. Technology

As the country achieves the economical goals the country invests more over technology. With the development in technology, there is a shift from natural goods to the artificial ones those are generally non-biodegradable by nature such as plastic, DDT etc. these non-biodegradable products are largely responsible for causing solid waste pollution because they persist unchanged in disposal operations such as landfills.

6.6. EFFECTS OF SOLID WASTE

1. Hazardous solid wastes are injurious to human health; some have acute effects while others pose a health hazard after prolonged period of exposure. Improper disposal of such wastes has resulted in the death of human and animals through contamination of crops and water supplies.
2. Solid waste disposal increases the rate of breeding of disease vectors, primarily files and rats.
3. There is danger of water pollution when the leachate from a solid waste dump enters surface water or ground water resource.

4. In addition, uncontrolled burning of open dumps can cause objectionable odour and air pollution.
5. Improper handling of solid waste causes damage to the environment. The environmental damage caused by solid waste is mostly un-aesthetic in nature. Uncontrolled dumping of urban waste destroys the beauty of the country-site.

6.7. MANAGEMENT OF SOLID WASTE

Some of the waste can be recovered and reprocesses a procedure commonly known as recycling. However, before the waste can be disposed of effectively, they must be collected efficiently. These activities, *i.e.*, collection, disposal and/or recovery are known as solid waste management. Four R's (Refuse, Reuse, Recycle, and Reduce) to be followed for waste management

1. Refuse

Instead of buying new containers from the market, use the ones that are in the house. Refuse to buy new items though you may think they are prettier than the ones you already have.

2. Reuse

Do not throw away the soft drink cans or the bottles; cover them with homemade paper or paint on them and use them as pencil stands or small vases.

3. Recycle

Use shopping bags made of cloth or jute, which can be used over and over again [will this come under recycle or reduce?]. Segregate your waste to make sure that it is collected and taken for recycling.

4. Reduce

Reduce the generation of unnecessary waste, e.g. carry your own shopping bag when you go to the market and put all your purchases directly into it.

6.7.1. Control Measures

Solid waste management is a manifold task involving many activities like-

Collecting of Solid Waste

The essential parts of the overall solid waste management programme are effective collection and transportation. The basic mode of collection is from communal storage point. The waste is delivered to fixed storage bins usually built from concrete blocks to the location from where the collection vehicle will ultimately transport it to the site of disposal. Daily collection is essential because the organic matter in the waste tends to decompose rapidly in the hot climate.

Disposal of Solid Waste

As cities are growing in size with a rise in the population, the amount of waste generated is increasing becoming unmanageable. The local corporations have adapted different methods for the disposal of waste – open dumps, landfills, sanitary landfills, composting and incineration plants.

Open Dumps

Open dumping of solid waste is done in low lying areas and outskirts of the towns and cities. Being comparatively cheaper and required no planning, this method of disposal is used extensively in India.

Open dumps refer to uncovered areas that are used to dump solid waste of all kinds. The waste is untreated, uncovered, and not segregated. It is the breeding ground for flies, rats, and other insects that spread disease. The rainwater run-off from these dumps contaminates nearby land and water thereby spreading disease. They also become a source of objectionable odours and cause air pollution when the wastes are burned I order to reduce their volume and conserve space.

Landfills

Landfills are generally located in urban areas where a large amount of waste is generated and has to be dumped in a common place. Unlike an open dump, it is a pit that is dug in the ground. The garbage is dumped and the pit is covered thus preventing the breeding of flies and rats. At the end of each day, a layer of soil is scattered on top of it and some mechanism, usually earth-moving equipment is used to compress the garbage, which now forms a cell. Thus, every day, garbage is dumped and becomes a cell. After the landfill is full, the final layer is covered by final cover of about one meter of earth to prevent rodents from burring into the refuse and scattering and the site can thereafter be developed as a parking lot or a park.

Landfills have many problems as all types of waste are dumped in landfills and when water seeps through them it gets contaminated and in turn pollutes the surrounding area. This contamination of groundwater and soil through landfills is known as leaching.

Sanitary Landfills

Sanitary land filling involves the disposal of municipal wastes on or in the upper layers of the earth's mantle especially in degraded area in the need of restoration. An alternative to landfills which will solve the problem of leaching to some extent is a sanitary landfill which is more hygienic and built in a methodical manner. These are lined with materials that are impermeable such as plastics and clay, and are also built over impermeable soil. The advantages of a sanitary landfill, as opposed to an open dump are:

1. The public health problems are minimized because flies, rats and other pests are unable to breed in the covered refuse.
2. There is no air pollution from burning.
3. Fire hazards are minimal.

Constructing sanitary landfills is very costly and they are having their own problems. Some authorities claim that often the plastic liner develops cracks as it reacts with various chemical solvents present in the waste.

The rate of decomposition in sanitary landfills is also extremely variable. This can be due to the fact that less oxygen is available as the garbage is compressed very tightly. It has also been observed that some biodegradable materials do not decompose in a landfill. Another major problem is the development of methane gas, which occurs when little oxygen is present, i.e. during anaerobic decomposition. In some countries, the methane being produced from sanitary landfills is tapped and sold as fuel.

Incineration Plants

This process of burning waste in large furnaces is known as incineration. In these plants the recyclable material is segregated and the highly combustible wastes like plastics, cardboard, paper, rubber and combustible wastes like cartons, wood scrap, floor sweepings, food wastes etc. are subjected to incineration i.e. burning at very high temperatures. At the end of the process all that is left behind is ash. During the process some of the ash floats out with the hot air. This is called fly ash. Both the fly ash and the ash that is left in the furnace after burning have high concentrations of dangerous toxins such as dioxins and heavy metals. Disposing of this ash is a problem. The ash that is buried at the landfills leaches the area and cause severe contamination.

Burning garbage is not a clean process as it produces tones of toxic ash and pollutes the air and water. A large amount of the waste that is burnt here can be recovered and recycled. In fact, at present, incineration is kept as the last resort and is used mainly for treating the infectious waste.

The advantages of incineration include wide range ability for handling varying loads and small space requirement for ultimate disposal. However, the method requires fairly high level of maintenance and the operation of sanitary landfill.

Composting

Organic matter constitutes 35%–40% of the municipal solid waste generated in India. This waste can be recycled by the method of composting, one of the oldest forms of disposal. It is the natural process of decomposition of organic waste that yields manure or compost, which is very rich in nutrients. Composting is a biological process in which micro-organisms, mainly fungi and bacteria, convert degradable organic waste into humus like substance. This finished product, which looks like soil, is high in carbon and nitrogen and is an excellent medium for growing plants. It recycles the nutrients and returns them to the soil as nutrients.

Apart from being clean, cheap, and safe, composting can significantly reduce the amount of disposable garbage. The organic fertilizer can be used instead of chemical fertilizers and is better specially when used for vegetables. It increases the soil's ability to hold water and makes the soil easier to cultivate. It helped the soil retain more of the plant nutrients.

Composting : Some benefits

1. Compost allows the soil to retain more plant nutrients over a longer period.
2. It supplies part of the 16 essential elements needed by the plants.
3. It helps reduce the adverse effects of excessive alkalinity, acidity, or the excessive use of chemical fertilizer.
4. It makes soil easier to cultivate.
5. It helps keep the soil cool in summer and warm in winter.
6. It aids in preventing soil erosion by keeping the soil covered.
7. It helps in controlling the growth of weeds in the garden.

Pyrolsis or Destructive Distillation

In this disposal method, the solid wastes are heated under anerobic conditions *i.e.* burning without oxygen. The organic components of the solid wastes split up into gaseous liquid and gaseous fractions (CO, CO_2, CH_4, tar, charred carbon). Unlike the highly exothermic process of combustion, pyrolysis is a highly endothermic process and that is why it is also called destructive distillation

Land Farming

The organic wastes are either applied on top of the land or injected below the soil surface with suitable equipment, where they undergo bacterial and chemical decomposition. At frequent intervals, the land farming sites can be reused without any adverse effects provided the land farming site is properly managed.

Waste Utilization

A developing country can not afford wastage. By proper utilization of solid waste a developing counter like India can avail of many advantages. Many solid wastes generated by industries can be utilized directly. Fly ash and bottom ash from power plants can be used commercially, largely as cement substitute. New uses are developed for fly ash, *e.g.* to make bricks, to dewater industrial wastewater sludge, as a land cover etc. Recycling involves the collection of used and discarded materials processing these materials and making them into new products. It reduces the amount of waste that is thrown into the community dustbins thereby making the environment cleaner and the air fresher to breathe.

6.8. CONSUMERISM AND WASTE PRODUCT

Consumerism is a term used to describe the effects of equating personal happiness with purchasing material possessions and consumption. It is often associated with criticisms of consumption starting with Karl Marx and Thorstein Veblen, but can actually be traced back to the first human civilizations. In economics, *consumerism* can also refer to economic policies that place an emphasis on consumption, and, in an abstract sense, the belief that the free choice of consumers should dictate the economic structure of a society.

Today's consumption is undermining the environmental resource base. It is exacerbating inequalities. And the dynamics of the consumption-poverty-inequality-environment nexus are accelerating. If the trends continue without change—not redistributing from high-income to low-income consumers, not shifting from polluting to cleaner goods and production technologies, not promoting goods that empower poor producers, not shifting priority from consumption for conspicuous display to meeting basic needs—today's problems of consumption and human development will worsen.

The real issue is not consumption itself but its patterns and effects.

Inequalities in consumption are stark. Globally, the 20% of the world's people in the highest-income countries account for 86% of total private consumption expenditures—the poorest 20% a minuscule 1.3%. **More specifically, the richest fifth:**

- Consume 45% of all meat and fish, the poorest fifth 5%
- Consume 58% of total energy, the poorest fifth less than 4%
- Have 74% of all telephone lines, the poorest fifth 1.5%
- Consume 84% of all paper, the poorest fifth 1.1%
- Own 87% of the world's vehicle fleet, the poorest fifth less than 1%

We consume a variety of resources and products today having moved beyond basic needs to include luxury items and technological innovations to try to improve efficiency. Such consumption beyond minimal and basic needs is not necessarily a bad thing in and of itself, as throughout history we have always sought to find ways to make our lives a bit easier to live. However, increasingly, there are important issues around consumerism that need to be understood. For example:

- How are the products and resources we consume actually produced?
- What are the impacts of that process of production on the environment, society, on individuals?
- What are the impacts of certain forms of consumption on the environment, on society, on individuals?
- Which factors influence our choices of consumption?
- Which factors influence how and why things are produced or not?
- What is a necessity and what is a luxury?
- How do demands on items affect the requirements placed upon the environment?

- How do consumption habits change as societies change?
- Businesses and advertising are major engines in promoting the consumption of products so that they may survive. How much of what we consume is influenced by their needs versus our needs?
- Also influential is the very culture of today in many countries, as well as the media and the political institutions themselves. What is the impact on poorer nations and people on the demands of the wealthier nations and people that are able to afford to consume more?
- How do material values influence our relationships with other people?
- What impact does that have on our personal values?

and so on.

Just from these questions, we can likely think of numerous others as well. We can additionally, see that consumerism and consumption are at the core of many, if not most societies. The impacts of consumerism, positive and negative are very significant to all aspects of our lives, as well as our earth (Environment). But equally important to bear in mind in discussing consumption patterns is the underlying system that promotes certain types of consumption and not other types.

"Over" population is usually blamed as the major cause of environmental degradation, but the above statistics strongly suggests otherwise. As we will see, consumption patterns today are not to meet everyone's needs. The system that drives these consumption patterns also contribute to inequality of consumption patterns too.

6.8.1. Effects of Consumerism

Because consumption is so central to many economies, and even to the current forms of globalization, its effects therefore are also seen around the world. How we consume, and for what purposes drives how we extract resources, create products and produce pollution and waste. Issues relating to consumption hence also affect environmental degradation, poverty, hunger, and even the rise in obesity that is nearing levels similar to the "official" global poverty levels. Political and economic systems that are currently promoted and pushed around the world in part to increase consumption also lead to immense poverty and exploitation. Much of the world cannot and do not consume at the levels that the wealthier in the world do.

Consumerism has been related both to the increase in the population size as well as increase in our demand due to change in life-style. In the modern society our needs have multiplied and so consumerism of resources has also multiplied.

We are beginning to get just a hint of how wasteful our societies are. Sugar, beef, and bananas are just the tip of the iceberg in terms of examples of wasted industry and waste structured within the current system. Not only are certain wasteful job functions unnecessary as a result, but the capital that employs this labor is therefore a wasteful use of capital. As a result, we see waste and misuse of the environment, as well as social and environmental degradation increasing. Our industries may be efficient for accumulating capital and making profits, but that does not automatically mean that it is efficient for society.

6.9. MANAGEMENT AND LEADERSHIP STANDARDS

ISO standards that provide requirements or give guidance on **good management practice** are among the best known of ISO's offering. Many, although not all, are modelled on the management system structure of ISO 9001 and ISO 14001. In addition, the management standards are divided into those which can be used for certification, such as ISO 9001 and ISO 14001, and those which are not certifiable, requirements standards, but provide guidance, such as ISO 26000 and ISO 31000.

ISO's management standards are now joined by ISO 26000, which gives **guidance on social responsibility.** It does not follow the management system model and **is not a certification standard.** Therefore, any claim by an organization to be certified to ISO 26000 is false and organizations should beware of any offers to certify them to the standard.

Published in 2009, ISO 31000 is experiencing a rapid take-up by the global market, while it adapts aspects of the management system approach, it **is not a certification standard.**

ISO 9001:2008 which gives the requirements for **quality management systems,** is now firmly established as the globally implemented standard for providing assurance about the ability to satisfy quality requirements and to enhance customer satisfaction in supplier-customer relationships.

ISO 14001:2004 – Environmental management which gives the requirements for **environmental management systems**, confirms its global relevance for organizations wishing to operate in an environmentally sustainable manner.

6.9.1. ISO 14000

The **ISO 14000** environmental management standards exist to help organizations (*a*) minimize how their operations (processes etc.) negatively affect the environment (*i.e.* cause adverse changes to air, water, or land); (*b*) comply with applicable laws, regulations, and other environmentally oriented requirements, and (*c*) continually improve in the above.

ISO 14000 is similar to ISO 9000 quality management in that both pertain to the process of how a product is produced, rather than to the product itself. As with ISO 9000, certification is performed by third-party organizations rather than being awarded by ISO directly. The ISO 19011 audit standard applies when auditing for both 9000 and 14000 compliance at once.

The concept of an environmental management system evolved in the early nineties and its origin can be traced back to 1972, when the United Nations organised a Conference on the Human Environment in Stockholm and the United Nations Environment Programme (UNEP) was launched (Corbett & Kirsch, 2001). These early initiatives led to the establishment of the World Commission on Environment and Development (WCED) and the adoption of the Montreal Protocol and Basel Convention.

In 1992, the first Earth Summit was held in Rio-de-Janeiro (Jiang & Bansal, 2001), which served to generate a global commitment to the environment (RMIT University). In the same year, BSI Group published the world's first environmental management systems standard, BS 7750. This supplied the template for the development of the **ISO 14000** series in 1996, by the International Organization for Standardization, which has representation from committees all over the world (ISO) (Clements 1996, Brorson & Larsson, 1999). As of 2010, ISO 14001 is now used by at least 223 149 organizations in 159 countries and economies.

6.9.2. Development of the ISO 14000 Series

The ISO 14000 family includes most notably the ISO 14001 standard, which represents the core set of standards used by organizations for designing and implementing an effective environmental management system. Other standards included in this series are ISO 14004, which gives additional guidelines for a good environmental management system, and more specialized standards dealing with specific aspects of environmental management. The major objective of the ISO 14000 series of norms is "to promote more effective and efficient environmental management in organizations and to provide useful and usable tools - ones that are cost effective, system-based, flexible and reflect the best organizations and the best organizational practices available for gathering, interpreting and communicating environmentally relevant information".

6.9.3. List of ISO 14000 Series Standards

- **ISO 14001** Environmental management systems—Requirements with guidance for use
- **ISO 14004** Environmental management systems—General guidelines on principles, systems and support techniques
- **ISO 14015** Environmental assessment of sites and organizations
- **ISO 14020** series (14020 to 14025) Environmental labels and declarations
- **ISO 14031** Environmental performance evaluation—Guidelines
- **ISO 14040** series (14040 to 14049), Life Cycle Assessment, LCA, discusses pre-production planning and environment goal setting.
- **ISO 14050** terms and definitions.
- **ISO 14062** discusses making improvements to environmental impact goals.
- **ISO 14063** Environmental communication—Guidelines and examples
- **ISO 14064** Measuring, quantifying, and reducing Greenhouse Gas emissions.

- **ISO 19011** which specifies one audit protocol for both 14000 and 9000 series standards together. This replaces **ISO 14011** meta-evaluation—how to tell if your intended regulatory tools worked. 19011 is now the only recommended way to determine this.

6.9.4. ISO 14001

ISO 14001 can be integrated with other management functions and assists companies in meeting their environmental and economic goals.

ISO 14001 is known as a generic management system standard, meaning that it is applicable to any size and type of organisation, product or service, in any sector of activity and can accommodate diverse socio-cultural and geographic conditions (Standards Australia/Standards New Zealand 2004). All standards are periodically reviewed by ISO and new ones issued (Standards Australia/ Standards New Zealand 2004).

ISO 14001 is based on the Plan-Do-Check-Act methodology which has been expanded to include 17 elements, grouped into five phases that relate to Plan-Do-Check-Act:

- Environmental Policy
- Planning
- Implementation and Operation
- Checking and Corrective Action
- Management Review

Organizations complying with the ISO 14001:2004 standard are recognizable by the ISO 14001 logo given by the certification body.

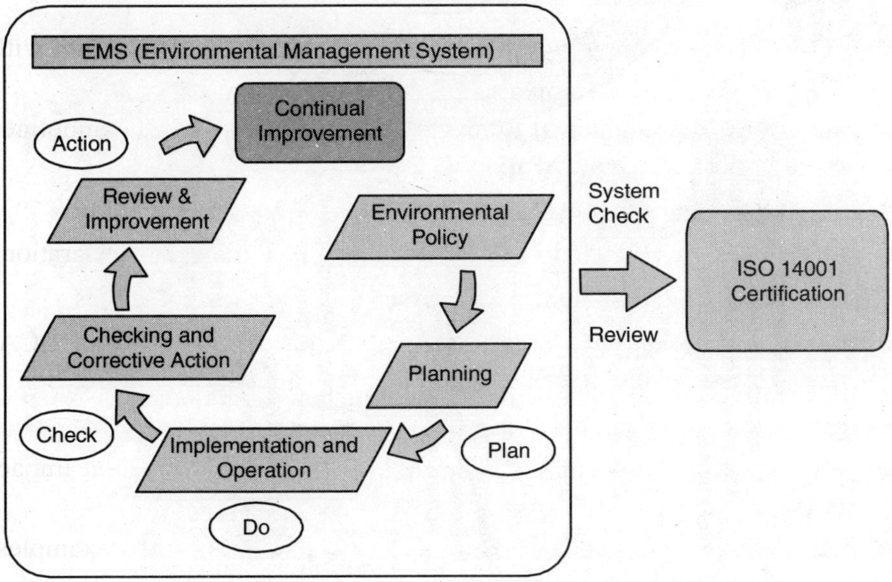

Fig. 6.8. Plan-Do-Check-Act Methodology

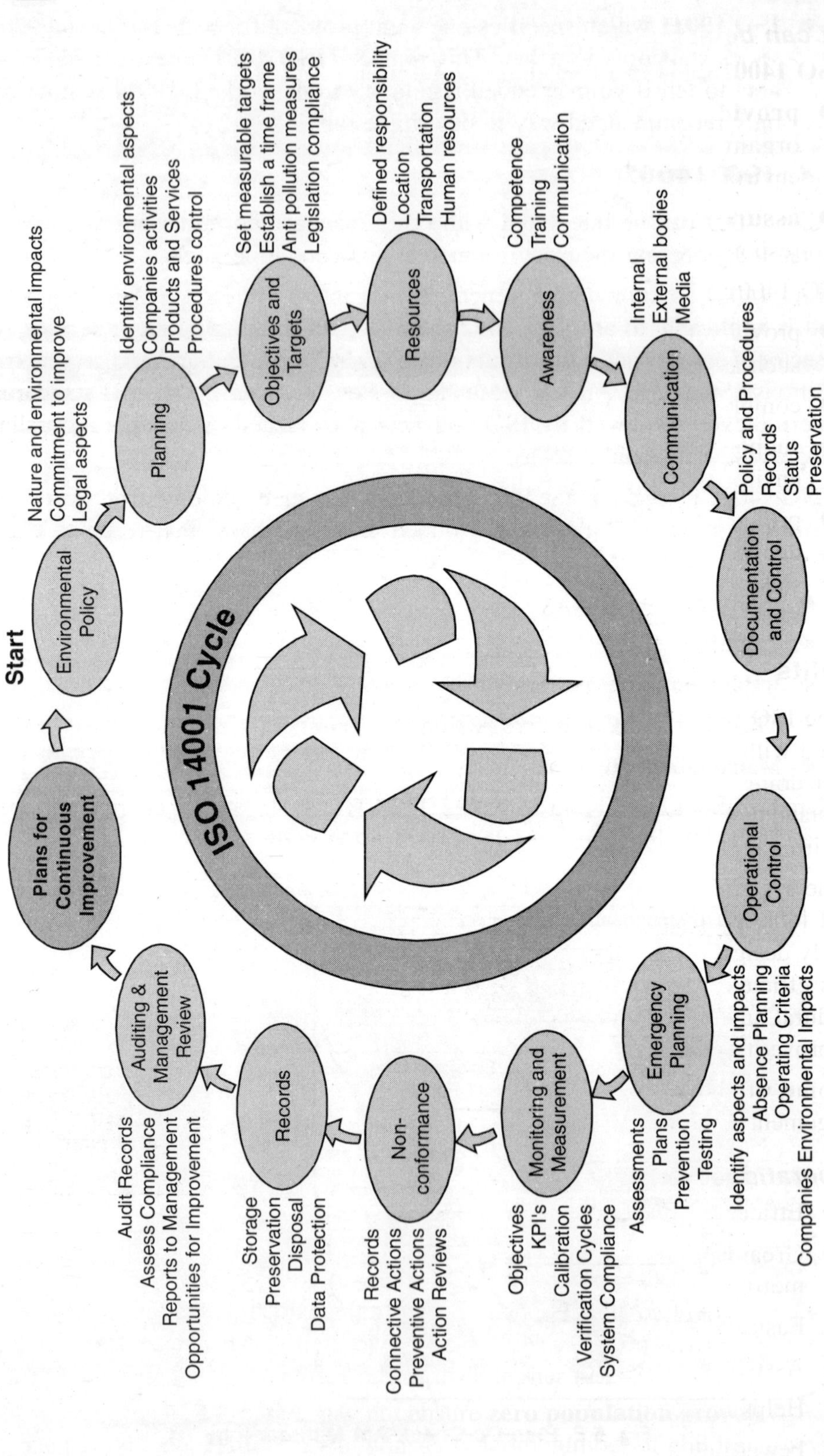

Fig. 6.9. ISO 14001 Process

What can be achieved

ISO 14001:2004 is a tool that can be used to meet internal objectives:

- **provide assurance to management** that it is in control of the organizational processes and activities having an impact on the environment
- **assure employees** that they are working for an environmentally responsible organization.

ISO 14001:2004 can also be used to meet **external objectives:**

- provide assurance on environmental issues to **external stakeholders** – such as customers, the community and regulatory agencies
- comply with environmental regulations
- support the organization's **claims and communication** about its own environmental policies, plans and actions
- provides a framework for **demonstrating conformity** via suppliers' declarations of conformity, assessment of conformity by an external stakeholder - such as a business client - and for certification of conformity by an independent certification body.

Benefits

The ISO 14000 standards are practical tools for organizations who are not satisfied with mere compliance with legislation – which may be perceived as a cost of doing business. They are useful tools for proactive organizations who understand that implementing a strategic approach can bring return on investment in environment-related measures.

The systematic ISO 14001:2004 approach requires the organization to take a hard look at all areas where its activities have an environmental impact. A properly designed ISO 14001:2004 Environmental Management System (EMS) allows efficient identification of opportunities for cost savings. It can trigger procedural and/or technological changes that reduce the total cost of a product or improve its value.

Some of the benefits of implementing an ISO 14000 Environmental Management System (EMS) in accordance with the ISO 14000 standards include:

1. Operational Benefits

- Efficiency, discipline and operational integration with ISO 9000
- Greater employee involvement in business operations with a more motivated workforce
- Easier to obtain operational permits and authorizations
- Assists in developing and transferring technology within the company
- Helps reduce pollution
- Fewer operating costs

- Savings from safer workplace conditions
- Reduction of costs associated with emissions, discharges, waste handling, transport and disposal
- Improvements in the product as a result of process changes
- Safer products

2. Environmental Benefits

- Minimizes hazardous and non-hazardous waste
- Conserves natural resources—electricity, gas, space and water with resultant cost savings
- Prevents pollution and reduces wastage

3. Marketing Benefits

- Demonstrates to customers that the firm has met environmental expectations
- Meets potential national and international government purchasing requirements
- Delivers profits from marketing "green" products
- Provides a competitive marketing tool
- Improves international competitiveness

4. Financial Benefits

- Improves the organization's relationship with insurance companies
- Elimination of costs associated with conformance to conflicting national standards
- Process cost savings by reduction of material and energy input
- Satisfying investor/shareholder criteria
- Helps reduce liability and risk
- Improved access to capital

6.9.5. Certification

Once a business has fully developed and implemented their ISO 14001 compliant environmental management system, they can choose to apply for certification (also referred to as registration). Certification involves evaluation of the company's EMS system, including a comprehensive on-site audit, to determine whether it meets the ISO 14001 requirements (Martin 1998). If the company conforms to the ISO standard it is issued with a certificate which is generally valid for a period of three years (ISO14001.com.au 2010). In some countries, certification can only be granted by an ISO-accredited Certification Body: ANSI-ASQ National Accreditation Board in the USA, the United Kingdom Accreditation Service in the UK, or the National Accreditation Board in Ireland.

Certification auditors need to be accredited by the International Registrar of Certification Auditors. Some countries however allow businesses to self-certify (Martin 1998).

How it Works

ISO 14001:2004 does not specify levels of environmental performance. If it specified levels of environmental performance, they would have to be specific to each business activity and this would require a specific EMS standard for each business. That is not the intention.

ISO has many other standards dealing with specific environmental issues. The intention of ISO 14001:2004 is to provide a **framework for a holistic, strategic approach** to the organization's environmental policy, plans and actions.

ISO 14001:2004 gives the **generic requirements** for an environmental management system. The underlying philosophy is that whatever the organization's activity, the requirements of an effective EMS are the same.

This has the effect of establishing a **common reference** for communicating about environmental management issues between organizations and their customers, regulators, the public and other stakeholders.

Because ISO 14001:2004 does not lay down levels of environmental performance, the standard can to be implemented by a **wide variety of organizations,** whatever their current level of environmental maturity. However, a **commitment to compliance** with applicable environmental legislation and regulations is required, along with a commitment to continual improvement – for which the EMS provides the framework.

6.10. ENVIRONMENTAL LEGISLATION

India is the first country in the world to have provisions for the protection and conservation of environment in its constitution. On 5 June 1972, environment war first discussed as an item of international agenda in the UN Conference on Human Environment in Stockholm and thereafter 5th June is celebrated all over the world as World Environmental Day.

The provisions for environmental protection in the constitution were made in 1976 through 42 amendment under the article 48-A and Article 51-A (g). This constitution includes environmental protection and conservation as one of our fundamental duties. Some of the important Acts passes by the Government of India are discussed here:

Wildlife (Protection) Act, 1972

(53 of 1972)
[9th September, 1972]

An Act to provide for the protection of wild animals, birds and plants and for matters connected therewith. It extends to the whole of India except the State of Jammu and Kashmir. It shall come into force in a State or Union Territory to which it extends, on such date as the Central Government may, by notification, appoint, and different dates may be appointed for different provisions of this Act or for different States or Union territories. The major activities and provisions in the Act can be summed up as follows

1. It provides for the appointment of wildlife advisory board.
2. Protection of some endangered species
3. The Chief Wild Life Warden may permit any person to hunt such animal, if any wild animal has become dangerous to human life or is so disabled or diseased as to be beyond recovery in writing and stating the reasons and any wild animal killed or wounded in defense of any person shall be Government property
4. Chief Wild Life Warden may grant a permit to any person in writing stating the reasons on payment of such fee that is prescribed for the purpose, of education, scientific research, scientific Management, collection of specimens, collection or preparation of snake-venom for the manufacture of life-saving drugs
5. No person shall willfully pick, uproot, damage, destroy, acquire or collect any specified plant from any forest land and any area specified or notification by the Central Government.
6. No person shall cultivate a specified plant except under and in accordance with a license granted by the Chief Wild Life Warden or any other officer authorized by the State Government in this behalf;
7. Every licence granted under this section shall specify the area in which and the conditions, if any, subject to which the licensee shall cultivate a specified plant.
8. The Act provides the declaration of sanctuary and National Parks
9. There is restriction on entry in sanctuary, The Chief Wild Life Warden may, on application, grant to any person a permit to enter or reside in a sanctuary for all or any of the following purposes, namely investigation or study of wild life, photography, scientific research, tourism and transaction of lawful business with any person residing in the sanctuary.
10. It provides for legal powers to officers and punishment to offenders. Any person who contravenes any provision of this Act or any rule or order made who commits a breach of any of the conditions of any license or permit granted under this Act, shall be guilty of an offence against this Act, and shall, on conviction, be punishable with imprisonment for a term which may punishable with imprisonment for a term which shall not be less than one year but which may extend to seven years and also with fine which shall not be less than five thousand
11. This Act gives the power of Central Government and State Government to make rules

Forest (Conservation) Act, 1980 with Amendments Made in 1988

An Act to provide for the conservation of forests and for matters connected therewith or ancillary or incidental thereto. It extends to the whole of India except the State of Jammu and Kashmir. It shall be deemed to have come into force on the 25th October, 1980.

Under section 2: Restriction on the dereservation of forests or use of forest land for non-forest purpose.

Notwithstanding anything contained in any other law for the time being in force in a State, no State Government or other authority shall make, except with the prior approval of the Central Government, any order directing-

(*i*) that any reserved forest (within the meaning of the expression "reserved forest" in any law for the time being in force in that State) or any portion thereof, shall cease to be reserved;

(*ii*) that any forest land or any portion thereof may be used for any non-forest purpose;

(*iii*) that any forest land or any portion thereof may be assigned by way of lease or otherwise to any private person or to any authority, corporation, agency or any other organization not owned, managed or controlled by Government;

(*iv*) that any forest land or any portion thereof may be cleared of trees which have grown naturally in that land or portion, for the purpose of using it for re-afforestation.

Explanation : For the purpose of this section, "non-forest purpose" means the breaking up or clearing of any forest land or portion thereof for :

(*a*) the cultivation of tea, coffee, spices, rubber, palms, oil-bearing plants, horticultural crops or medicinal plants;

(*b*) any purpose other than re-afforestation;

but does not include any work relating or ancillary to conservation, development and management of forests and wildlife, namely, the establishment of check-posts, fire lines, wireless communications and construction of fencing, bridges and culverts, dams, waterholes, trench marks, boundary marks, pipelines or other like purposes.

Under section 3: Constitution of Advisory Committee

The Central Government may constitute a Committee consisting of such number of persons as he may deem fit to advise that Government with regard to-

(*i*) the grant of approval, under Section 2; and

(*ii*) any other matter connected with the conservation of forests which may be referred to h by the Central Government.

Under section 3A: Penalty for contravention of the provisions of the Act

Whoever contravenes or abets the contravention of any of the provisions of Section 2, shall be punishable with simple imprisonment for a period which may extend to fifteen days.

Under section 3B: Offences by the Authorities and Government Departments

(1) Where any offence under this Act has been committed -

(*a*) by any department of Government, the head of the department; or

(*b*) by any authority, every person who, at the time the offence was committed, was directly in charge of, and was responsible to, the authority for the conduct of the business of the authority as well as the authority;

shall be deemed to be guilty of the offence and shall be liable to be proceeded against and punished accordingly.

Under section 4: Power to make rules

(1) The Central Government may, by notification in the Official Gazette, makes rules for carrying out the provisions of this Act.

Under section 5: Repeal and saving

(1) The Forest (Conservation) Ordinance, 1980 is hereby replaced.

(2) Notwithstanding such repeal, anything done or any action taken under the provisions of the said Ordinance shall be deemed to have been done or taken under the corresponding provisions of this Act.

Some amendments are made in forest Act in 1992 :

- Cultivation of tea, coffee, rubber are included under non-forestry activity
- Exploration and survey of National park and wildlife sanctuaries are totally prohibited.

The Water (Prevention and Control of Pollution) Act, 1974

[Act No. 6 of 1974]

An Act to provide for the prevention and control of water pollution and the maintaining or restoring of wholesomeness of water for the establishment, with a view to carrying out the purposes aforesaid, of Boards for the prevention and control of water pollution, for conferring on and assigning to such Boards powers and functions relating thereto and for matters connected therewith

It applies in the first instance to the whole of the States of Assam, Bihar, Gujarat, Haryana, Himachal Pradesh, Jammu and Kashmir, Karnataka, Kerala, Madhya Pradesh, Rajasthan, Tripura and West Bengal and the Union Territories; and it shall apply to such other State which adopts this Act by resolution passed in that behalf under clause (1) of article 252 of the Constitution.

It shall come into force at once in the States of Assam, Bihar, Gujarat, Haryana, Himachal Pradesh, Jammu and Kashmir, Karnataka, Kerala, Madhya Pradesh, Rajasthan, Tripura and West Bengal and in the Union Territories, and in any other State which adopts this Act under clause (1) of article 252 of the Constitution on the date of such adoption and any reference in this Act to the commencement of this Act shall, in relation to any State or Union Territory, means the date on which this Act comes into force in such State or Union Territory.

The Water Act was enacted by Parliament Act, 1974 purpose to provide for the prevention of control of water pollution and the maintaining or restoring of wholesomeness of water. Water pollution is defined as 'such contamination of water, or such alteration of physical, chemical or biological properties of water, or such discharge as is likely to cause a nuisance or render the water harmful or injurious to public health and safety or harmful for any other use or to aquatic plants and other organism or animal life'. As on day, it is applicable in all the states of India. The relevant provisions of this and are given as below :

Under section 16: Functions of Central Board

(1) Subject to the provisions of this Act, the main function of the Central Board shall be to promote cleanliness of streams and wells in different areas of the States.

(2) In particular and without prejudice to the generality of the foregoing function, the Central Board may perform all or any of the following functions, namely:

(*a*) Advise the Central Government on any matter concerning the prevention and control of water pollution;

(*b*) Co-ordinate the activities of the State Boards and resolve disputes among them;

(*c*) Provide technical assistance and guidance to the State Boards, carry out and sponsor investigations and research relating to problems of water pollution and prevention, control or abatement of water pollution;

(*d*) Plan and organize the training of persons engaged or to be engaged in programs for the prevention, control or abatement of water pollution on such terms and conditions as the Central Board may specify;

(*e*) Organize through mass media a comprehensive programme regarding the prevention and control of water pollution;

(*f*) Collect, compile and publish technical and statistical data relating to water pollution and the measures devised for its effective prevention and control and prepare manuals, codes or guides relating to treatment and disposal of sewage and trade effluents and disseminate information connected therewith;

(*g*) Lay down, modify or annul, in consultation with the State Government concerned, the standards for a stream or well: Provided that different standards may be laid down for the same stream or well or for different streams or wells, having regard to the quality of water, flow characteristics of the stream or well and the nature of the use of the water in such stream or well or streams or wells;

(*h*) Plan and cause to be executed a nation-wide programme for the prevention, control or abatement of water pollution;

(*i*) Perform such other functions as may be prescribed.

(3) The Board may establish or recognize a laboratory or laboratories to enable the Board to perform its functions under this section efficiently, including the analysis of samples of water from any stream or well or of samples of any sewage or trade effluents.

Under section 17: Functions of State Board

(1) Subject to the provisions of this Act, the functions of a State Board shall be :

(*a*) To plan a comprehensive programme for the prevention, control or abatement of pollution of streams and wells in the State and to secure the execution thereof;

(*b*) To advise the State Government on any matter concerning the prevention, control or abatement of water pollution;

(*c*) To collect and disseminate information relating to water pollution and the prevention, control or abatement thereof;

(*d*) To encourage, conduct and participate in investigations and research relating to problems of water pollution and prevention, control or abatement of water pollution;

(*e*) To collaborate with the Central Board in organizing the training of persons engaged or to be engaged in programmes relating to prevention, control or abatement of water pollution and to organize mass education programmes relating thereto;

(*f*) To inspect sewage or trade effluents, works and plants for the treatment or sewage and trade effluents and to review plans, specifications or other data relating to plants set up for the treatment of water, works for the purification thereof and the system for the disposal of sewage or trade effluents or in connection with the grant of any consent as required by this Act;

(*g*) Lay down, modify or annul effluent standards for the sewage and trade effluents and for the quality of receiving waters (not being water in an inter-State stream) resulting from the discharge of effluents and to classify waters of the State;

(*h*) To evolve economical and reliable methods of treatment of sewage and trade effluents, having regard to the peculiar conditions of soils, climate and water resources of different regions and more specially the prevailing flow characteristics of water in streams and wells which render it impossible to attain even the minimum degree of dilution;

(*i*) to evolve methods of utilization of sewage and suitable trade effluents in agriculture;

(*j*) to evolve efficient methods of disposal of sewage and trade effluents on land, as are necessary on account of the predominant conditions of scant stream flows that do not provide for major part of the year the minimum degree of dilution;

(*k*) to lay down standards of treatment of sewage and trade effluents to be discharged into any particular stream taking into account the minimum fair weather dilution available in that stream and the tolerance limits of pollution permissible in the water of the stream, after the discharge of such effluents;

(*l*) to make, vary or revoke any order :

(*i*) For the prevention, control or abatement of discharge of waste into streams or wells;

(*ii*) Requiring any person concerned to construct new systems for the disposal of sewage and trade effluents or to modify, alter or extend any such existing system or to adopt such remedial measures as are necessary to prevent control or abate water pollution;

(*m*) to lay down effluent standards to be complied with by persons while causing discharge of sewage or sullage or both and to lay down, modify or annul effluent standards for the sewage and trade effluents;

(*n*) to advice the State Government with respect to the location of any industry the carrying on of which is likely to pollute a stream or well;

(*o*) to perform such other functions as may be prescribed or as may, from time to time be entrusted to it by the Central Board or the State Government.

(2) The Board may establish or recognize a laboratory or laboratories to enable the Board to perform its functions under this section efficiently, including the analysis of samples of water from any stream or well or of samples of any sewage or trade effluents.

Under section 18: Powers to give directions

In the performance of its functions under this Act :

(*a*) the Central Board shall be bound by such directions in writing the Central Government may give to it; and

(*b*) every State Board shall be bound by such directions in writing as the Central Government or the State Government may give to it:

Under section 19

The entire National Capital Territory of Delhi has been declared as water pollution prevention control area.

Under section 21

Officials of DPCC can take samples of the water effluent from any industry stream or well or sewage sample for the purpose of analysis

Under section 23

Officials of the state boards can enter any premises for the purpose of examining any plant, record, register etc. or any of the functions of the Board entrusted to him.

Under section 24

No person shall discharge any poisonous, noxious or any polluting matter into any stream, or well or sewer or on land.

Under section 25

No person shall without the previous consent to establish shall

(*a*) Establish or take any step to establish any industry, operation or process or any treatment and disposal system for any extension or addition thereto, which is likely to discharge sewage or trade effluent into a stream or well or sewer or on land or

(*b*) Bring into use any new or altered outlet for the discharge of sewage or

(*c*) Begin to make any new discharge of sewage.

Under this section the state board may grant consent to the industry after satisfying itself on pollution control measures taken by the unit or refuse such consent for reasons to be recorded in writing.

Under section 27

A state board may from time to time review any condition imposed by it on the person under section 25 and 26 and may vary or revoke that condition.

Under section 28

Any person aggrieved by the order made by the State Board under Section 25, 26 or section 27 may within thirty days from the date on which the order is communicated to hi, prefer an appeal to such authority (referred to as the appellate authority) as the State Govt. may think fit to constitute (in case of NCT of Delhi Appellate authority under this section is Financial Commissioner, Delhi Administration).

Under section 33

The State Board can direct any person who is likely to cause or has cause the pollution of water in street or well to desist from taking such action as is likely to cause its pollution or to remove such matters as specified by the Board through court.

Under section 33A

DPCC can issue any directions to any person, officer or authority, and such person, officer or authority shall be bound to comply with such directions. The direction includes the power to direct:-

(*i*) The closure, prohibition of any industry.

(*ii*) Stoppage or regulations of supply of electricity, water or any other services.

Under section 43

Whoever contravenes the provisions of Section 24 shall be punishable with imprisonment for a term which shall not be less than one year & six months but which may extend to six years with fine.

Under section 45

If any who has been convicted of any offence under section 24, or Section 25 or Section 26 is again found guilty of an offence involving a contravention of the same proviso shall be on the second and on every subsequent conviction be punishable with imprisonment for a term which shall not less than two years but which may extend to seven years with fine.

Under section 45A

Whoever contravenes any of the provisions of this act or fails to comply with any order or direction given under this act, for which no penalty has been elsewhere provided in this Act, shall be punishable with imprisonment which may extend to three months or with fine which may extend to ten thousand rupees or with both.

The Air (Prevention and Control of Pollution) Act, 1981

[Act No. 14 of 1981]

An Act to provide for the prevention, control and abatement of air pollution, for the establishment, with a view to carrying out the aforesaid purposes, of Boards, for conferring on and assigning to such Boards, powers and functions relating thereto and for matters connected therewith

Whereas decisions were taken at the United Nations Conference on the Human Environment held in Stockholm in June, 1972, in which India participated, to take appropriate steps for the preservation of the natural resources of the earth which, among other things, include the preservation of the quality of air and control of air pollution; And whereas it is considered necessary to implement the decisions aforesaid insofar as they relate to the preservation of the quality of air and control of air pollution.

In the Act, air pollution has been defined as 'the presence of any solid, liquid or gaseous substance (including noise) in the atmosphere in such concentration as may be or tend to be harmful; to human beings or any other living creatures or plant or property or environment.' Noise pollution has been inserted and pollution in the Act in 1987.

The population in the Air (Prevention and Control of Pollution) Act divided in the following components :

1. Short Title, Extent, Commencement and Definitions
2. Components of Central and State boards for the prevention and control of air pollution
3. Powers and functions of boards: the Functions of Central Board, Functions of State Boards and Power to give Directions are given under the section 16, 17 and 18 respectively.
4. Prevention and Control of Air Pollution: The Power to declare Air Pollution control areas, Power to give Instructions for ensuring Standards for Emission from automobiles and Power to put the restrictions on use

of certain industrial plants are mentioned under the section 19, 20 and 21 respectively.

5. Fund, accounts and Audit
6. Penalties and Procedure
7. Miscellaneous: the power of state government to supersede state board, power of central government to make rules and power of state government to make rules describes under the section 47, section 53 and section 54 respectively

Under section 16: Functions of Central Board

(1) Subject to the provisions of this Act, and without prejudice to the performance of its functions under the Water (Prevention and Control of Pollution) Act, 1974 (6 of 1974), the main functions of the Central Board shall be to improve the quality of air and to prevent, control or abate air pollution in the country.

(2) In particular and without prejudice to the generality of the foregoing functions, the Central Board may :

(*a*) advise the Central Government on any matter concerning the improvement of the quality of air and the prevention, control or abatement of air pollution;

(*b*) plan and cause to be executed a nation-wide programme for the prevention, control or abatement of air pollution;

(*c*) co-ordinate the activities of the State Boards and resolve disputes among them;

(*d*) provide technical assistance and guidance to the State Boards, carry out and sponsor investigations and research relating to problems of air pollution and prevention, control or abatement of air pollution;

(*e*) plan and organize the training of persons engaged or to be engaged in programmes for the prevention, control or abatement of air pollution on such terms and conditions as the Central Board may specify;

(*f*) organize through mass media a comprehensive programme regarding the prevention, control or abatement of air pollution;

(*g*) collect, compile and publish technical and statistical data relating to air pollution and the measures devised for its effective prevention, control or abatement and prepare manuals, codes or guides relating to prevention, control or abatement of air pollution;

(*h*) lay down standards for the quality of air;

(*i*) collect and disseminate information in respect of matters relating to air pollution;

(*j*) perform such other functions as may be prescribed.

(3) The Central Board may establish or recognise a laboratory or laboratories to enable the Central Board to perform its functions under this section efficiently.

(4) The Central Board may :

(*a*) delegate any of its functions under this Act generally or specially to any of the committees appointed by it;

(*b*) do such other things and perform such other acts as it may think necessary for the proper discharge of its functions and generally for the purposes of carrying into effect the purposes of this Act.

Under section 17: Functions of State Boards

(1) Subject to the provisions of this Act, and without prejudice to the performance of its functions, if any, under the Water (Prevention and Control of Pollution) Act, 1974 (6 of 1974), the functions of a State Board shall be-

(*a*) to plan a comprehensive programme for the prevention, control or abatement of air pollution and to secure the execution thereof;

(*b*) to advise the State Government on any matter concerning the prevention, control or abatement of air pollution;

(*c*) to collect and disseminate information relating to air pollution;

(*d*) to collaborate with the Central Board in organizing the training of persons engaged or to be engaged in programmes relating to prevention, control or abatement of air pollution and to organize mass-education programme relating thereto;

(*e*) to inspect, at all reasonable times, any control equipment, industrial plant or manufacturing process and to give, by order, such directions to such persons as it may consider necessary to take steps for the prevention, control or abatement of air pollution;

(*f*) to inspect air pollution control areas at such intervals as it may think necessary, assess the quality of air therein and take steps for the prevention, control or abatement of air pollution in such areas;

(*g*) to lay down, in consultation with the Central Board and having regard to the standards for the quality of air laid down by the Central Board, standards for emission of air pollutants into the atmosphere from industrial plants and automobiles or for the discharge of any air pollutant into the atmosphere from any other source whatsoever not being a ship or an aircraft:

PROVIDED that different standards for emission may be laid down under this clause for different industrial plants having regard to the quantity and composition of emission of air pollutants into the atmosphere from such industrial plants;

(*h*) to advise the State Government with respect to the suitability of any premises or location for carrying on any industry which is likely to cause air pollution;

(*i*) to perform such other functions as may be prescribed or as may, from time to time, be entrusted to it by the Central Board or the State Government;

(*j*) to do such other things and to perform such other acts as it may think necessary for the proper discharge of its functions and generally for the purpose of carrying into effect the purposes of this Act.

(2) A State Board may establish or recognize a laboratory or laboratories to enable the State Board to perform its functions under this section efficiently.

Under section 18: Power to give directions

In the performance of its functions under this Act :

(*a*) the Central Board shall be bound by such directions in writing as the Central Government may give to it; and

(*b*) every State Board shall be bound by such directions in writing as the Central Board or the State Government may give to it:

Provided that where a direction given by the State Government is inconsistent with the direction given by the Central Board, the matter shall be referred to the Central Government for its decision.

Under section 19

The State Government may, after consultation with the State Board, by notification in the Official Gazette, declare in such manner as may be prescribed, any area or areas within the State as air pollution control area or areas for the purposes of this Act.

Under section 20

With a view to ensuring that the standards for emission of air pollutants from automobiles laid down by the State Board. the State Government shall, in consultation with the State Board, give such instructions as may be deemed necessary to the concerned authority in charge of registration of motor vehicles under the Motor Vehicles Act, 1939(4 of 1939), and such authority shall contained in that Act or the rules made there under be bound to comply with such instructions.

Under section 21

Subject to the provisions of this section, no person shall, without the previous consent of the State Board, establish or operate any industrial plant in an air pollution control area

Under section 24

Subject to the provisions of this section, any person empowered by a State Board shall have a right to enter at all reasonable times in any place for the purpose of performing any of the functions of the State Board entrusted to him or for the purpose of determining whether any such functions are to be performed or whether any provisions of this Act or the rules made there under or any notice, order, direction or authorization served, made, given or granted under this Act is being or has been complied with;

Under section 25

For the purposes of carrying out the functions entrusted to it, the State Board or any officer empowered to get any information (including information regarding the types of air pollutants emitted into the atmosphere and the level of the emission of such air pollutants) from the occupier or any other person carrying on any industry or operating any control equipment or industrial plant and for the purpose of verifying the correctness of such information, the State Board or such officer shall have the right to inspect the premises where such industry, control equipment or industrial plant is being carried on or operated.

The Environmental (Protection) Act, 1986

[23rd May, 1986]

The Act comes into force on Nov. 19, 1986, the birth anniversary of our late Prime Minister Mrs. Indra Gandhi, who was a pioneer of environmental protection issues in our country.

An Act to provide for the protection and improvement of environment and for matters connected there with:

WHEREAS the decisions were taken at the United Nations Conference on the Human Environment held at Stockholm in June, 1972, in which India participated, to take appropriate steps for the protection and improvement of human environment;

AND WHEREAS it is considered necessary further to implement the decisions aforesaid in so far as they relate to the protection and improvement of environment and the prevention of hazards to human beings, other living creatures, plants and property;

It extends to the whole of India. It shall come into force on such date as the Central Government may, by notification in the Official Gazette, appoint and different dates may be appointed for different provisions of this Act and for different areas. This Act gives the definitions of

(*a*) "environment" includes water, air and land and the inter- relationship which exists among and between water, air and land, and human beings, other living creatures, plants, micro-organism and property;

(*b*) "environmental pollutant" means any solid, liquid or gaseous substance present in such concentration as may be, or tend to be, injurious to environment;

(*c*) "environmental pollution" means the presence in the environment of any environmental pollutant;

(*d*) "hazardous substance" means any substance or preparation which, by reason of its chemical or physico-chemical properties or handling, is liable to cause harm to human beings, other living creatures, plant, micro-organism, property or the environment;

Silent Features of EPA, 1986

1. Subject to the provisions of this Act, the Central Government shall have the power to take all such measures as it deems necessary or expedient for the purpose of protecting and improving the quality of the environment and preventing controlling and abating environmental pollution. Central Government have the power for such measures may include measures with respect to all or any of the following matters, namely :

   (*a*) planning and execution of a nation-wide programme for the prevention, control and abatement of environmental pollution;

   (*b*) laying down standards for the quality of environment in its various aspects;

   (*c*) laying down standards for emission or discharge of environmental pollutants from various sources whatsoever: provided that different standards for emission or discharge may be laid down under this clause from different sources having regard to the quality or composition of the emission or discharge of environmental pollutants from such sources;

   (*d*) restriction of areas in which any industries, operations or processes or class of industries, operations or processes shall not be carried out or shall be carried out subject to certain safeguards;

   (*e*) laying down procedures and safeguards for the prevention of accidents which may cause environmental pollution and remedial measures for such accidents;

   (*f*) laying down procedures and safeguards for the handling of hazardous substances; examination of such manufacturing processes, materials and substances as are likely to cause environmental pollution;

   (*g*) carrying out and sponsoring investigations and research relating to problems of environmental pollution;

   (*h*) inspection of any premises, plant, equipment, machinery, manufacturing or other processes, materials or substances and giving, by order, of such directions to such authorities, officers or persons as it may consider necessary to take steps for the prevention, control and abatement of environmental pollution;

   (*i*) establishment or recognition of environmental laboratories and institutes to carry out the functions entrusted to such environmental laboratories and institutes under this Act;

(*j*) collection and dissemination of information in respect of matters relating to environmental pollution;

(*k*) preparation of manuals, codes or guides relating to the prevention, control and abatement of environmental pollution;

(*l*) such other matters as the Central Government deems necessary or expedient for the purpose of securing the effective implementation of the provisions of this Act.

2. The Central Government may appoint officers with such designation as it thinks fit for the purposes of this Act and may entrust to them such of the powers and functions under this Act as it may deem fit. The officers appointed shall be subject to the general control and direction of the Central Government or, if so directed by that Government, also of the authority or authorities.
3. Notwithstanding anything contained in any other law but subject to the provisions of this Act, the Central Government may, in the exercise of its powers and performance of its functions under this Act, issue directions in writing to any person, officer or any authority.
4. The Central Government may, by notification in the Official Gazette, make rules in respect of all or to protect and improve environment
5. In particular, and without prejudice to the generality of the foregoing power, such rules may provide for all or any of the following matters, namely :

(*a*) the standards of quality of air, water or soil for various areas and purposes;

(*b*) the maximum allowable limits of concentration of various environmental pollutants (including noise) for different areas;

(*c*) the procedures and safeguards for the handling of hazardous substances;

(*d*) the prohibition and restrictions on the handling of hazardous substances in different areas;

(*e*) the prohibition and restriction on the location of industries and the carrying on process and operations in different areas;

(*f*) the procedures and safeguards for the prevention of accidents which may cause environmental pollution and for providing for remedial measures for such accidents.

6. No person carrying on any industry, operation or process shall discharge or emit or permit to be discharged or emitted any environmental pollutants in excess of such standards as may be prescribed.
7. No person shall handle or cause to be handled any hazardous substance except in accordance with such procedure and after complying with such safeguards as may be prescribed.

8. Where the discharge of any environmental pollutant in excess of the prescribed standards occurs or is apprehended to occur due to any accident or other unforeseen act or event, the person responsible for such discharge and the person in charge of the place at which such discharge occurs or is apprehended to occur shall be bound to prevent or mitigate the environmental pollution caused as a result of such discharge.
9. Subject to the provisions of this section, any person empowered by the Central Government in this behalf shall have a right to enter, at all reasonable times with such assistance as he considers necessary, any place for the purpose of performing any of the functions of the Central Government entrusted to him;
10. The Central Government or any officer empowered by it in this behalf shall have power to take, for the purpose of analysis, samples of air, water, soil or other substance from any factory, premises or other place in such manner as may be prescribed.

Environmental (Protection) Rule, 1986

For the purpose of protecting and improving the quality of the environment and preventing the abating pollution, standards have been specified under Schedule 1-IV of Environmental (Protection) Rule, 1986 for emission of gaseous pollutant and discharge of effluent/waste.

Under the Environmental (Protection) Rule, 1986, an amendment was made in 1994 for Environmental Impact Assessment (EIA) of various development projects. There are 29 types of project which required clearance form the Central government before establishing.

Other requires clearance form the SPCB. The proposed project or expansion activity proponent has to provide EIA report, risk assessment report, No Objection Certificate (NOC) from SPCB, commitment regarding availability of water and electricity.

EPA, 1984 has also made provision for environmental audit as a means of checking whether or not a company is complying with the environmental laws and regulations.

Issues involved in the Enforcement of Environmental Legislations

There are certain drawbacks of the Acts due to which the Government has faces different problems in their enforcement. They are discussed as below:

- The offenders of the Wild life (Protection) Act are not subjected to vary harsh punishment. The offender can be fined of Rs. 25,000 or imprisoned for 3 years or both which is not enough to deal with the poachers and wild life traders.

- The Jammu & Kashmir state has its own Wildlife Act which does not follow the Central Wildlife Act and hunting and trading of several endangered species which are prohibited in other states are allowed in J & K. As a result the wildlife traders can get wildlife articles illegally from other state.
- In Wild life (Protection) Act the ownership certificates for animal articles (like tiger, leopard skin etc.) are permissible which very often serve as a tool for illegal trading.
- Since the Chairmen and Members of the Board are often drawn either from forest services or administrative services, they don't have technical background in pollution control and they find difficulties in providing proper guidance to their subordinate.
- To enforce the punishment to the polluters the State Pollution Control Board has to file a case in the lower court and due to insufficient availability of time and over work load the lower courts are not able to solve the cases on time and it result a number of cases pending for years.
- Due to lack of the sufficient equipments the Pollution Control Boards are not able to investigating, suggesting for remedial measure and monitoring the implementation wherever required.
- The pollution Control Boards are expected to function as statutory autonomous bodies, but due to over dependence on the Government for their existence they can not function properly.
- Due to insufficient funds the Pollution Control Boards are not able to pursue the legal actions.

6.11. PUBLIC AWARENESS

Education has been the utmost priority of any civilized society since the ages and will remain same for all eternity. The rise and fall of the human race depends largely on the quality of education being provided through the available modes and agencies. Environmental Education is a process of recognizing values and clarifying concepts in order to develop skills and added tools necessary to understand and appreciate the inter-relationship among man, his culture and his bio-physical surrounding. It creates an overall perspective, which acknowledges the fact that natural environment and man-made environment are interdependent. Its main objective is to improve public awareness and understanding of environmental issues with a view to promote the conservation and wide use of nature and natural resources.

The Ministry of Environment and Forests (MoEF), Govt. of India, has been interacting with the Ministry of Human Resource Development (MHRD), NCERT, State Departments of Education etc. to ensure that environmental components are adequately covered at the school levels by infusion into the school curricula at various levels. The Ministry has also taken an initiative to educate people through a structural course on environment education and it presently working out a frame work for environmental appreciation courses in consultation with IGNOU. (Source: MoEF).

Environmental Education, Awareness and Training plays a significant role in encouraging and enhancing people's participation in activities aimed at conservation, protection and management of the environment, essential for achieving sustainable development.

Case Study

In Punjab, the environment education at the Primary school in the state is adequate, where the syllabus designed by Punjab State Education Board is based on NCERT syllabus. For the primary classes, a project "Environment Education in school system" has been undertaken by Punjab education Department, which is been implemented by the Ministry of Environment and Forests, Govt. of India with the financial aid from the World Bank. Various activities are being undertaken to strengthen the Environment Education, which includes the development of Educational material like Greening of Books of classes VI, VII and VIII (Hindi, Punjabi, Science & Social studies),development of Environment Education programs & also Trainings are being organized for educational administrators, Teachers & NGO's.

Non formal education in the state is also being prompted by the department of environment through PSCST and Punjab Pollution Control Board. Under this Program, important environment days are celebrated from time to time, environmental workshop are conducted and competition, seminars, lectures and Public camps on Environment awareness are organized. A number of NGOs of the state have been actively participating in National Environment Awareness campaign organized by MoEF, GoI.

6.12. HUMAN VALUES

Human values are the values of the human beings for the human beings and by the human beings. Morals and values are the foundation of human values. Just as a building has a foundation of concrete, the building of human values rests on morals and values. Human values can also be defined as values of life.

One's ethical standards are also greatly influenced by values and morals. People who place a high value on money and material possessions may not have strong ethical standards regarding behaviors that facilitate the accumulation of that wealth. People who value the quality of life enjoyed by all living creatures will probably have strong ethics with respect to the proper treatment of others.

India is said to be country of multi-cultures. As in olden days means of transport and communication were very poor; large number of languages, habits, values developed which continue even today. These facts of history made India a multi culture nation but the core values are same throughout the country. Thus, there is unity in diversity.

Human Values = Morals + Ethics + Values

6.12.1. Concept of Values

The word Value is derived from the Latin word "Valerie" which means "to be strong and vigorous." The German Philosopher Friedrich Nietzsche first used the word 'Values' in plural to denote moral beliefs and attitudes in 1880.

In the modem democratic society, the word 'Values' has come to be used in plurals for over a century. Thus, the universal values like Truth, Beauty, Goodness, Charity, Equality, Compassion, Responsibility, and Justice cannot be replaced by any subjective values.

As per dictionary, value gives the following meaning: relative worth, utility or importance, degree of excellence, quality, intrinsically valuable. Value literally means something that has a price, something precious, dear and worthwhile, hence something; one is ready to suffer and sacrifice for, if necessary one is ready, even to die for it or quality of being useful or worthwhile or important

The Encyclopedia of social sciences refers to values, as "interest, pleasure, duties, likes, preferences, moral obligations, desires, wants, needs and many modalities of social orientation".

Edward Spranger defines the values "as the constellation of likes, dislikes, view points inner inclinations, rational and irrational judgments, prejudices, and association pattern that determine a person's view of the world."

From a philosophical point of view values are those standards or codes for moral behavior conditioned by one's cultural trends and guarded by conscience, according to which a human being is supposed to conduct himself and shape his life patterns by integrating his beliefs, ideas and attitudes to realize cherished ideals and aims of life. Thus from philosophical point of view, value is directly related to what one believes or thinks.

Intellectually, value may be defined as a "concept which is accepted by the sub-conscious mind, is understood by all and perceived by the individual."

In other words, values represent basic convictions or enduring belief that "a – specific mode of conduct or end-state of existence is personally or socially preferable to an opposite or a converse mode of conduct or end-state of existence."

Values are a Set of principles or standards of behavior that are regarded as desirable, important and held in high esteem by society .They are based on the moral norms or standards. Value is the importance or usefulness of something. Since we give importance, we try to protect values. We value money and jewellery, so keep it under lock. Values reflect personal attitudes and judgments, decisions and choices, behavior and relationships, dreams and vision. They give rise to beliefs, thoughts, feelings and actions.

Giving respect to others, good listening, sincerity, admiring good actions of others, understanding the responsibility, expressing views, projection of individuality are the Values which help any individual to come up in life.

To know about value, we have to understand its meaning contribute to the welfare of the larger social unit such as the family, the community and the nation of which the individual is a member.

Value may have two possible forms: The first, which is simpler provides a list of values such as courage, beauty, honesty and compassion. The second tries to answer what exactly is known by beauty or courage or friendship. It may seem to be relatively easy to compile a list of different values, but difficult to answer the second part of the question because, the list of values may change from culture to culture and time to time. We cannot estimate whether the particular values are useful or not, if not one knows the nature of value in general.

Values are taken from life, environment, from self, society and culture, and, beyond all, from the ideal, transcended dimension of human existence and experience. As per the opinions of psychologists and social scientists, values are mere preferences and aversions, as, desirable goals, emotions and interests.

Value makes actions, characters, traits and objects of anyone good or bad. Evaluating the moral qualities of people or actions and their non-moral characters enviably raises the qualities of the nature or source of those values

The unity and transcendence of the value system, grounded in both human actualities and human possibilities, are seldom envisaged by the sciences of man, society and culture. Values and the process of valuation and development of the value system are analyzed in strikingly divergent and piecemeal manner by the various psychological, social and philosophical studies according to their image of man and their conception of human nature and destiny.

With the aid of values, man delays his satisfactions and sets his mind and behavior to distant and sometimes unrealizable goals, strivings and ideals. Values guide us to do the right things and be morally sound. His value judgment enables him to select between alternative courses of behavior, and solve chronic inner tensions and conflicts by accepting standards and demands that control him from beyond.

Human values emerge due to two factors:

- The impingement of society and its meanings and norms on the fulfillment of the individual's needs or drives.
- The introduction of his own awareness, choice and judgment in need fulfillment.

These two processes are interdependent. Values are essentially social products, and at the same time involve the individual's assumption of certain common goals and purposes of the social milieu that have become a part of him.

Thus values may be defined as stable, regulative, future-oriented patterns of expression of formalized social life in its relations to man, society and world.

The process of valuation starts at the biological dimension of health, efficiency, well-being and security. Here the generic values are life-maintenance and enhancement and the values are wealth, status, equality and justice. It then reaches out and fastens upon potential wholeness, balance and serenity and the realization of being at the ontological or a transcendent dimension and the values here are truth, beauty, love, harmony and holiness. The principal branches of human learning are biology, psychology, social science, ethics and metaphysics which can show multi-dimensional frames of human values. Their findings and explorations now need to be organized into a coherent, general theory of values for the guidance of man, society and civilization.

Values are indispensable, encountered by man everywhere and in all thoughts, relations and actions, no matter what he thinks or believes or how he responds to his fellowman, society and cosmos. Values are subjective, laden with feelings, and flexible and at the same time objective, rational and regulative.

We have values about work, equality, authority self respect, freedom, honesty, obedience and other matters. These values provide the basic foundation for understanding a person's personality, perceptions and attitudes.

6.12.2. Characteristics of Values

From the concept of value and definitions discussed above the following characteristics of Human values can be formulated. These characteristics can help in understanding the nature of human values:

- All values are not equal in merit. They are usually organized in terms of hierarchy and are related to each other both internally as well as externally.
- Values are at the core of personality and are a powerful force affecting behavior.
- Values are integrated experiences that touch simultaneously all dimensions of human adaptation, organic, social and cultural and transcend them in all their propriety towards orientation. Values mold the human nature in our social environment.
- Values contain a judgmental element in that they carry an individual's ideas as to what is right, or desirable.
- Values have both content and intensity attributes. The content attribute indicates that a mode of conduct or end-state of existence is important. The intensity attribute specifics how important it is.
- A value is always an experience, never a thing or an object. Psychologically every individual is unique. Thus values vary from person to person that is no two individuals can have same value system. However, all men everywhere possess the same values in different degrees and levels.
- Values are ideas, abstractions and conceptions of the desirable that are relevant to selective behavior. Values are a special kind of attitudes functioning as standards by which choices are evaluated. Values are something that we consider worthwhile.

- Values cannot be directly observed, seen or perceived but have to be inferred from some value indicates like, what people say and do when they value 'economic justice', or loyalty. The behavioral action and statements of people give clues about their values.
- Values are not fixed, but they change over time.
- Many values are relatively stable and enduring. This is because of the way in which they are originally learned.
- Values can also derive their origin from a variety of sources like literature, philosophy, religion, science, social customs and the constitution.
- Values are the desirables and states, which act as a guide to human endeavor, or the most general statements of legitimate ends, which guide social action.
- Every individual takes care of things, which they value. Thus values have emotional dimensions, a powerful emotional commitment, a strong liking for something. A value can be defined compared, analyzed and generalized. They can be used as standards to judge the worth of things.
- Values are normative standards by which human beings are influenced in their choice among the alternative courses of action which they perceive.
- Values influence our thoughts, feelings and action. They guide us to do the right things.
- Values don't exists as objects in space and time, but are established by judgment by judging things, qualities, events or actions from a personal point of view.
- Every individual have values but they may or may not be aware of what values are. Values exist in minds of individual, which act as guiding principles of life.
- All of us have a hierarchy of values that forms our value system. But everyone does not hold the same values.

6.12.3. Types of Values

There are several methods to use to categorize the value. Each plays an important role in lives of people, but these values don't all play equal roles in the formation of moral standards and moral norms. These are the most common categories that value are put into:

Universal value : These are value that there is nearly unanimous agreement as to the importance of them. Something is of universal value if it has the same value or worth for all, or almost all, people. The five Universal Values are truth, righteous conduct, peace, love and non-violence.

Preference value : The expression of preference is the expression of some value people hold. When people say that they prefer to play sports, they are saying that we value that activity.

Instrumental value : These are personal value that can be used to get something else. In other words the value is an instrument which allows you to get some other things. In fact, it has been argued that values represent a major influence on how we process information how we feel about, Issues, and how we behave. Examples of these would include ambition, courage, persistence, politeness etc. They are not the end but a mean of achieving terminal values. Instrumental values are those values concerning the way we approach end states. These relate to means for achieving desired results. That is, do we believe in ambition. Cleanliness, honesty or obedience, courage, etc. Below are listed several Instrumental Values:

- Assertiveness; standing up for yourself
- Being helpful or caring toward others
- Dependability; being counted upon by others
- Education and intellectual pursuits
- Hard work and achievement
- Obedience; following the wishes of others
- Open-mindedness; receptivity to new ideas
- Self-sufficiency; independence
- Truthfulness; honesty
- Being well-mannered and courteous toward others

Terminal values are those end-state goals that we praise such as comfortable life, a sense of accomplishment, equality among all people, self-respect, family security. A list of Terminal Values are listed below :

- Happiness; satisfaction in life
- Knowledge and wisdom
- Peace and harmony in the world
- Pride in accomplishment
- Prosperity; wealth
- Lasting friendships
- Recognition from peers
- Salvation; finding eternal life
- Security; freedom from threat

Intrinsic value : Something which has intrinsic value is valued purely for itself - it isn't used simply as a means to some other end and it isn't simply "preferred" above other possible options. Something has intrinsic worth simply because of what it is and not necessarily what it will lead to or because of its acceptance. Some possible examples of intrinsic value would include beauty, artistic expression, and happiness. We value them because they are an important aspect of life.

Prerequisite value : These are value that are necessary before you can get to some bigger goal. It is similar to the prerequisite course that you must take in order to get to the more advanced course. Some good examples of this type of value include safety (which is needed before people can even think about having anything else), Justice (which is needed before we can move onto equality), or the common good (which must be honored if we can ever get to a state of peace).

Paramount values : Think of this type of value like one think of Paramount Studios with the large mountain. It is the value which is above all other things. Some examples of this might include freedom (which many people have given up their lives for and see as essential to a decent life) or sanctity of life (which if we do not value or have renders everything else worthless).

Operative values : This type of values are the ways that we make judgments on how to live the rest of our lives. We use these values as the overarching and guiding principles which tell us what is always right and wrong. These are things such as Integrity, Honesty, and Loyalty.

The basis for values is morals. There are different types of values like economic value, time value, psychological and moral or philosophical values. Values can also be classified as personal values, social values, and behavioral values, moral and spiritual values. Broadly values are of various kinds and named according to their specifications. The values may be categorized as in the following manner:

(1) Personal Values

Personal values refer to those values which are desired and cherished by the individual irrespective of his or her social relationship, i.e. What is profitable and good for the welfare of the individual. Personal values developed very early in life may be resistant to change. Personal values denote a sense of right or wrong, good or bad and other judgmental criteria based on our strong sense of what the ideal ought to be. They may be derived from those of particular groups or systems, such as culture, religion, and political party. However, personal values are not universal; one's family, nation, generation and historical environment help determine one's personal values. The individual determines his own standards of achievement and attains these goals without explicit interaction with any other person. Examples: Courage, Creativity, Contentment, Determination, Ambition, Diligence, Excellence, Honesty, Cleanliness, Hope, Dignity of labor, Maturity, Regularity, Self-confidence, Self-concept, Self-motivation, Punctuality, Simplicity, etc. According to Jones, E.E. and Gerard (1967) anything "for which the individual strives or approaches extols, embraces, voluntarily consumes, incurs expense to acquire a positive value. Anything that the individual avoids, escapes from, deplores, rejects, or attacks a negative value."

(a) Negative Values

- Anger
- Meanness

- Arrogance
- Crookedness
- Greed
- Lust

They generate negative thoughts.

(b) Positive Values

- Integrity, honesty
- Truthfulness
- Kind heartiness, humility
- Friendliness
- Faith
- Self-respect
- Open mindedness
- Creativity
- Civil sense
- Simplicity
- Forgiveness
- Poise
- Detachment, etc.

They generate positive thoughts.

(2) Social Values

Social values abstract standards or empirical variables in social life which are believed to be important and/or desirable. These refer to those values, which are oriented towards others. These values are developed and practiced because of our association with others. The social values are always practiced in relation to our neighbor, society, nation and the world. Social values are local and temporary in application. These values change with the change of social circumstances. Examples: Brotherhood, concern for environment protection and conservation, courtesy, forgiveness, justice, love, freedom, friendship, team spirit, sympathy, service sharing, hospitality, patience, tolerance, repentance, etc.

(3) Moral Values

The moral values are those values which are related to an individual's character conforming to what is right and virtuous. They reveal a person's self-control. In other words these values are relating to ethics. Examples: honesty, integrity, sense of responsibility, etc. It is observed that the list of moral values includes both social values and personal values.

(4) Cultural Values

Cultural values commonly held standards of what is acceptable or unacceptable, important or unimportant, right or wrong, workable or unworkable, etc., in a community or society. They are the values which involve the survival of the culture.

(5) Essential Values

These involve the basic nature of man himself.

(6) Institutional Values

Which include the political and establishment values.

6.12.4. Principles of Values

Though all values are at par with each other, the following principles may help an individual for preference of values:

1. Intrinsic values are to be preferred to those that are extrinsic.
2. Values that are productive and relatively permanent are to be preferred to those that are less productive and less permanent. Some values like the economic values tend to be used up in the activities of life. Thus the economic value while necessary for life is not a permanent value. The social, aesthetic, intellectual and religious values tend to provide more permanent satisfaction than material values.
3. One ought to select values on the basis of self-chosen ideals or ends. The values we seek ought to be our values and these values ought to be consistent with each other and with the demands life makes on us.

Core Values

Though there are wide variations in various cultural variables in the country, there are some core values which are common to all. These widely held beliefs interests, feelings, relationships has brought similarity in diversity (variations) and unified the country. Let us take few of the core human values and understand what it is and how important they are.

Integrity

Integrity is consistency of actions, values, methods, measures, principles, expectations and outcome. As a holistic concept, it judges the quality of a system in terms of its ability to achieve its own goals

The dictionary meaning of Integrity is the quality of being honest and morally upright. It is the state of being whole or unified. Thus integrity is an act of honesty. There is a close relationship between integrity and honesty. The consistent honest behavior develops integrity in an individual. The moral of honesty results in integrity.

Integrity may be seen as the quality of having a sense of honesty and truthfulness in regard to the motivations for one's actions. Integrity helps the other person to predict your behavior and action. If there is consistency in what you say and what you do, people will have belief in you. Lack of integrity, results in unpredictable behavior and action. This value will help the engineers to gain confidence from management and help him to move up ladder. All the organizations are looking for this value in employees.

Ethic

Ethics is a branch of philosophy which seeks to address questions about morality, such as what the fundamental semantic, ontological, and epistemic nature of ethics or morality is, how moral values should be determined, how a moral outcome can be achieved in specific situations, how moral capacity or moral agency develops and what its nature is, moral values people actually abide by. This value will help the engineer to work hard and discipline himself and build team in an organization.

Service Learning

Service-learning is a method of teaching, learning and reflecting that combines academic classroom curriculum with meaningful service, frequently youth service, throughout the community. As a teaching methodology, it falls under the philosophy of experiential education. More specifically, it integrates meaningful community service with instruction and reflection to enrich the learning experience, teach civic responsibility, encourage lifelong civic engagement, and strengthen communities for the common good.

The National and Community Service Act of 1990, defines service-learning as: "a method under which students or participants learn and develop through active participation in thoughtfully organized service that is conducted in and meets the needs of a community; is coordinated with an elementary school, secondary school, institution of higher education, or community service program, and with the community; and helps foster civic responsibility; and that is integrated into and enhances the academic curriculum of the students, or the educational components of the community service program in which the participants are enrolled; and provides structured time for the students or participants to reflect on the service experience."

Civic Virtue

Civic virtues are personal habits and attitudes that are conducive to social harmony and group well-being. The identification of the character traits that constitute civic virtue has been a major concern of political philosophy. The term civility refers to behavior between persons and groups that conforms to a social mode (that is, in accordance with the civil society), as itself being a foundational principle of society and law.

Respects for Others

Respect is not a word. It is a way of life, and takes self-awareness and discipline to practice. People can't demand respect, although some falsely believe they can. We earn respect when we stand up for ourself, but also respect others.

Caring

Caring is loving feeling or showing care or exhibiting concern and empathy and/or compassion for others; having or displaying warmth or affection;

Sharing

Sharing is the joint use of a resource or space. In its narrow sense, it refers to joint or alternating use of an inherently finite good, such as a common pasture or a shared residence. It is also the process of dividing and distributing. Apart from obvious instances, which we can observe in human activity, we can also find many examples of this happening naturally in nature. When an organism takes in nutrition or oxygen for instance, its internal organs are designed to divide and distribute the energy taken in, to supply parts of its body that need it. Flowers divide and distribute their seeds. In a broader sense, it can also include the free granting of use rights to a good that is capable of being treated as a nonrival good, such as information. Still more loosely, "sharing" can actually mean giving something as an outright gift: for example, to "share" one's food really means to give some of it as a gift.

Honesty

Honesty is the best policy, we always hear from our elders. Honesty is the value of speaking truth and creating trust in minds of others. This includes all varieties of communication, both verbal and non-verbal. Honesty implies a lack of deceit. A statement can be strictly true and still be dishonest if the intention of the statement is to deceive its audience. Similarly, a falsehood can be spoken honestly if the speaker actually believes it to be true.

Honesty is typically considered virtuous behavior, and has strong positive connotations in most situations. A principal reason for this may be that honesty simplifies communication, in that honest statements can be trusted at face value, not necessarily as true, but as genuinely believed. Additionally, honesty helps to form bonds of trust in human relationships.

Conversely, dishonesty can be defined simply as behavior that is performed with intent to deceive. Lying, lying by omission, fraud, and plagiarism are all examples of this sort of behavior. Other examples can be doing one thing and telling the other, as if you are hiding something.

While there are a great many moral systems, generally speaking, honesty is considered moral and dishonesty is considered immoral. There are several exceptions, such as hedonism, which values honesty only insofar as it improves ones own sense of pleasure, and moral nihilism, which denies the existence of objective morality outright. Additionally, even in moral systems which approve in general of honesty over dishonesty, there are situations in which dishonesty may be preferable.

Courage

Courage, also known as bravery, will, intrepidity, and fortitude, is the ability to confront fear, pain, risk/danger, uncertainty, or intimidation. "Physical courage" is courage in the face of physical pain, hardship, or threat of death, while "moral courage" is the ability to act rightly in the face of popular opposition, shame, scandal, or discouragement.

Collegiality

Collegiality is the relationship between colleagues. Colleagues are those explicitly united in a common purpose and respecting each other's abilities to work toward that purpose. A colleague is an associate in a profession or in a civil or ecclesiastical office. Thus, the word collegiality can connote respect for another's commitment to the common purpose and ability to work toward it.

Empathy

Empathy is the capability to share and understand another's emotions and feelings. It is often characterized as the ability to "put oneself into another's shoes," Empathy does not necessarily imply compassion, sympathy, or empathic concern because this capacity can be present in context of compassionate or cruel behavior.

Self-confidence

The socio-psychological concept of self-confidence relates to self-assuredness in one's personal judgment, ability, power, etc., sometimes manifested excessively. Lack of self-confidence is called timidness or timidity. Being afraid of failure is a sign of this.

Character

Character is an evaluation of a particular individual's moral qualities. The concept of character can imply a variety of attributes including the existence or lack of virtues such as integrity, courage, fortitude, honesty, and loyalty, or of good behaviors or habits. The word "character" is derived from the Greek word charaktêr, which was originally used of a mark impressed upon a coin. Later and more generally, it came to mean a point by which one thing was told apart from others (Timpe 2007). There are two approaches when dealing with moral character: Normative ethics involve moral standards that exhibit right and wrong conduct. It is a test of proper behavior and determining what is right and wrong. Applied ethics involve specific and controversial issues along with a moral choice, and tend to involve situations where people are either for or against the issue (Timpe 2007).

Character primarily refers to the assemblage of qualities that distinguish one individual from another - although on a cultural level, the set of moral behaviors to which a social group adheres can be said to unite and define it culturally as distinct from others. Psychologist Lawrence Pervin defines character as "a disposition to behave expressing itself in consistent patterns of functioning across a range of situations" (Pervin 1994, p. 108).

Spirituality

Spirituality is matters of the spirit, a concept often but not necessarily tied to to a spirit world, a multidimensional reality and one or more deities. Spiritual matters regard humankind's ultimate nature and purpose, not as material biological organisms, but as spirits or energy with an eternal relationship beyond the bodily senses, time and the material world.

The spiritual is contrasted with the physical, matter and the temporary. A sense of connection is central of spirituality — connection to a reality beyond than the physical world and oneself, which may include an emotional experience of awe and reverence. Spirituality may also include the development of the individual's inner life through practices such as meditation and prayer, including the search for God, the supernatural, a divine influence, or information about the afterlife. Spirituality is the personal, subjective aspect of religion, mysticism, magic and occult.

Tolerance

Tolerance are terms used in social, cultural and religious contexts to describe attitudes which are "tolerant" (or moderately respectful) of practices or group memberships that may be disapproved of by those in the majority. In practice, "tolerance" indicates support for practices that prohibit ethnic and religious discrimination. Conversely, 'intolerance' may be used to refer to the discriminatory practices sought to be prohibited. The co-existence of various religions, cultures, languages has been possible due to tolerance. The co-existence of various religions, cultures and languages has been possible due to tolerance of people.

Liberty and Equality

Liberty is a concept of political philosophy and identifies the condition in which an individual has the right to act according to his or her own will. Individualist and classical liberal conceptions of liberty relate to the freedom of the individual from outside compulsion or coercion.

Social equality is a social state of affairs in which all people within a specific society or isolated group have the same status in a certain respect. At the very least, social equality includes equal rights under the law, such as security, voting rights, freedom of speech and assembly, and the extent of property rights. However, it also includes access to education, health care and other social securities. It also includes equal opportunities and obligations, and so involves the whole society. The principle of equality of sex, castes, religions, voting right and employment have been provided in various articles of the Indian constitution.

6.13. ETHICS

'Ethics' is defined in the Oxford English Dictionary as the 'science of morals; moral principles or code'. Ethics are considered the moral standards by which people judge behavior. Ethics are often summed up in what is considered the "golden rule"—do unto others as you would have them do unto you. While this makes sense as a general rule of thumb, it does not truly offer much in terms of helping one to make moral decisions in daily life.

Ethics are about making choices that may not always feel good or seem like they benefit you but are the 'right' choices to make. They are the choices that are examples of 'model citizen' and are the examples of 'golden rules' like; don't hurt, don't steal, don't be dishonest, don't lie. But if we take Ethics as a subjective philosophy then what will happen to these golden rules. Especially at the time of facing any ethical dilemma how one should decide - what is ethical and what's unethical? Organization provides rules, regulations, code of conduct, protocols which provide guidelines to work, it shows how to walk, but it does not show the correct path to walk on. *Ethical dilemmas* faced by managers are often more real to life and highly complex with no clear guidelines, whether in law or often in religion.

Ethics is the part of philosophy that talks about good and evil but today rarely any business school would have left which does not provide lessons on Ethics Management. Some philosophers call ethics the "science of morality", morality is what someone thinks or feels is good or bad. However, other philosophers believe that ethics is subjective.

In its simplest form, a code of ethics is nothing but a kind of mission statement for human beings to live by. It flows out of the overriding necessity to ensure that any given task is undertaken and completed in conformity with the principles of accountability, equity, propriety, prudence, probity, rectitude, honesty, integrity and transparency, and paying due regard to the imperatives of national security, public safety, health and sanitation, environmental protection and ecological balance. These fundamental requirements admit of no compromise.

Ethics is concerned with how a moral person *should* behave. Ethical values are beliefs concerning what is morally right and proper as opposed to what is simply correct or effective. An individual may personally believe that drinking is immoral. However, drinking is not, in and of itself, unethical. Game shows are a sophisticated way of gambling thus immoral but not unethical. Further, it is unethical to impose your personal moral values on another.

Ethical values go above cultural, religious, or ethnic differences. Ethical values embrace a more universal worldview.

Albert Einstein said, "A man's ethical behavior should be based effectually on sympathy, education, social ties; no religious basis is necessary."

Sometimes, we have to do things which are legally correct but not morally, or go ethically right but not morally. At workplace we have to keep a proper balance of Morality, Ethics and Law. The safest place is where your action covers all the three aspects (EML). From the organization's point of view we must not cross the boundaries of law *i.e.* perform within EL, EM, L or EML. Further, An employee may not always follow the law sometime he could listen to his ethics but one should always have the answer of breaking the law of Worklife.

In the words of great *Abraham Lincon – "When I do good, I feel good; when I do bad, I feel bad. That's my religion."*

Society often places certain standards of conduct upon the people which are governed together. Groups of people who agree about what constitutes major unethical acts, such as stealing or murder, work to have their viewpoint written into laws that govern the entire body of people. However, not all ethical areas are governed by law, such as lying to a parent or friend. Other areas of law, such as the debate on abortion or stem cell research, are considered ethical judgments made by governments on behalf of the people for the greater good.

Religious and political groups have often been criticized for trying to impose their own ethics and morals upon the larger society.

Companies and professional societies will also have ethical standards, often referred to as a code of conduct or rules of conduct, by which they and their employees abide. Each group determines their ethical stance based upon different guidelines, whether it is established principals, religious book, or group agreement. While members of the group may follow the ethical guidelines as a principle by which they believe everyone should act, they might also have a set of personal ethics which also governs their own behaviors.

Areas that ethical codes include:

- Honesty
- Objectivity
- Integrity
- Carefulness
- Openness
- Respect for intellectual property
- Confidentiality
- Responsible publication
- Responsible mentoring
- Respect for colleagues
- Social responsibility
- Non-discrimination
- Competence
- Legality
- Human subjects protection

Ethics in whatever profession or avocation does not stand by itself. A human being too cannot be ethical in certain respects and unethical in certain others. He is either ethical or he is not. That is why it becomes necessary to adopt a holistic approach to it.

The field of ethics (or moral philosophy) involves systematizing, defending, and recommending concepts of right and wrong behavior. Philosophers today usually divide ethical theories into three general subject areas: metaethics, normative ethics, and applied ethics.

The term "meta" means after or beyond, and, consequently, the notion of metaethics involves a removed, or bird's eye view of the entire project of ethics. We may define metaethics as the study of the origin and meaning of ethical concepts. When compared to normative ethics and applied ethics, the field of metaethics is the least precisely defined area of moral philosophy. Metaethics investigates where our ethical principles come from, and what they mean. Metaethical answers to these questions focus on the issues of universal truths, the will of God, the role of reason in ethical judgments, and the meaning of ethical terms themselves.

Normative ethics involves arriving at moral standards that regulate right and wrong conduct. In a sense, it is a search for an ideal litmus test of proper behavior. The Golden Rule is a classic example of a normative principle. Normative ethics takes on a more practical task, which is to arrive at moral standards that regulate right and wrong conduct. This may involve articulating the good habits that we should acquire, the duties that we should follow, or the consequences of our behavior on others. The key assumption in normative ethics is that there is only one ultimate criterion of moral conduct, whether it is a single rule or a set of principles.

Finally, applied ethics is the branch of ethics which consists of the analysis of specific, controversial moral issues such as abortion, animal rights, euthanasia, infanticide, animal rights, environmental concerns, homosexuality, capital punishment, or nuclear war. In recent years applied ethical issues have been subdivided into convenient groups such as medical ethics, business ethics, environmental ethics, and sexual ethics.

Morality

Again, what morally correct is not always ethically also. Morals and the expression, "moral values" are generally associated with a personal view of values. Personal morals tend to reflect beliefs relating to drinking, gambling, etc. They can reflect the influence of religion, culture, family and friends. These concepts and beliefs about right and wrong are often generalized and codified by a culture or group, and thus serve to regulate the behavior of its members. Conformity to such codification may also be called *morality*, and the group may depend on widespread conformity to such codes for its continued existence. The systematic study of morality is a branch of philosophy called ethics.

Law

When we talk about *Law at work place*, we discuss the set of rules imposed by authority. In other words, law is a rule or body of rules of conduct inherent in organization and essential to or binding upon all the employees.

6.13.1. Characteristics of an Ethical Person

1. Giving of one's best

The first and foremost distinguishing feature of an ethical personality is the spirit of service itself. It basically comprises seven ingredients:

- Excellence is the state or quality of excelling. Achieving a little more than what is strictly expected. Particularly in the field of business and organizations, excellence is considered to be an important value, and a goal to be pursued.
- Sensitivity is the quality of sensing intensely, instinctively and instantly the feelings, emotions and reactions of other human beings, and avoiding anything that will alienate people, and going out of the way to anticipate and help.
- Integrity is consistency of actions, values, methods, measures, principles, expectations and outcome. As a holistic concept, it judges the quality of a system in terms of its ability to achieve its own goals.
- Responsiveness is being receptive to ideas, suggestions and requests in a positive way and with "will do" spirit.
- Vigour is being energetic, tireless and dynamic in performance;
- Commitment is being passionately dedicated to cause or task and total identification with it.
- Empathy is the capability to share and understand another's emotions and feelings. It is often characterized as the ability to "put oneself into another's shoes".

All of these fall within the definition of being true to oneself, by maintaining the highest standards of conduct and behavior.

2. Loyalty to team associates

Second important characteristic is building up the self-esteem, competence and capabilities of one's associates. The first and foremost aim of everyone in any field of endeavor is to be an inspiration and an exemplar. No one who aspires to lead or manage can afford to lose sight of the fact that human beings are the most precious assets of any organization. Harmonious team work, strict discipline and total involvement are the mainsprings of achievement.

3. Loyalty to customers

Third is putting in the maximum efforts to live up to the essence of what has come to be known as 'customer delight'. Mahatma Gandhi defined this ethical imperative very concisely and to the point. His tribute to the greatness of the customer is quoted: "A customer is the most important visitor on our premises. He is not dependent on us. We are dependent on him. He is not an interruption on our work He is the purpose of it. He is not an outsider on our business. He is a part of it. We are not doing him afavour by serving him. He is doing us afar our by giving us an opportunity to do so."

Sensitivity makes for loyal customers who help enlarge the market share by attracting new ones by their endorsement of the quality of the product or service. As such, it is worthwhile taking due note of the framework for strengthening relations with customers.

4. Social responsibility

Last is the sense of social responsibility. This distinguishes a human being or organization from animals. The justification for this rests on a number of important considerations. Tangibly, social responsibility also makes business sense, from the standpoint of corporate entities in any sector, as it is an essential pathway to good public relations and general acceptance.

A marketing manager's social responsibility includes developing marketing programs and increasing the awareness and acceptability of social ideas and practices. An often associated concept is not-for-profit marketing, which champions a cause for the betterment of individuals and society rather than for financial profits.

6.13.2. Ethical Behavior

Ethical behavior refers to that conduct that is beyond reproach and is in accordance to the laid down standards of a society, organization or institution.

Ethical Behavior in Society

This varies from place to place. The behaviors considered unethical in one society can be acceptable in another. For instance, it is an unethical social behavior to engage in favoritism in the public. Refusing to relate with others on the basis of their age, sex and race are also unethical social behaviors.

Ethical Behavior Within the Work Place

This refers to conduct that organizations expect their employees to hold while at work. Most organizations have formulated documents referred to as 'codes of conduct', that set out the accepted behaviors within the work place. They have set out the rules and regulations that need to be adhered to by those in their employ. This is mainly because an individual's conduct can affect the relationships within an organization, thereby taking a negative toil on the overall performance of an organization.

Employee's Unethical Behavior

In any organization, it is unethical of an employee to be corrupt. One is not supposed to accept or give out bribes in exchange for services or goods. An employee who engages in corruption is said to have acted in an unethical manner. Taking excessive breaks or repeated sick leave are also unethical behaviors. This should only happen when there is a serious need to do so. Mild headaches do not count. Again, taking office supplies home is extremely unethical. Office property should be kept separately from one's personal property. In some organizations, this conduct can be translated to theft.

Employer's Unethical Behavior

It is not only employees who can behave in unethical ways. Employers also have a common code of conduct and are required to act in a given manner. For instance, an employer is not supposed to use the company money for his personal expenses.

Ethical Behavior in Academic Institutions

For students within learning institutions it is unethical behavior to cheat during examinations. Ethical behavior is to be cultivated within an individual from an early age. Without strong moral convictions, one can be easily swayed to engage in contrary behaviors.

6.13.3. Professional Ethics

Professional ethics concerns the moral issues that arise because of the specialist knowledge that professionals attain, and how the use of this knowledge should be governed when providing a service to the public. The definition of professional ethics endorsed by the working party is: 'giving of one's best to ensure that clients' interests are properly cared for, but in doing so the wider public interest is also recognized and respected'. This is taken from *'Professional Ethics and the Rules of Conduct of the Royal Institution of Chartered Surveyors'*, by Howard Land FRICS, CPD Study Pack, College of Estate Management.

Professional ethics are moral principles and standards of conduct guiding professionals in performing their functions. Codes of professional ethics are established by organizations of professionals, such as the American Institute of professionals and state societies. A violation of professional ethics will make the accountant subject to disciplinary action. The accountant's professional ethics affect the reputation of the profession and the confidence of the public. Some of the professional ethic requirements follow. The professionals should be *independent* of clients served so that objectivity and integrity are maintained. The professional must be professionally *competent*, exercise *due professional care*, and properly *plan and supervise* the engagement. The professionals must keep confidential information obtained from the client in the course of a professional engagement. A professional in public practice shall not engage in an incompatible occupation creating a conflict of interest.

Professional Responsibility

The professional carries additional moral responsibilities to those held by the population in general. This is because professionals are capable of making and acting on an informed decision in situations that the general public cannot, because they have not received the relevant training. For example, a layman member of the public could not be held responsible for failing to act to save a car crash victim because they could not give an emergency tracheotomy. This is because they do not have the relevant knowledge. In contrast, a fully trained doctor (with the correct equipment) would be capable of making the correct diagnosis and carrying out the procedure and we would think it wrong if they stood by and failed to help in this situation. We cannot be held accountable for failing to do something that we do not have the ability to do.

This additional knowledge also comes with authority and power. The client places trust in the professional on the basis that the service provided will be of benefit to them. It would be quite possible for the professional to use his authority to exploit the client. An obvious example is that of the dentist who carries out unneeded dental work on his paticnts in order to gain more money. It is likely that the patient will not have sufficient knowledge to question what is being done, and so will undergo and pay for the treatment.

The concept of Professional Ethics is partly comprised of what a professional should or should not do in the work place. It also encompasses a much greater part of the professionals life. If a professional is to have ethics then that person needs to adopt that conduct in all of his dealings. Things that are included are concepts like: professional respect, avoidance of dishonest or fraudulent activity such as plagiarism and the professional development of the individual. Another aspect of this is the enhancement of the profession and the industry within which the professional works. This concerns a professional's conduct and behaviour while carrying out their professional work. This then, is work for the good of the community and mankind.

Five basic assumptions underpin the understanding of ethics often covered by professional organizations:

Professional ethics is a process. Ethics are not a fixed text to be learnt once. It is a way of reviewing behavior against constantly changing standards. What may be ethical today, or in a particular society, may be viewed differently by others or at another time.

Human behavior is caused. There is a motive for all human behavior e.g., financial gain, power, compassion.

Actions have consequences. The equivalent of Newton's Third Law of Physics – 'Every force has an equal and opposite reaction'.

What is ethical depends on the individual's point of view. This is influenced by a variety of factors including published codes and statements.

Good ethical business practice rests on mutual vulnerability. We are each susceptible to the actions of others, and the way we are treated depends on how we treat others. Respect is not a right and must be earned.

The need for professional ethics is based upon the vulnerability of others. The client (and where relevant, opposing negotiators) must be protected from exploitation in a situation in which they are unable to protect themselves because they lack the relevant knowledge to do so.

Codes of Practice

Questions arise as to the ethical limits of the professional's responsibility and how power and authority should be used in service to the client and society. Most professions have internally enforced codes of practice that members of the profession must follow, to prevent exploitation of the client and preserve the integrity of the profession. This is not only to the benefit of the client but to the benefit of those belonging to the profession. For example, an American business may approach an engineer to certify the safety of a project which is not safe. Whilst one engineer may refuse to certify the project on moral grounds, the business may find a less scrupulous engineer who will be prepared to certify the project for a bribe, thus saving the business the expense of redesigning. Disciplinary codes allow the profession to draw a standard of conduct and ensure that individual practitioners meet this standard, by disciplining them from the professional body if they do not practice accordingly. This allows those professionals who act with conscience to practice in the knowledge that they will not be undermined commercially by those who have less ethical qualms. It also maintains the public's trust in the profession, meaning that the public will continue to seek their services.

Things to Do

- **Do** return value to your customer (internal and external) in all business decisions.
- **Do** return value to your community locally and globally
- **Do** deliver quality in a timely fashion
- **Do** be honest in your work by telling the client, customer, or boss that the task or project you are working on will not meet the target date.
- **Do** ask for help in order to meet the project or task deadline. A professional will not feel slighted if he or she acknowledges that he or she needs help.
- When you accept an assignment, **Do** start using words like *we*, *us* and *ours*. A professional never works at cross-purpose with the employer.
- If the employer wants respect from employees, he/she should treat all employees as professionals in their own right. Remember, if you treat someone as a professional, they will (hopefully) treat you like one too.
- Promote your profession i.e. Information Technology
- Do things for the good of yourself, the customer and the profession. You are not a true professional if you don't deliver outcomes that satisfy all three of these areas.
- Do provide respect to others. This means truly thinking of their beliefs and desires, and the contribution they make to the organization, however small.

Things to Not Do

- **Do not** tell the client, customer, or boss that you can do something when you cannot.
- When you accept an assignment, **Do not** use words like *me*, *mine*, *you* and *yours*.
- Do not steal from your employer.
- Do not underestimate your capabilities.

Difference Between Professional Ethics and Personal Ethics

Ethics seeks to answer questions of right and wrong that every person faces in life. However, even more important is choosing to do what is right. Ethics may also be the basis by which people judge others, although what may be right for one person is not always right for another.

Professional ethical codes are generally established by a business or a professional organization to govern behavior within it. A personal ethical code is how one person will govern his life based on his own beliefs about what is right or wrong.

Personal ethics depends highly on personal beliefs and loyalties, and results in behavior in a person's life in his relationships, family and friends. We bind ourselves by our personal ethics as they define who we are, and give us the pride and confidence in our lives. Personal ethics are often deep within us, but not written. Personal ethics can also change depending on major events in our lives.

Professional ethics depends highly on the requirements of the profession, looking out for the interests of the organization as well as the customers. We are often bound legally by our professional ethics, such as confidentiality and due diligence. Professional ethics are usually written and signed, and rarely change over time.

A professional code of ethics generally only applies in our professional lives. It is meant to govern the rights and wrongs related to work. However, a personal code of ethics is an established system by which one determines what is right and wrong in her everyday life.

Professional ethical codes within a business are generally statements that express the values of the organization. The code may express the commitment that a company is making to its employees, stockholders and communities. Personal ethical codes are usually written in the first person expressing the things that a person feels are important to him, such as honesty and loyalty.

Personally all people are equal. Generally, we all have the capacity for knowing the difference between what is right and wrong and conducting one's self accordingly or suffering the consequences. However, professionally, not all people are equal. A manager will have more responsibility than a subordinate to do what is right according to the organization's standards.

There are often legal ramifications to violating a professional code of ethics. Violating a professional code of ethics such as confidentiality or due diligence may result in sanctions and legal penalties. However, the violation of one's personal ethical code will generally only result in feelings of guilt.

In summary, the similarity ends at the fact that we bind our action to the ethical rules expected of us. Differences are many. While professional ethics apply only while we are working in our professional, we check our actions by our personal ethics wherever we go.

6.13.4. Ethics in Engineering

Engineering ethics is the field of applied ethics which examines and sets standards for engineers' obligations to the public, their clients, employers and the profession.

Engineering does not have a single uniform system, or standard, of ethical conduct across the entire profession. Ethical approaches vary somewhat by discipline and jurisdiction, but are most influenced by whether the engineers are independently providing professional services to clients, or the public if employed in government service; or if they are employees of an enterprise creating products for sale.

Current Codes of Ethics

Many American engineering professional societies have prepared codes of ethics. Some go back to the early decades of the twentieth century. These have been incorporated to a greater or lesser degree into the regulatory laws of several states.

The Institution of Civil Engineers (ICE) in the UK has a code of ethics incorporated into its standards of conduct. The Canadian societies of Professional engineers likewise have as well. These codes of ethics share many similarities.

General Principles

Codes of engineering ethics identify a specific precedence with respect to the engineer's consideration for the public, clients, employers, and the profession.

This is an example from the American Society of Civil Engineers (ASCE):

Fundamental Canons

1. Engineers shall hold paramount the safety, health and welfare of the public and shall strive to comply with the principles of sustainable development in the performance of their professional duties.
2. Engineers shall perform services only in areas of their competence.
3. Engineers shall issue public statements only in an objective and truthful manner.
4. Engineers shall act in professional matters for each employer or client as faithful agents or trustees, and shall avoid conflicts of interest.
5. Engineers shall build their professional reputation on the merit of their services and shall not compete unfairly with others.
6. Engineers shall act in such a manner as to uphold and enhance the honor, integrity, and dignity of the engineering profession and shall act with zero-tolerance for bribery, fraud, and corruption.
7. Engineers shall continue their professional development throughout their careers, and shall provide opportunities for the professional development of those engineers under their supervision."

Like virtually all professional societies and chartering authorities, ASCE expands upon these and publishes specific guidance.

A basic ethical dilemma is that an engineer has the duty to report to the appropriate authority a possible risk to others from a client or employer failing to follow the engineer's directions. According to first principles, this duty overrides the duty to a client and/or employer. An engineer may be disciplined, or have their license revoked, even if the failure to report such a danger does not result in the loss of life or health.

In many cases, this duty can be discharged by advising the client of the consequences in a forthright matter, and assuring the client takes the engineer's advice. However, the engineer must ensure that the remedial steps are taken and, if they are not, the situation must be reported to the appropriate authority. In very rare cases, where even a governmental authority may not take appropriate action, the engineer can only discharge the duty by making the situation public. As a result, whistleblowing by professional engineers is not an unusual event, and courts have often sided with engineers in such cases, overruling duties to employers and confidentiality considerations that otherwise would have prevented the engineer from speaking out.

6.13.5. Ethical Issues

There are several other ethical issues that engineers may face. Some have to do with technical practice, but many others have to do with broader considerations of business conduct. These include:

- Relationships with clients, consultants, competitors, and contractors
- Ensuring legal compliance by clients, client's contractors, and others
- Conflict of interest
- Bribery and kickbacks, which also may include:
- Gifts, meals, services, and entertainment
- Treatment of confidential or proprietary information
- Consideration of the employer's assets
- Outside employment/activities (Moonlighting)

Some engineering societies are addressing environmental protection as a stand-alone question of ethics.

The field of business ethics often overlaps and informs ethical decision making for engineers.

6.13.6. Work Ethics

Work ethic is a set of values based on hard work and diligence. It is also a belief in the moral benefit of work and its ability to enhance character. An example would be the Protestant work ethic. A work ethic may include being reliable, having initiative or maintaining social skills.

Work ethics include not only how one feels about their job, career or vocation, but also how one does his/her job or responsibilities. This involves attitude, behavior, respect, communication, and interaction; how one gets along with others. Work ethics demonstrate many things about whom and how a person is.

Work ethics involve such characteristics as honesty and accountability. Essentially, work ethics break down to what one does or would do in a particular situation. The begging question in a situation involves what is right and acceptable, and above board, versus what is wrong, underhanded, and under the table.

Work ethics, such as honesty (not lying, cheating, and stealing), doing a job well, valuing what one does, having a sense of purpose and feeling/being a part of a greater vision or plan is vital. Philosophically, if one does not have proper work ethics, a person's conscience may be bothered. People for the most part have good work ethic(s); we should not only want to do, but desire to do the proper thing in a given situation.

Work ethics are intrinsic; they come from within. A question may involve where they came from, if they come from within. Philosophically, this may lead to various perspectives; Work ethics come from God the creator. God made humans in His image, and His word proclaims these various work ethics — honesty, integrity, doing a job well, keeping things above board, and accountability factors.

There are two fundament work ethics — humility and the treatment of others. Humility is being humble, no task is too demeaning. Humility involves servitude, which emphasizes placing other peoples need before ones own. Treating others with decency and respect equate to the *golden rule*. The treatment of others involves loving your neighbor, loving our enemy, doing good to those who dislike us. It involves valuing others, and knowing they have worth.

Workers exhibiting a good work ethic in theory (and ideally in practice) should be selected for better positions, more responsibility and ultimately promotion. Workers who fail to exhibit a good work ethic may be regarded as failing to provide fair value for the wage the employer is paying them and should not be promoted or placed in positions of greater responsibility.

The idea of a meritocracy is based somewhat on the work ethic, in that under a meritocracy, workers who possess a good work ethic (work hard and play by the rules) are to be rewarded and move ahead, and workers who do not have a good work ethic are to be punished, or not be rewarded.

Criticism of Work Ethic Concept

Slacker and hippie cultures have challenged these values in recent times. In the 19th century, the Arts and Crafts movement of William Morris in the UK and Elbert Hubbard in the US noted how "alienation" of workers from ownership of the tools of production and their work product was destructive of the work ethic because in the expanding firms of that era, the workers saw no point in doing more than the minimum. The industrial engineer Frederick Taylor revised the notion of work ethic to include giving up control over the work process to management so that the latter could study and "rationalize" the work process, and the notion of work ethic thereafter included acknowledgment of management control.

Marxists, and most non-Marxist sociologists, make short shrift of "work ethic" as a useful sociological concept. They argue having a "work ethic" in excess of management's control doesn't appear rational in any mature industry where the employee can't rationally hope to become more than a manager whose fate still depends on the owner's decisions. Sociology prefers to renarrate excess work ethic as a form of alienation from truer needs for family and community connections, and twentieth century "critical theory" sees the "work ethic" as a unilateral demand which evolved from a mass confusion between Max Weber's "Protestant work ethic" of company founders, and a sociologically uninteresting phenomenon (rare enough to not register on a mass radar screen) which in fact produces deviance (of interest to the sociologist) in the form of addiction and family neglect.

In fact, an excess "work ethic" is often fueled (according again to sociology) by addictions such as the widespread addiction to crystal meth in American rural communities, which has been shown to result from excessive work schedules. In other environments, it creates a sharp divide between work and play, and neglect in the latter of any recreation (time with family, volunteer work, or cultural pursuits) which reminds the workaholic of "work."

The French Leftist philosopher André Gorz wrote: "The work ethic has become obsolete. It is no longer true that producing more means working more, or that producing more will lead to a better way of life. The connection between more and better has been broken; our needs for many products and services are already more than adequately met, and many of our as-yet- unsatisfied needs will be met not by producing more, but by producing differently, producing other things, or even producing less. This is especially true as regards our needs for air, water, space, silence, beauty, time and human contact. Neither is it true any longer that the more each individual works, the better off everyone will be. The present crisis has stimulated technological change of an unprecedented scale and speed: `the micro-chip revolution'. The object and indeed the effect of this revolution has been to make rapidly increasing savings in labour, in the industrial, administrative and service sectors. Increasing production is secured in these sectors by decreasing amounts of labour. As a result, the social process of production no longer needs everyone to work in it on a full-time basis. The work ethic ceases to be viable in such a situation and workbased society is thrown into crisis." André Gorz, Critique of Economic Reason,Gallilé, 1989.

Working with Ethics

HP, Johnson & Johnson, Ben and Jerry are some of the well known names in the field of '*Working with Ethics*'. These companies have adopted some guidelines to cultivate and maintain ethics in work life. Some of the guidelines listed below may be followed by a company for the said purpose:

1. A company and its employees should always believe that managing ethics is a continuous process and it is inherent like any other management function in the organization.
2. The ground of ethics management is relationship and behavior is the root, thus without fair and just behavior an organization cannot make their employees follow ethical practices.
3. As the Mc Gregor's *Red hot stove rule* says, the administration of discipline should be more like touching a hot stove i.e. immediate, impersonal, consistent, and foreseeable; in the same way the act of breaking ethical norms/ code must be rooted out in its very initial stage. The codes must be applicable to each and every employee without discrepancy.
4. Value forgiveness. It's better to try operating ethically and committing few mistakes than not trying at all.
5. Participation of employees in such programs must be sought out. Employee must feel that working with ethics is as much necessary for the good health for the organization as fresh air is for human being.

A company may adopt different ways and methods to bring in work ethics in work life. Further, the other side of the coin is "*Good people do not need laws to tell them to act responsibly, while bad people will find a way around the laws": Pluto.*

Whether the circumstance is business or life, ethical values should be ground-rules for behavior. When we live by these values we are demonstrating that we are worthy of trust.

6.13.7. Environmental Ethics

Environmental ethics is the discipline that studies the moral relationship of human beings to, and also the value and moral status of, the environment and its nonhuman contents. It all depends on how do we think and act.

Environmental ethics or environmental philosophy considers the relationship between human beings and the natural environment. It exerts influence on a large range of disciplines including law, sociology, theology, economics and geography. There are many ethical decisions that human beings make with respect to the environment. For example:

- Should we continue to clear cut the forests for the sake of human consumption?
- Should we continue to make gasoline powered vehicles, depleting fossil fuel resources while the technology exists to create zero-emission vehicles?
- What environmental obligations do we need to keep for future generations?

- Is it right for humans to knowingly cause the extinction of a species for the (perceived or real) convenience of humanity?

We have to follow Eco-centric view *i.e.* 'Nature has provided us with all the resources for leading a beautiful life and she nourishes us like a mother, we should respect her and nurture her' instance of Anthropocentric view, which says 'Man is all powerful and the supreme creature on this earth and man is the master of nature and can harness it as his will'.

Environmental ethics can provide us the guidelines for putting our beliefs into action and help us decide what to do when faced with crucial situation. Some important ethical guidelines known as earth ethics or environmental ethics are as follows :

- The Earth should exist for the benefit of humanity and we should love and respect the earth.
- Humans should have any ethical obligations with respect to the natural world. We should celebrate the turning of its seasons.
- We should not have the right to take all the Earth's resources for our own use.
- We should not hold our-self above other living things and have no right to drive them to extinction. Other species have an intrinsic right to exist.
- We should believe on various religions and should follow what they say about humanity's relationship to the rest of the living world.
- We should be grateful to the plants and animals which nourishes us by giving us food.
- We should consume the material goods in moderate amount so that all may share the earth's precious treasure of resources. We should not waste our resources on destruction weapons
- We should not steal from future generation their right to live in a clean and safe planet by polluting it.

6.14. ENVIRONMENTAL IMPACT ASSESSMENT (EIA)

Environmental Impact Assessment (EIA) is an important management tool for ensuring optimal use of natural resources for sustainable development. A beginning in this direction was made in our country with the impact assessment of river valley projects in 1978-79 and the scope has subsequently been enhanced to cover other developmental sectors such as industries, thermal power projects, mining schemes etc. To facilitate collection of environmental data and preparation of management plans, guidelines have been evolved and circulated to the concerned Central and State Government Departments. EIA has now been made mandatory under the Environmental (Protection Act, 1986 for 29 categories of developmental activities involving investments of Rs. 50 crores and above.

An **environmental impact assessment (EIA)** is an assessment of the possible impact—positive or negative—that a proposed project may have on the environment, together consisting of the natural, social and economic aspects.

The purpose of the assessment is to ensure that decision makers consider the ensuing environmental impacts when deciding whether to proceed with a project. The International Association for Impact Assessment (IAIA) defines an environmental impact assessment as "the process of identifying, predicting, evaluating and mitigating the biophysical, social, and other relevant effects of development proposals prior to major decisions being taken and commitments made."

After an EIA, the precautionary and polluter pays principles may be applied to prevent, limit, or require strict liability or insurance coverage to a project, based on its likely harms. Environmental impact assessments are sometimes controversial.

Environmental Impact Assessment (EIA) studies need a significant amount of primary and secondary environmental data. The primary data are those which need to be collected in the field to define the status of environment (like air quality data, water quality data etc.). The secondary data are those data which have been collected over the years and can be used to understand the existing environmental scenario of the study area. The environmental impact assessment (EIA) studies are conducted over a short period of time and therefore the understanding the environmental trends based on few months of primary data has its own limitations. Ideally, the primary data has to be considered along with the secondary data for complete understanding of the existing environmental status of the area. In many EIA studies, the secondary data needs could be as high as 80% of the total data requirement. EIC is the repository of one stop secondary data source for environmental impact assessment in India.

The Environmental Impact Assessment (EIA) experience in India indicates that the lack of timely availability of reliable and authentic environmental data has been a major bottle neck in achieving the full benefits of EIA. The environment being a multi-disciplinary subject, a multitude of agencies is involved in collection of environmental data. However, there is no single organization in India which tracks the data available amongst these agencies and makes it available in one place, in a form and manner required by practitioners in the field of environmental impact assessment in India. Further, the environmental data is not available in value added forms that can enhance the quality of the EIA. This in turn adversely affects the time and efforts required for conducting the environmental impact assessments (EIAs) by project proponents and also timely environmental clearances by the regulators.

6.14.1. Stages of EIA

The EIA process makes sure that environmental issues are raised when a project or plan is first discussed and that all concerns are addressed as a project gains momentum through to implementation. Recommendations made by the

EIA may necessitate the redesign of some project components, require further studies, suggest changes which alter the economic viability of the project or cause a delay in project implementation. To be of most benefit it is essential that an environmental assessment is carried out to determine significant impacts early in the project cycle so that recommendations can be built into the design and cost-benefit analysis without causing major delays or increased design costs. To be effective once implementation has commenced, the EIA should lead to a mechanism whereby adequate monitoring is undertaken to realize environmental management. An important output from the EIA process should be the delineation of enabling mechanisms for such effective management.

The way in which an EIA is carried out is not rigid: it is a process comprising a series of steps. The main steps in the EIA process are:

- **Screening** often results in a categorization of the project and from this a decision is made on whether or not a full EIA is to be carried out.
- **Scoping** is the process of determining which are the most critical issues to study and will involve community participation to some degree. It is at this early stage that EIA can most strongly influence the outline proposal.
- Detailed **prediction and mitigation** studies follow scoping and are carried out in parallel with feasibility studies.
- The main output report is called an *Environmental Impact Statement*, and contains a detailed plan for managing and **monitoring environmental** impacts both during and after implementation.
- Finally, an **audit** of the EIA process is carried out some time after implementation. The audit serves a useful feedback and learning function.

Figure 6.10 shows a general flow diagram of the EIA process, how it fits in with parallel technical and economic studies and the role of public participation. In some cases, such as small-scale irrigation schemes, the transition from identification through to detailed design may be rapid and some steps in the EIA procedure may be omitted.

The result of an EIA is assembled in a document known as an Environmental Impact Statement (EIS) which looks at all the positive and negative effects of a particular project on the environment. This report is just one component of the information required to aid decision makers in making their ultimate choices about a project.

EIA can be considered as a mechanism which maximises the efficient use of natural and human resources. It can also reduce costs and time taken to reach a decision by ensuring that subjectivity and duplication of effort are minimised, as well as identifying and attempting to evaluate the primary and secondary consequences which might require the introduction of expensive pollution control equipment or compensation and other costs at a later date.

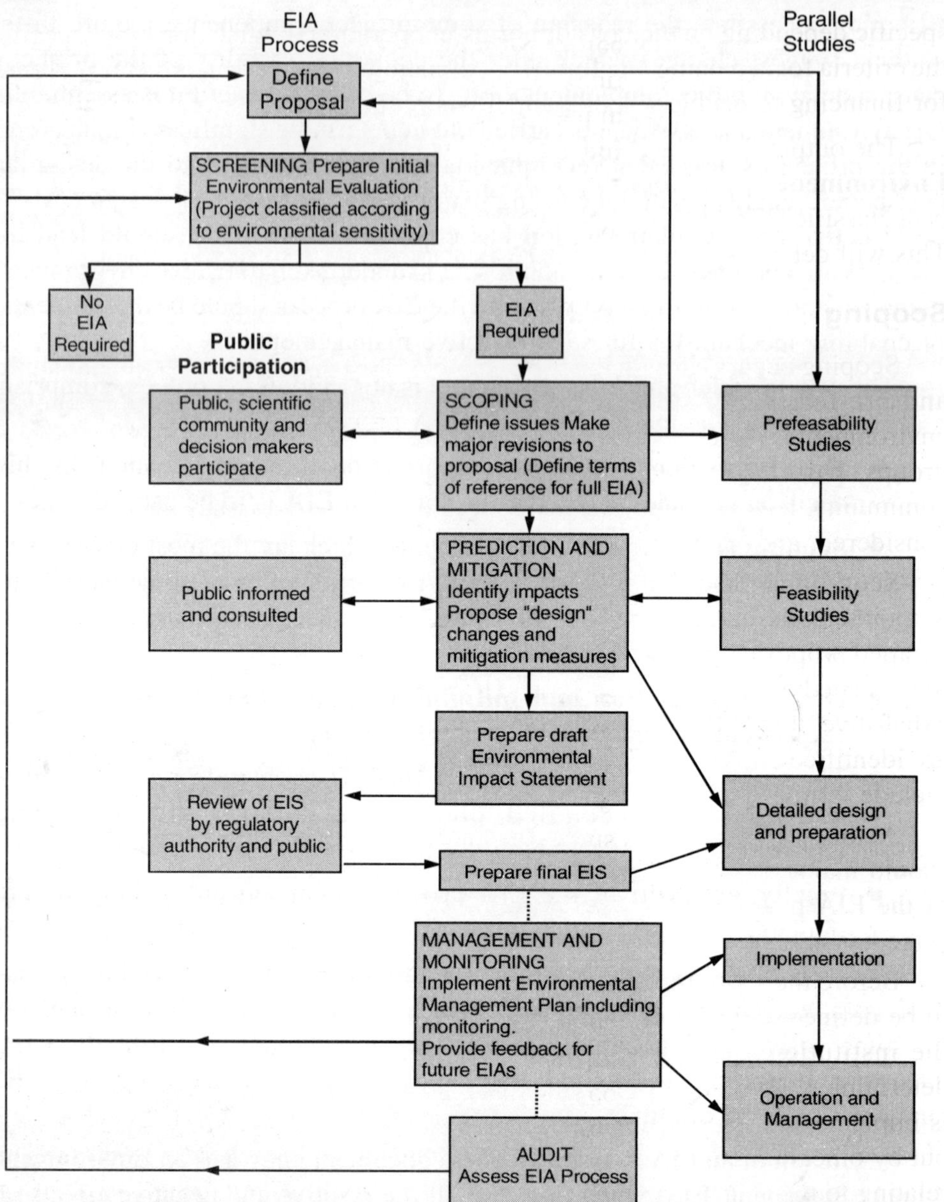

Fig. 6.10. Flow Diagram of the EIA Process and Parallel Studies

Screening

Screening is the process of deciding on whether an EIA is required. This may be determined by size (*e.g.* greater than a predetermined surface area of irrigated land that would be affected, more than a certain percentage or flow to be diverted or more than a certain capital expenditure). Alternatively it may be based on site-specific information. For example, the repair of a recently destroyed diversion structure is unlikely to require an EIA whilst a major new headwork structure may. Guidelines for whether or not an EIA is required will be country specific depending on the laws or norms in operation. Legislation often specifies the criteria for screening and full EIA. All major donors screen projects presented for financing to decide whether an EIA is required.

The output from the screening process is often a document called an **Initial Environmental Examination or Evaluation (IEE).** The main conclusion will be a classification of the project according to its likely environmental sensitivity. This will determine whether an EIA is needed and if so to what detail.

Scoping

Scoping occurs early in the project cycle at the same time as outline planning and pre-feasibility studies. Scoping is the process of identifying the key environmental issues and is perhaps the most important step in an EIA. Several groups, particularly decision makers, the local population and the scientific community, have an interest in helping to deliberate the issues which should be considered, and scoping is designed to canvass their views, (Wathern 1988).

Scoping is important for two reasons. First, so that problems can be pinpointed early allowing mitigating design changes to be made before expensive detailed work is carried out. Second, to ensure that detailed prediction work is only carried out for important issues. It is not the purpose of an EIA to carry out exhaustive studies on all environmental impacts for all projects. If key issues are identified and a full scale EIA considered necessary then the scoping should include terms of reference for these further studies.

At this stage the option exists for cancelling or drastically revising the project should major environmental problems be identified. Equally it may be the end of the EIA process should the impacts be found to be insignificant. Once this stage has passed, the opportunity for major changes to the project is restricted.

Before the scoping exercise can be fully started, the remit of the study needs to be defined and agreed by the relevant parties. These will vary depending on the institutional structure. At a minimum, those who should contribute to determining the remit will include those who decide whether a policy or project is implemented, those carrying out the EIA (or responsible for having it carried out by others) and those carrying out parallel engineering and economic studies relating to the proposal. A critical issue to determine is the breadth of the study. For example, if a proposed project is to increase the area of irrigated agriculture in a region by 10%, is the remit of the EIA to study the proposal only or also to consider options that would have the same effect on production?

A major activity of scoping is to identify key interest groups, both governmental and non-governmental, and to establish good lines of communication. People who are affected by the project need to hear about it as soon as possible. Their knowledge and perspectives may have a major bearing on the focus of the EIA. Rapid rural appraisal techniques provide a means of assessing the needs and views of the affected population.

The main EIA techniques used in scoping are *baseline studies, checklists, matrices* and *network diagrams*. These techniques collect and present knowledge and information in a straightforward way so that logical decisions can be made about which impacts are most significant. *Risk and uncertainty* are discussed further in the section *Managing uncertainty*.

Prediction and Mitigation

Once the scoping exercise is complete and the major impacts to be studied have been identified, prediction work can start. This stage forms the central part of an EIA. Several major options are likely to have been proposed either at the scoping stage or before and each option may require separate prediction studies. Realistic and affordable mitigating measures cannot be proposed without first estimating the scope of the impacts, which should be in monetary terms wherever possible. It then becomes important to quantify the impact of the suggested improvements by further prediction work. Clearly, options need to be discarded as soon as their unsuitability can be proved or alternatives shown to be superior in environmental or economic terms, or both. It is also important to test the "without project" scenario.

An important outcome of this stage will be recommendations for mitigating measures. This would be contained in the Environmental Impact Statement. Clearly the aim will be to introduce measures which minimize any identified adverse impacts and enhance positive impacts. Formal and informal communication links need to be established with teams carrying out feasibility studies so that their work can take proposals into account. Similarly, feasibility studies may indicate that some options are technically or economically unacceptable and thus environmental prediction work for these options will not be required.

Many mitigating measures do not define physical changes but require management or institutional changes or additional investment, such as for health services. Mitigating measures may also be procedural changes, for example, the introduction of, or increase in, irrigation service fees to promote efficiency and water conservation.

By the time prediction and mitigation are undertaken, the project preparation will be advanced and a decision will most likely have been made to proceed with the project. Considerable expenditure may have already been made and budgets allocated for the implementation of the project. Major changes could be disruptive to project processing and only accepted if prediction shows that impacts will be considerably worse than originally identified at the scoping stage. For example, an acceptable measure might be to alter the mode of operation of a reservoir to protect downstream fisheries, but a measure proposing an alternative to dam construction could be highly contentious at this stage. To avoid conflict it is important that the EIA process commences early in the project cycle.

This phase of an EIA will require good management of a wide range of technical specialists with particular emphasis on:

- prediction methods;
- interpretation of predictions, with and without mitigating measures;
- assessment of comparisons.

It is important to assess the required level of accuracy of predictions. *Mathematical modelling* is a valuable technique, but care must be taken to choose models that suit the available data. Because of the level of available knowledge and the complexity of the systems, physical systems are modelled more successfully than ecological systems which in turn are more successfully modelled than social systems. Social studies (including institutional capacity studies) will probably produce output in non-numerical terms. *Expert advice*, particularly from experts familiar with the locality, can provide quantification of impacts that cannot be modelled. Various techniques are available to remove the bias of individual opinion.

Checklists, matrices, networks diagrams, graphical comparisons and *overlays*, are all techniques developed to help carry out an EIA and present the results of an EIA in a format useful for comparing options. The main quantifiable methods of comparing options are by applying *weightings*, to environmental impacts or using economic cost-benefit analysis or a combination of the two. Numerical values, or weightings, can be applied to different environmental impacts to (subjectively) define their relative importance. Assigning economic values to all environmental impacts is not recommended as the issues are obscured by the single, final answer. However, *economic techniques*, can provide insight into comparative importance where different environmental impacts are to be compared, such as either losing more wetlands or resettling a greater number of people.

When comparing a range of proposals or a variety of mitigation or enhancement activities, a number of characteristics of different impacts need to be highlighted. The relative importance of impacts needs agreeing, usually following a method of reaching a consensus but including economic considerations. The uncertainty in predicting the impact should be clearly noted. Finally, the time frame in which the impact will occur should be indicated, including whether or not the impact is irreversible.

Management and Monitoring

The part of the EIS covering monitoring and management is often referred to as the **Environmental Action Plan** or Environmental Management Plan. This section not only sets out the mitigation measures needed for environmental management, both in the short and long term, but also the institutional requirements for implementation. The term 'institutional' is used here in its broadest context to encompass relationships:

- established by law between individuals and government;
- between individuals and groups involved in economic transactions;

- developed to articulate legal, financial and administrative links among public agencies;
- motivated by socio-psychological stimuli among groups and individuals (Craine, 1971).

The above list highlights the breadth of options available for environmental management, namely: changes in law; changes in prices; changes in governmental institutions; and, changes in culture which may be influenced by education and information dissemination. All the management proposals need to be clearly defined and costed. One of the more straightforward and effective changes is to set-up a monitoring programme with clear definition as to which agencies are responsible for data collection, collation, interpretation and implementation of management measures.

The purpose of monitoring is to compare predicted and actual impacts, particularly if the impacts are either very important or the scale of the impact cannot be very accurately predicted. The results of monitoring can be used to manage the environment, particularly to highlight problems early so that action can be taken. The range of parameters requiring monitoring may be broad or narrow and will be dictated by the 'prediction and mitigation' stage of the EIA. Typical areas of concern where monitoring is weak are: water quality, both inflow and outflow; stress in sensitive ecosystems; soil fertility, particularly salinization problems; water related health hazards; equity of water distributions; groundwater levels.

The use of satellite imagery to monitor changes in land use and the 'health' of the land and sea is becoming more common and can prove a cost-effective tool, particularly in areas with poor access. Remotely sensed data have the advantage of not being constrained by political and administrative boundaries. They can be used as one particular overlay in a GIS. However, authorization is needed for their use, which may be linked to national security issues, and may thus be hampered by reluctant governments.

Monitoring should not be seen as an open-ended commitment to collect data. If the need for monitoring ceases, data collection should cease. Conversely, monitoring may reveal the need for more intensive study and the institutional infrastructure must be sufficiently flexible to adapt to changing demands. The information obtained from monitoring and management can be extremely useful for future EIAs, making them both more accurate and more efficient.

The Environmental Management Plan needs to not only include clear recommendations for action and the procedures for their implementation but must also define a programme and costs. It must be quite clear exactly how management and mitigation methods are phased with project implementation and when costs will be incurred. Mitigation and management measures will not be adopted unless they can be shown to be practicable and good value for money. The plan should also stipulate that if, during project implementation, major changes are introduced, or if the project is aborted, the EIA procedures will be re-started to evaluate the effect of such actions.

Auditing

In order to capitalise on the experience and knowledge gained, the last stage of an EIA is to carry out an Environmental Audit some time after completion of the project or implementation of a programme. It will therefore usually be done by a separate team of specialists to that working on the bulk of the EIA. The audit should include an analysis of the technical, procedural and decision-making aspects of the EIA. Technical aspects include: the adequacy of the baseline studies, the accuracy of predictions and the suitability of mitigation measures. Procedural aspects include: the efficiency of the procedure, the fairness of the public involvement measures and the degree of coordination of roles and responsibilities. Decision-making aspects include: the utility of the process for decision making and the implications for development, (adapted from Sadler in Wathern, 1988). The audit will determine whether recommendations and requirements made by the earlier EIA steps were incorporated successfully into project implementation. Lessons learnt and formally described in an audit can greatly assist in future EIAs and build up the expertise and efficiency of the concerned institutions.

Public Participation

Projects or programmes have significant impacts on the local population. Whilst the aim is to improve the well being of the population, a lack of understanding of the people and their society may result in development that has considerable negative consequences. More significantly, there may be divergence between national economic interests and those of the local population. For example, the need to increase local rice production to satisfy increasing consumption in the urban area may differ from the needs as perceived by the local farmers. To allow for this, public participation in the planning process is essential. The EIA provides an ideal forum for checking that the affected public have been adequately consulted and their views taken into account in project preparation.

The level of consultation will vary depending on the type of plan or project. New projects involving resettlement or displacement will require the most extensive public participation. As stated before, the purpose of an EIA is to improve projects and this, to some extent, can only be achieved by involving those people directly or indirectly affected. The value of environmental amenities is not absolute and consensus is one way of establishing values. Public consultation will reveal new information, improve understanding and enable better choices to be made. Without consultation, legitimate issues may not be heard, leading to conflict and unsustainability.

The community should not only be consulted they should be actively involved in environmental matters. The International Union for the Conservation of Nature, IUCN promotes the concept of Primary Environmental Care whereby farmers, for example, with assistance from extension services, are directly involved in environmental management. The earlier the public are involved, the better. Ideally this will be before a development proposal is fully defined. It is an essential feature of successful scoping, at which stage feedback will have the maximum influence. Openness about uncertainty should be a significant feature of this process. As the EIA progresses, public consultation is likely to be decreased though it is important to disseminate information. The publication of the draft *Environmental Impact Statement* (EIS), will normally be accompanied by some sort of public hearing that needs to be chaired by a person with good communication skills. He/she may not be a member of the EIA team.

There are no clear rules about how to involve the public and it is important that the process remains innovative and flexible. In practice, the views of people affected by the plan are likely to be heard through some form of representation rather than directly. It is therefore important to understand how decisions are made locally and what are the methods of communication, including available government extension services. The range of groups outside the formal structure with relevant information are likely to include: technical and scientific societies; Water User Groups; NGOs; experts on local culture; and religious groups. However, it is important to find out which groups are under-represented and which ones are responsible for access to natural resources, namely: grazing, water, fishing and forest products. The views of racial minorities, women, religious minorities, political minorities and lower cast groups are commonly overlooked, (World Bank, 1991).

There has been an enormous increase in the number of environmental NGOs and "Green" pressure groups throughout the world. Such organizations often bring environmental issues to the attention of the local press. However, this should not deter consultation with such organizations as the approach to EIA should be open and positive with the aim of making improvements. Relevant NGOs should be identified and their experience and technical capacity put to good use.

In some countries, open public meetings are the most common technique to enable public participation. However, the sort of open debate engendered at such meetings is often both culturally alien and unacceptable. Alternative techniques must be used. Surveys, workshops, small group meetings and interviews with key groups and individuals are all techniques that may be useful. Tools such as maps, models and posters can help to illustrate points and improve communication. Where resettlement is proposed, extensive public participation must be allowed which will, at a minimum, involve an experienced anthropologist or sociologist who speaks the local language. He/she can expect to spend months, rather than weeks, in the field.

Information dissemination can be achieved using a number of mechanisms including the broadcasting media, in particular newspapers and radio. Posters and leaflets are also useful and need to be distributed widely to such locations as schools, clinics, post offices, community centres, religious buildings, bus stops, shops etc. The EIA process must be seen to be fair.

The public participation/consultation and information dissemination activities need to be planned and budgeted. The social scientist team member should define how and when activities take place and also the strategy: extensive field work is expensive. It is important to note that public participation activities are often reported as a separate section of the final EIA. Where experience of managing community involvement is limited, training is highly recommended. Further reading on public participation can be obtained from: Ahmed L and G K Sammy (1988) and on Rapid Rural Appraisal from Chambers R (1981). Rapid Rural Appraisal techniques may be an appropriate and cost effective method of assessment.

History of EIA in India

The Indian experience with Environmental Impact Assessment began over 20 years back. It started in 1976-77 when the Planning Commission asked the

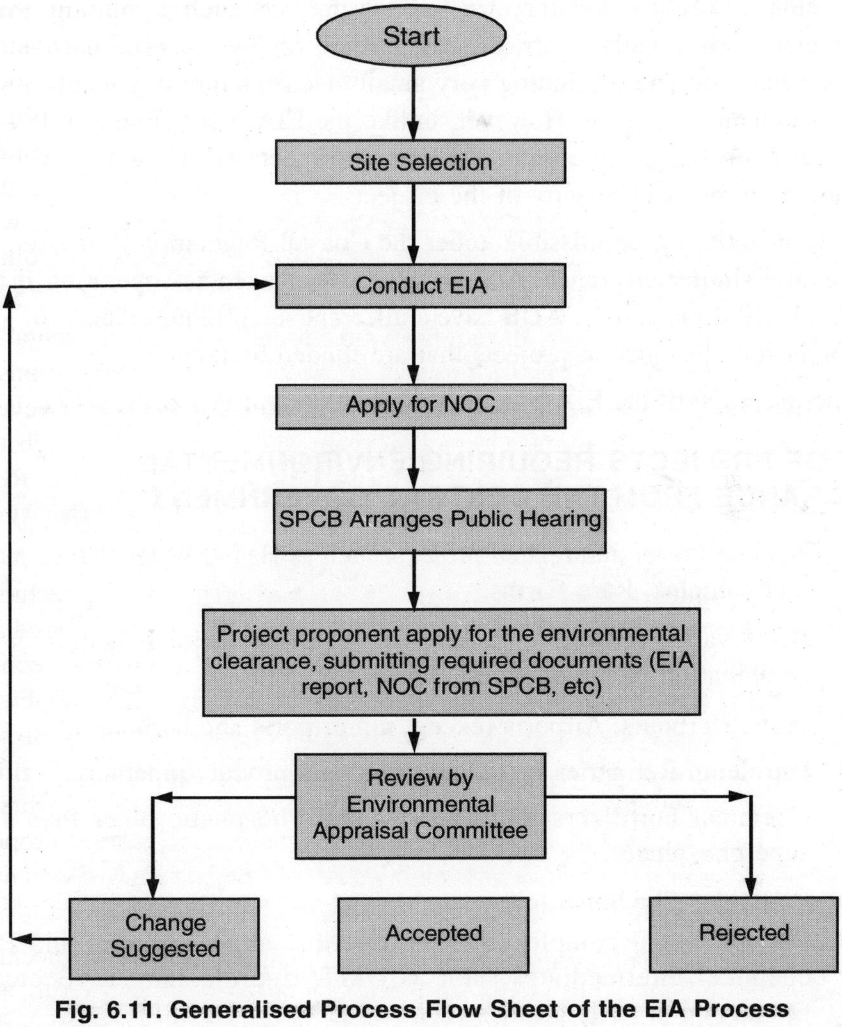

Fig. 6.11. Generalised Process Flow Sheet of the EIA Process

Source: The manual in perspective, EIA Training Resource Manual, United Nations Environment Programme, 2002

Department of Science and Technology to examine the river-valley projects from an environmental angle. This was subsequently extended to cover those projects, which required the approval of the Public Investment Board. Till 1994, environmental clearance from the Central Government was an administrative decision and lacked legislative support.

On 27 January 1994, the Union Ministry of Environment and Forests (MEF), Government of India, under the Environmental (Protection) Act 1986, promulgated an EIA notification making Environmental Clearance (EC) mandatory for expansion or modernisation of any activity or for setting up new projects listed in Schedule 1 of the notification. Since then there have been 12 amendments made in the EIA notification of 1994.

The MoEF recently notified new EIA legislation in September 2006. The notification makes it mandatory for various projects such as mining, thermal power plants, river valley, infrastructure (road, highway, ports, harbours and airports) and industries including very small electroplating or foundry units to get environment clearance. However, unlike the EIA Notification of 1994, the new legislation has put the onus of clearing projects on the state government depending on the size/capacity of the project.

Certain activities permissible under the Coastal Regulation Zone Act, 1991 also require similar clearance. Additionally, donor agencies operating in India like the World Bank and the ADB have a different set of requirements for giving environmental clearance to projects that are funded by them.

The overview of the EIA process is represented in Fig. 6.11.

LIST OF PROJECTS REQUIRING ENVIRONMENTAL CLEARANCE FROM THE CENTRAL GOVERNMENT

1. Nuclear Power and related projects such as Heavy Water Plants, nuclear fuel complex, Rare Earths.
2. River Valley projects including hydel power, major Irrigation & their combination including flood control.
3. Ports, Harbours, Airports (except minor ports and harbours).
4. Petroleum Refineries including crude and product pipelines.
5. Chemical Fertilizers (Nitrogenous and Phosphatic other than single superphosphate).
6. Pesticides (Technical).
7. Petrochemical complexes (Both Olefinic and Aromatic) and Petrochemical intermediates such as DMT, Caprolactam, LAB etc. and production of basic plastics such as LLDPE, HDPE, PP, PVC.
8. Bulk drugs and pharmaceuticals.

9. Exploration for oil and gas and their production, transportation and storage.
10. Synthetic Rubber.
11. Asbestos and Asbestos products.
12. Hydrocyanic acid and its derivatives.
13. (*a*) Primary metallurgical industries (such as production of Iron and Steel, Aluminium, Copper, Zinc, Lead and Ferro Alloys).

    (*b*) Electric arc furnaces (Mini Steel Plants).
14. Chlor alkali industry.
15. Integrated paint complex including manufacture of resins and basic raw materials required in the manufacture of paints.
16. Viscose Staple fibre and filament yarn.
17. Storage batteries integrated with manufacture of oxides of lead and lead antimony alloys.
18. All tourism projects between 200m—500 metres of High Water Line and at locations with an elevation of more than 1000 metres with investment of more than Rs. 5 crores.
19. Thermal Power Plants.
20. Mining projects *(major minerals)* with leases more than 5 hectares.
21. Highway Projects **except projects relating to improvement work including widening and strengthening of roads with marginal land acquisition along the existing alignments provided it does not pass through ecologically sensitive areas such as National Parks, Sanctuaries, Tiger Reserves, Reserve Forests**
22. Tarred Roads in the Himalayas and or Forest areas.
23. Distilleries.
24. Raw Skins and Hides
25. Pulp, paper and newsprint.
26. Dyes.
27. Cement.
28. Foundries (individual)
29. Electroplating
30. Meta amino phenol

Comparative Review of EIA Procedures and Practices

Developed countries	EIA in developing countries	EIA in India
Well-framed EIA legislation in place. For instance, in Canada, Canadian Environmental Assessment Act regulates EIA while EU countries are guided by Directive on EIA (1985).	Lack of formal EIA legislation in many developing countries. For instance, EIA is not mandatory in many African countries	Formal legislation for EIA. It has been enacted by making an amendment in the Environment Protection Act 1986.
In developed countries, active involvement of all participants including competent authority, government agencies and affected people at early stages of the EIA. This makes the process more robust and gives a fair idea of issues, which need to be addressed in the initial phase of EIA.	Limited involvement of public and government agencies in the initial phases. This often results in poor representation of the issues and impacts in the report, adversely affecting the quality of the report.	Limited involvement of public and government agencies in the initial phases.
Integrated approach to EIA followed. All aspects including social and health taken into account.	Mainly environmental aspects considered. Poor on social or health aspects.	No provision in place to cover landscape and visual impacts in the Indian EIA regulations
Proper consideration of alternatives in EIA	The consideration of alternatives in developing countries is more or less absent.	Same as developing countries
The process of screening is well defined. For instance, in EU countries competent authorities decide whether EIA is required after seeking advice from developer, NGO and statutory consultees. In Japan, screening decision is made by the authorizing agency with respect to certain criteria. In Canada, federal authority determines whether an environmental assessment is required or not.	In developing countries, screening practice in EIA is weak. In most cases, there is a list of activities that require EIA but without any threshold values.	Screening done on the basis of a defined list. Threshold values on the size of the project has been used to decide whether the project will be cleared by the state government or the central government.
Scoping process is comprehensive and involves consultation with all the stakeholders. In many countries like US, Netherlands, Canada and	Scoping process in most developing countries is very poorly defined. In many countries including China, Pakistan, etc. there is no provision for	Earlier scoping was done by consultant or proponent with an inclination towards meeting pollution control requirements, rather than

Europe, the involvement of the public and their concern are addressed in the scoping exercise. Besides this, funding organisations such as World Bank, ADB and ERDB have provision for consultation with the affected people and NGOs during identification of issues in scoping exercise.	scoping. In some countries like in Nigeria and Indonesia, a term of reference is followed for scoping while in some countries like Ghana, Taiwan and Chile, a general checklist is followed. In countries where it is undertaken, there is no public consultation during scoping. Moreover, in most developing countries, scoping is often directed towards meeting pollution control requirements, rather than addressing the full range of potential environmental impacts from a proposed development.	addressing the full range of potential environmental impacts from a proposed development. However, the new notification has put the onus of scoping on the expert committee based on the information provided by the proponent. Consultation with public is optional and depends on the discretion of the expert committee.
Most reports in local language	Most reports in English and not in the local language.	Most reports in English and not in the local language. In some case, executive summary is translated into local language.
A multi-disciplinary approach. Involvement of expert with expertise in different areas.	Lack of trained EIA professionals often leads to the preparation of inadequate and irrelevant EIA reports in developing countries	Same in India. Preparation of EIA is done by consultants. Therefore, the selection criterion for the organisation is fees/cost rather than the expertise of EIA team.
Two tier of EIA review, One conducted after the completion of EIA to check the adequacy and effectiveness of EIA and the second done before decision-making.	Poor review or monitoring.	In India too, EIA review is not upto the marks. The review agency called Impact Assessment Agency (IAA) lacks inter-disciplinary capacity. No representation of NGO in IAA, which is a violation of the EIA notification.
Expertise in EIA: The International Association for Impact Assessment (AIA) and other organisations demonstrate that there are a large number of individuals with the capability to design, conduct, review and evaluate EIAs from countries of the North. The major portion of teaching about environmental assessment also takes place in industrial countries.	The expertise in EIA is slowly developing. In most cases, students from the developing countries go to the developed countries to gain knowledge of the subject.	Expertise in this area is developing.

Source: Compiled by Industry & Environment Unit, Centre for Science & Environment, 2006

Summary

- All solid and semi-solid waste arising from human and animal activities, except human-excrete and sullage (liquid wastes from bathroom and kitchen etc.) are discarded as useless or unwanted, is called solid wastes.
- The disposal of solid waste by least pollution the environment is known as solid waste management.
- Solid waste can be disposed by any of the following method depending upon the nature of waste as open dumps, landfills, sanitary landfills, incineration plants and composting.
- Consumerism refers to economic policies that place an emphasis on consumption, and the belief that the free choice of consumers should dictate the economic structure of a society.
- Consumerism has been related both to the increase in the population size as well as increase in our demand due to change in life-style.
- In the modern society our needs have multiplied and so consumerism of resources has also multiplied.
- India is the first country in the world to have provisions for the protection and conservation of environment in its constitution.
- On 5 June 1972, environment war first discussed as an item of international agenda in the UN Conference on Human Environment in Stockholm and thereafter 5th June is celebrated all over the world as World Environmental Day.
- The provisions for environmental protection in the constitution were made in 1976 through 42 amendment under the article 48-A and Article 51-A (g).
- Some of the important Acts passes by the Government of India are Wildlife (protection) Act, 1972, Forest (Conservation) Act, 1980, The Water (Prevention and Control of Pollution) Act, 1974, The Air (Prevention and Control of Pollution) Act, 1981 and The Environmental (Protection) Act, 1986.
- Environmental ethics are the relationship between human beings and the natural environment.
- Environmental ethics can provide us the guidelines for putting our beliefs into action and help us decide what to do when faced with crucial situation.
- Ethics are about making choices that may not always feel good or seem like they benefit you but are the 'right' choices to make.
- Moral values are generally associated with a personal view of values.
- Personal morals tend to reflect beliefs relating to drinking, gambling, etc.
- Morals can reflect the influence of religion, culture, family and friends.

- Law is a rule or body of rules of conduct inherent in organization and essential to or binding upon all the employees.
- Environmental Impact Assessment (EIA) is an important management tool for ensuring optimal use of natural resources for sustainable development.

Descriptive Type Questions

1. How industrial, domestic, agricultural and other wastes affect the fertility of the soil? Discuss the measures to check it.
2. Briefly describe the idea of 'solid waste management' stating suitable examples. Distinguish between the incineration and composting methods of solid waste disposal.
3. What is solid waste? Explain the sources and effects of solid waste.
4. Classify the composting techniques based on oxygen use. What are the advantages of 'Vermicomposting' over conventional compositing? What are the advantages of solid waste incineration?
5. What are 'Solid Wastes'? Discuss their types, effects and name the various methods used to dispose solid wastes. Explain any one of them with merits and demerits.
6. Explain the various methods commonly employed for disposal of solid waste, with their advantages and disadvantages.
7. Write a note on solid waste management practices.
8. Discuss in brief, the salient features of the Air (Prevention and Control of Pollution) Act 1981.
9. Describe power of the State Government to declare air pollution control areas and restrictions on use of certain industrial plants as given in the Air (Prevention and Control of Pollution) Act 1981.
10. Briefly discuss the salient features of Environment (Protection) Act, 1986.
11. Briefly discuss the salient features of the Forest (Conservation) Act, 1980.
12. How do you define pollution as per Water (Prevention and Control of Pollution) Act 1974? What are the salient features of the Act?
13. Write down the major highlights of Wildlife Protection Act of India.

Short Answer Type Questions

1. Write short notes on Solid waste management practices.
2. What do you understand by consumerism and how it is related by waste product?
3. Differentiate ethics, moral and law.
4. Define environmental ethics. What do you mean by eco-friendly view and anthropocentric view?
5. Differentiate the Pyrolysis and composting.
6. Write a note on Environmental impact assessment.

Objective Type Questions

Multiple Choice Questions

1. Which of the following is a likely characteristic of hazardous waste?
   (a) Ignitability (b) Corrosivity
   (c) Reactivity (d) Any of the above
2. High level radioactive waste can be managed in which of the following ways?
   (a) Open dumping (b) Composting
   (c) Incineration (d) Dumping in sealed containers
3. Biomedical waste may be disposed off by?
   (a) Incineration (b) Autoclaving
   (c) Land filling (d) Both (b) and (c)
4. The highest heating value is of :
   (a) Garbage (b) Rubbish
   (c) Hospital waste (d) Industrial waste
5. The highest moisture content is of :
   (a) Garbage (b) Rubbish
   (c) Hospital waste (d) Agricultural waste
6. World Environmental Day is celebrated all over the world on :
   (a) May 1 (b) June 5
   (c) Nov 14 (d) Aug 15
7. The Environmental (Protection) Act comes into force on :
   (a) 1986 (b) 1981
   (c) 1974 (d) 1972
8. The Air (Prevention and Control of Pollution) Act comes into force on :
   (a) 1986 (b) 1981
   (c) 1974 (d) 1972
9. The Water (Prevention and Control of Pollution) Act comes into force on :
   (a) 1986 (b) 1981
   (c) 1974 (d) 1972
10. Wildlife (protection) Act comes into force on :
    (a) 1986 (b) 1981
    (c) 1974 (d) 1972
11. Forest (Conservation) Act, comes into force on :
    (a) 1980 (b) 1981
    (c) 1974 (d) 1972
12. The amendments made in Forest (Conservation) Act on :
    (a) 1980 (b) 1988
    (c) 1974 (d) 1972

13. The provisions for environmental protection in the constitution were made through :

(*a*) 42 amendment (*b*) 10 amendment
(*c*) 21 amendment (*d*) None of these

14. The provisions for environmental protection in the constitution were made under the :

(*a*) Article 48-A and Article 58
(*b*) Article 26 and Article 51
(*c*) Article 48-A and Article 51-A (g)
(*d*) None of these

15. It is concerned with how a moral person *should* behave :

(*a*) Ethics (*b*) Law
(*c*) Moral (*d*) None of these

16. The systematic study of morality is a branch of philosophy called :

(*a*) Ethics (*b*) Law
(*c*) Moral (*d*) None of these

17. A rule or body of rules of conduct inherent in organization and essential to or binding upon all the employees is :

(*a*) Ethics (*b*) Law
(*c*) Moral (*d*) None of these

18. EIA stands for :

(*a*) Ethical improved association
(*b*) Environmental impact assessment
(*c*) Ecological impact assessment
(*d*) None of these

19. EIA has now been made mandatory for :

(*a*) 29 categories of developmental activities
(*b*) 50 categories of developmental activities
(*c*) 10 categories of developmental activities
(*d*) 100 categories of developmental activities

20. EIA has now been made mandatory for developmental activities involving investments of :

(*a*) Rs. 5 crores (*b*) Rs. 15 crores
(*c*) Rs. 25 crores (*d*) Rs. 50 crores and above.

ANSWERS

Multiple Choice Questions

1. (*d*)	**2.** (*d*)	**3.** (*d*)	**4.** (*b*)	**5.** (*d*)
6. (*b*)	**7.** (*a*)	**8.** (*b*)	**9.** (*c*)	**10.** (*d*)
11. (*a*)	**12.** (*b*)	**13.** (*a*)	**14.** (*c*)	**15.** (*a*)
16. (*a*)	**17.** (*b*)	**18.** (*b*)	**19.** (*a*)	**20.** (*d*)

INDEX

A

B

C

D

E

F

G

H

O

P

R

S

❑❑❑

Win Prizes

ATTENTION READERS

Win Prizes

We request for your frank assessment regarding various aspects of the book as given below :

ENERGY, ENVIRONMENT, ECOLOGY AND SOCIETY

By : SMRITI SRIVASTAVA

Please Fill up the Sheet in Neat and Clean Writing

(i) What topics of your syllabus, important for your examination point of view, are not covered in this book ?

....

....

....

....

(ii) What are the chapters wherein the treatment of the subject matter is not systematic or organised or updated ?

....

....

....

....

(iii) Have you come across misprints/mistakes/factual inaccuracies in the book ? Please specify the chapters and the page numbers.

....

....

....

....

(iv) What other books on the same subject have you found/heard better than the present book ? Please specify in terms of price and quality.

....

....

....

....

(v) Further suggestions and comments for the improvement of this book.

....

....

....

....

(PS : If need be please attach a separate sheet to write your views)

OTHER DETAILS

(i) Who recommended you this book ? (Please tick (✓) in the box near the option relevant to you.

☐ Teacher ☐ Friends ☐ Bookseller

....

....

....

....

(ii) Name of the recommending teacher, his designation and address:

....

....

....

....

(iii) Name and address of the bookseller you purchased the book from :

....

....

....

....

(iv) Name and address of your institution (Please mention the University or Board as the case may be)

....

....

....

....

(v) Your name and complete address :

....

....

....

....

(vi) Course you are enrolled in :

....

....

....

....

The best assessment will be awarded each month : The award will be in the form of our publications as decided by the Editorial Board.

Please mail the filled up coupon at your earliest to :

S.K. KATARIA & SONS®
4760-61/23, Ansari Road,
Daryaganj, New Delhi-110002
☎ 23243489, 23269324
Fax : 011-23243489
email : katariabooks@yahoo.com